AF691414

In Memory of

Roger J. Rossiter, 1913–1976

An inspiring mentor

A pioneering neurochemist

An exuberant leader

A man for all seasons

Essentials of Human Metabolism

W.C. McMurray, Ph.D.

Professor, Department of Biochemistry
Faculties of Medicine and Dentistry
University of Western Ontario
London, Ontario

Essentials of Human Metabolism

THE RELATIONSHIP OF BIOCHEMISTRY TO HUMAN PHYSIOLOGY AND DISEASE

Second Edition

HARPER & ROW, PUBLISHERS

PHILADELPHIA

Cambridge
New York
Hagerstown
San Francisco

London
Mexico City
São Paulo
Sydney

1817

ı Burgower
Winters
. Smith

,harles W. Field
:aphics
nnelly & Sons Company

The author and publisher have exerted every effort to ensure that drug selection and dosage set forth in this text are in accord with current recommendations and practice at the time of publication. However, in view of ongoing research, changes in government regulations, and the constant flow of information relating to drug therapy and drug reactions, the reader is urged to check the package insert for each drug for any change in indications and dosage and for added warnings and precautions. This is particularly important when the recommended agent is a new or infrequently employed drug.

5 4 3 2 1

Library of Congress Cataloging in Publication Data

McMurray, W. C., Date
Essentials of human metabolism.

Bibliography: p.
Includes index.
1. Metabolism. 2. Metabolism—Disorders. I. Title.
[DNLM: 1. Biochemistry. 2. Metabolic diseases.
3. Metabolism. QU 120 M168e]
QP177.M33 1983 612'.39 83-280
ISBN 0-06-141643-6

Second Edition
Copyright © 1983, by Harper & Row, Publishers, Inc.
Copyright © 1977, by Harper & Row, Publishers, Inc. All rights reserved. No part of this book may be used or reproduced in any manner whatsoever without written permission except for brief quotations embodied in critical articles and reviews.
Printed in the United States of America. For information write Harper & Row, Publishers, Inc., East Washington Square, Philadelphia, Pennsylvania 19105.

Preface

Metabolism, the chemical transformations of molecules in the body, is the subject of change, and its study is greatly subject to change as well! In the five years since the first edition, much metabolic water has passed beneath the bridge. I have tried to show the more notable developments at the frontiers of study in the sequel to this book, *A Synopsis of Human Biochemistry*, which deals with the topics of blood, nutrition, hormone action, cell membranes, and molecular genetics. Since the *Synopsis* is intended as a companion to *Essentials*, in revising and updating the latter I have tried to make the contents of the two books complement one another with minimal redundancy. I have also added to or enlarged upon areas that time and readers' comments had indicated to me as being deficient in the original. However, the structure of the book, its intent to present the subject as concisely and simply as possible, and above all its lightness of tone and weight, I trust have not been compromised.

As with the first edition, I am grateful to colleagues in this department who have provided encouragement, advice, and corrections. I am also indebted to my students, and to interns and practitioners who wrote to tell me of strengths and weaknesses of the first edition. I would like to extend special thanks to Dr. Bob Murray, University of Toronto, who kindly sent me an extensive list of my sins of omission and commission and made useful suggestions for inclusions, several of which have been added.

One area of overlap that I have not eliminated in the two books concerns the molecular nature and expression of the gene. Several readers praised the concept of introducing the proteins in the first chapter of *Essentials* by an outline of the process whereby their sequences are predestined. Since the subject is considered in greater detail in the last chapter of *Synopsis*, it had been suggested that I eliminate the treatment in this book. On reflection I decided to retain this section. In many ways biochemistry is a circular subject; metabolism often goes by reversible or cyclical paths, and moreover, the most detailed descriptions of molecules and their actions generally return to first principles about their structures.

And so the journey's end is also its beginning. Once I nearly missed a trans-Atlantic flight in Copenhagen, arriving just in time to find the tourist section fully boarded and only the gangway left for first-class passengers. As I settled into my unaccustomed sumptuous surroundings, weakly protesting to the attendant, he explained with a quotation that probably suits this occasion equally, "You must remember what it says in the Bible—'the first shall be last; and the last shall be first!'."

W.C. McMurray, Ph.D.

Preface to the First Edition

A young man with the overburdened expression characteristic of medical students approaching examinations came to the desk of a medical librarian and inquired wearily, "Have you anything light on obstetrics?" This book was written having in mind his preclinical counterpart seeking something less ponderous than the standard biochemistry texts for medical students.

My intent is to explain the metabolic processes and their regulation in the human body, emphasizing the relationship of biochemistry to human physiology and disease. Treatment of specific tissues and organs is considered as an integral part of metabolism, not as an addendum.

I have sought to explain the transformations of chemical components in the human body while largely omitting detailed analyses of the structures and properties of these components. By concentrating upon the broader outlines of metabolism, as viewed from the wrong end of the telescope, I have hopefully obliterated individual trees, which should render the metabolic forest more passable to the aspiring clinician. Since many problems in comprehension by the student arise from the terminology of the subject, I have endeavoured to limit the vocabulary to essentials and to define each new term (italicized) where it first appears in the text. The definitions for most of the common biochemical terms (which appear in bold type) are collected in the glossary section which ends the text.

Beginning with a description of simple molecular structures, Chapter 1 outlines the principles involved in construction of the giant polymers and membrane structures of cells. Chapter 2 surveys the general properties of enzymes and the genetic and physiologic factors that modify their activities. The major energy-transducing systems involved in biologic oxidations and active transport processes are summarized in Chapter 3. The next three chapters discuss carbohydrates, lipids, and nitrogen-containing compounds from the standpoint of individual organ systems—intestinal tract, liver, muscle, adipose tissue, brain, kidney, and red blood cells—indicating processes of digestion, absorption, transport, and homeostasis of major nutrients. Specialization of pathways of metabolism and control mechanisms of individual tissues, as well as pathologic aberrations of metabolic diseases affecting the tissues, are elaborated on. The final Chapter 7 provides an integrative approach to the biochemical interrelationships between organ systems, including examples of metabolic imbalances and adaptive responses in the body during starvation and obesity.

Our knowledge and understanding of human metabolism increase daily through the efforts of clinical and basic science investigators. Major advances come often in unexpected ways, by study of mammals and vertebrates other than man, of insects and other invertebrate creatures, of yeasts, plants and microorganisms. We should be mindful and appreciative of the contributions to medical science made by these so-called lesser species of life. In focusing my attention upon man and his metabolism, in the interests of brevity I have omitted descriptions or attributions to individual workers and to the many forms of living organisms upon which our knowledge of life processes is founded. However, in paying tribute both to the investigators and the investigatees of biochemistry I must point out that this book is addressed primarily to an aspiring clinician rather than to a biochemist or an *Escherichia coli.*

Many of my associates have shown an interest in this book and have provided encouragement and advice about the presentation. In particular I would like to acknowledge with gratitude the contributions of two colleagues, Bill Magee and Valentina Donisch, who read the entire manuscript "stem to gudgeon" and spent a great deal of time and effort providing concrete suggestions for improvement. I am grateful to Ms. Doris Lesser for typing the manuscript, and to the staff of Harper and Row who have helped so much to fulfill the intent of the book.

W.C. McM.

Introduction

As a teacher and member of assorted bodies involved with curricula, I have been frequently questioned, at times assailed, by students concerning the relevance of biochemistry to their major studies of medicine, dentistry, pharmacy, home economics, and so forth. I must admit at the outset that the nature of my training and background equips me better to explain the subject matter of biochemistry than to act as an apologist. Indeed, it must be exceedingly difficult for anyone to set guidelines for the limits of relevant material from basic sciences to be assimilated in the training and education of those in various professions. However, the very persistent nature of the question demands that it be recognized and that teachers reconsider the goals of their courses in relation to career goals of their students.

Naturally there will be students within various professional fields for whom biochemistry or some other basic science will become a major interest, possibly leading to a specialized career of investigation. The majority of professional students, however, will be less enthusiastic about learning everything possible about the sciences and quite reasonably will wish to know the extent to which they are expected to master these subjects. If we accept the view that the mind of an average medical or dental student has a finite capacity for the absorption of facts and ideas within the limitations of available time, we must set some limits of subject material which are acceptable and sensible.

In establishing principles to outline the essence of biochemical knowledge, first I shall dispose of two related points; the declining retention of information of the professional person after leaving school and the rapid increase in scientific information, rendering some subject areas obsolescent even as they are taught. To my mind gaps of information are less important than gaps of awareness. A professor of biochemistry once awoke one of his napping students with the question "What is the mechanism of oxidative phosphorylation, Jones?", to which Jones, barely surfacing, replied, "Sorry Sir, I forget". The professor was understandably chagrined. "A tragedy," he moaned. "The only person who knew the mechanism of oxidative phosphorylation and he has forgotten!" Most of our students will lose less important fragments of their science subjects. The major consideration is that once a grasp of the fundamentals is acquired, though details may recede with time, a solid basis persists: the comprehension of those concepts necessary for refreshing areas of immediate application or of newer developments. This is my justification for the emphasis in subsequent sections upon the explanations of metabolic events at the expense of much of the detail.

"Exactly how much biochemistry will I find essential in my chosen profession?" It would be as easy to answer if you were a prospective sailor considering how much theory of navigation you require. If you plan to sail only inland on small lakes in a shallow draft dinghy, then simple chart reading, compass and rules of the road may suffice; if you venture into coastal waters or larger lakes, you will need in addition some pilotage, reading of tides and so forth; for ocean sailing you will need to manage sextant and astrolabe, know some theory and practice of celestial navigation, dead reckoning, drift and so on. Unfortunately, it is unlikely that either you or I can predict with great accuracy that your future nautical course will be confined to peregrinations upon a mill-pond amid balmy zephyrs. In matters of seamanship, as in careers of medicine, dentistry, and so forth, it is undoubtedly preferable to be somewhat overequipped and prepared for the eventuality that you are swept into deeper, more turbulent waters than anticipated at the outset. Even among the amateur sailors there will appear those like Chichester and Banting who possessed the flexibility, resourcefulness, and relentless exploitation of sound theoretic thinking to venture on the high seas and beat the pros at their own game.

My nautical analogy may seem inappropriate in the sense that it is the metropolitan clinician who will be most affected by and dependent upon cognizance and understanding of modern progress in scientific medicine, while his remoter colleague in the hinterlands without the full benefits of technologic progress would be expected to concern himself with more mundane approaches to the healing art. Nonetheless we must admit that a catalogue of human disease with a clear biochemical basis would show the higher incidence of molecular blood diseases such as sickle cell anaemia in West Africa, enzyme defects causing sensitivity to antimalarial drugs along the coastal Mediterranean areas, plus a host of other genetically determined metabolic diseases which concentrate among closed or in-bred populations of regions isolated from the rest of humanity. Moreover, the health problems of the world as a whole must be viewed in the context of those debilitating nutritional deficiencies of dietary protein, vitamins, and other essentials, which still afflict a large proportion of the people in areas we designate as remote and underdeveloped. Thus it may well be that those who practice in the outposts of civilization will be required to front most squarely the applications and implications of scientific principles in the solution of disease and health-related problems.

Let us now consider the goals of the study of biochemical phenomena, as related to the human species. For convenience these may be divided into theoretical or conceptual problems on the one hand and practical or applied ones on the other. There is a constant overlap and interplay between these areas. The spin-off leading to exploitation of basic principles in technologic developments and the equally exciting revelations gained by theoretical ramifications from pragmatic investigations have been thoroughly documented in biochemistry as in other areas of science. The following division is therefore somewhat arbitrary, and we should recognize that there is no hard line between the two sets of goals.

The theoretical objectives of biochemistry are to provide a description and understanding of living things, from the standpoints of their composition, form, functions, and requirements in health and disease, at the molecular level. Practical objectives of human biochemistry are *inter alia* to diagnose and follow the course of diseases by monitoring the chemical composition of blood, urine and other body fluids; to provide a scientific basis for chemotherapy in the treatment and management of disease by the selective destruction of invading cells, by the neutralization of toxic substances, and by the restoration of essential substances which are deficient; and to furnish rational approaches for dietary maintenance appropriate to stages of growth, differentiation, and so forth.

Quite apart from these rather specific educational objectives there are two additional philosophical reasons for presentation of a solid foundation in the biological sciences for those undertaking health-related professions. The first is based upon the premise that stressing the scientific study of man and his diseases will encourage a more rational, critical approach for later work in the clinic. The second is the knowledge that once initiated to the mystery and beauty of its intricate molecular makeup, the student will be conditioned in his future manipulations of the living organism by an enhanced respect and reverence for human life.

A final caution, and a challenge. In attempting to outline the events of human metabolism, too often we must fill the gaps in our knowledge by extrapolations, not only from the other species of life which may be more amenable to study, but also from the various *in vitro* preparations of cells and tissues which allow us to measure metabolic events in the test tube. In many cases our information about actual control systems of metabolism in the intact body are thus mere shadows of the true picture. Increasingly we must come to the realization that the only true subject for the study of metabolism in man is man himself, the whole man at that. Hopefully, some of those who read this book will seek to apply some of the basic concepts about metabolism to their future clinical studies, to fill in the gaps in our knowledge about human metabolism, and to elucidate the chemical changes in the body as they relate to development, function and behavior in health and disease.

SOME RECENT REFERENCE TEXTS

Bhagavan NV: Biochemistry, 2nd ed. Philadelphia, JB Lippincott Co, 1978

Devlin M (ed): Textbook of Biochemistry with Clinical Correlations. New York, John Wiley & Sons, 1982

Lehninger AL: Principles of Biochemistry. New York, Worth, 1982

Martin DW Jr, Mayer PA, Rodwell VW: Harper's Review of Biochemistry, 18th ed. Los Altos, CA, Lange Medical Publications, 1981

McGilvery RW: Biochemistry: A Functional Approach, 2nd ed. Philadelphia, WB Saunders, 1979

McMurray WC: A Synopsis of Human Biochemistry with Medical Applications. Philadelphia, Harper & Row, 1982

Orten JM, Neuhaus OW: Human Biochemistry, 10th ed. St. Louis, Mo, CV Mosby Co, 1982

Pasternak CA: An Introduction to Human Biochemistry. Oxford, Oxford University Press, 1979

Schwarz V: A Clinical Companion to Biochemical Studies. San Francisco, WH Freeman & Co, 1978

Stryer L: Biochemistry, 2nd ed. San Francisco, WH Freeman & Co, 1981

White A, Handler P, Smith EL: Principles of Biochemistry, 6th ed. New York, McGraw–Hill, 1978

Contents

Essentials of Human Metabolism

Chapter One Life as a Molecular Process

The story of life is the story of atoms and molecules reacting in ways fundamentally similar to the ways they react outside our bodies. The unique properties of the "molecular machine" that becomes a living cell lie in its complexity and in its organization. Analytic reasoning, which breaks down concepts into their component parts, can therefore only dimly perceive the nature of life. By analogy, a metallurgist's analytic description of a watch may be Fe, 59.82%; Cu, 11.97%; Zn, 7.21%, and so on. A comprehension of the machinery, however, requires an appreciation of the intricate structural features and mechanical principles that allow the components to work together as a whole. Hence, we need basic information about the construction both of the parts and of the whole mechanism. We need to know, obviously, what those relations are as a function of time; we need to ascertain what energy is imparted to the system and how the mechanism converts that energy to its own ends; finally, we must understand how the total functions of the parts intermesh, smoothly and in harmony, so that the actions of the whole machine are purposeful and regulated. This approach to the timepiece involves synthesis as well as analysis in our rational process of investigation. Similarly, to be meaningful, a study of the body demands both analytic and synthetic reasoning to characterize the parts of the living machine and to understand how the interrelations of those parts confer upon the whole a higher level of existence than each part could have in isolation. Therefore let us first analyze the components of the body, starting with the simplest, and hopefully we may, by degrees, thus comprehend the more complex and eventually arrive at a synthesis of the whole.

SIMPLE MOLECULES

All cells contain an abundance of inorganic substances: gaseous elements of O_2 and CO_2 participate in cellular oxidations. Soluble salts, particularly those of K^+ and Na^+, maintain the excitability and electrical properties of cells. Insoluble salts, mainly calcium phosphates, provide rigidity and strength to structures such as bone. Finally, and perhaps most importantly, the universal solvent—water—is the common medium for solution of the reactants within cells and for exchange of food and wastes outside the cells.

Of course, the characteristic molecules of life are those we term organic. All these contain carbon in high proportions, with varying amounts of hydrogen, oxygen, nitrogen and, to a lesser extent, phosphorus, sulfur, iron, and other metals. Carbon is the primary requisite for all living systems as we know them. The reason for this is the tremendous versatility with which this atom can share the eight electrons in its outer valence ring with four other atoms. These tetravalent structures may be highly reduced (containing hydrogen) or highly oxidized (containing oxygen); the attendant energy released by going from the former to the latter state provides the energy to sustain life. In later sections we will consider the categories of the carbohydrates, the fats and other lipids, and the nitrogen-containing substances of the body in the context of their changes and interconversions in the processes that are collectively known as ***metabolism.*** For the purposes of this discussion, we shall consider only a few representatives of the small, simple, organic substances. As we shall see, these become the units, or individual bricks, that go together to form the large, complex edifices of the cells and tissues.

Among the simpler organic compounds of the body are the sugars, the organic acids, the amino acids, and the nitrogen-ring compounds (Fig. 1-1). ***Sugars,*** such as glucose, serve as major fuels for cellular energy needs and are free to diffuse through body fluids from cell to cell. To be used inside the cells, they must be converted to esters of phosphate such as glucose-6-phosphate. ***Organic acids,*** such as acetic acid, act as intermediate forms in the breakdown and formation of myriad other substances. They may contain more than one carboxylic acid group, as in the case of succinic acid and citric acid, or longer chains of carbon atoms joined to the carboxyl group, as in the case of hydrocarbon groups of fats and other lipids. ***Amino acids,*** such as alanine, are the constituent parts of proteins in the body. All contain at least one amino group as well as the carboxylic acid group, but the carbon atom that joins these two may have many different groups attached; leucine, for example, has an aliphatic group and phenylalanine contains an aromatic substituent. ***Nitrogen-ring compounds,*** such as uracil and adenine, form the bases that compose the nucleic acids of the genetic apparatus. They are generally found linked to sugar phosphates to form what are known as ***nucleotides,*** with one, two, or three phosphate groups also interlinked. Among the multitude of com-

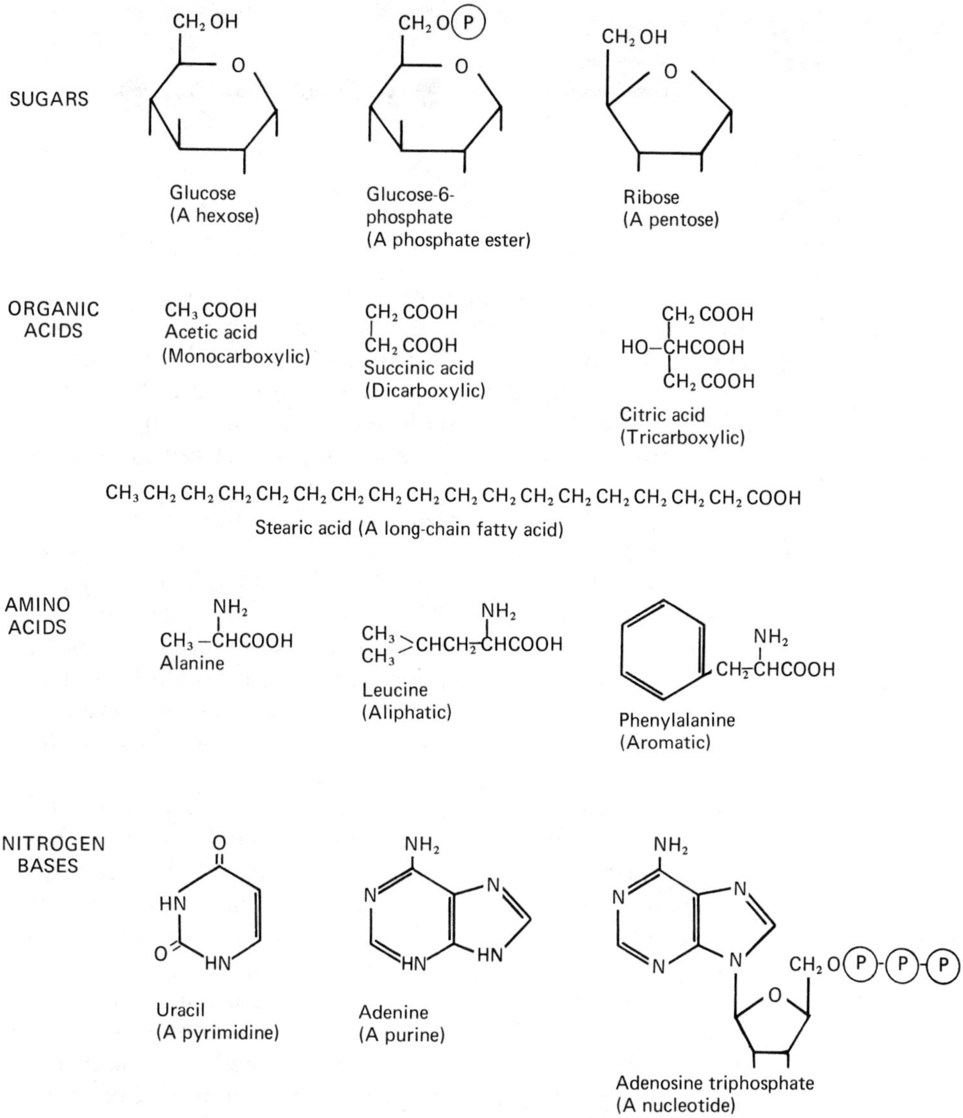

FIG. 1-1. Some of the major small organic molecules involved in metabolic processes.

plicated molecules in cells, virtually all are constructed from these few relatively simple starting materials.

MACROMOLECULES

An understanding of the nature and purposes of these complex giant molecules requires some consideration of the process of ***polymerization.*** This

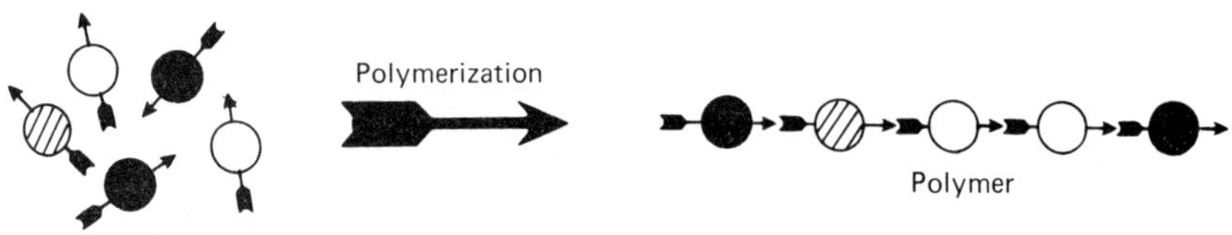

FIG. 1-2. Formation of polymers from small molecules.

refers to the joining of single molecules into long chains, analogous to the beads on a necklace (Fig. 1-2). The individual molecular beads are termed ***monomers,*** and they must be capable of forming linkages with one another head to tail to produce a linear sequence of many units, referred to as the ***polymer.*** Because of their very great molecular weights, which range as high as the millions, biologic polymers are frequently referred to as ***macromolecules.***

Monomers, the units for polymerization, include the sugars which form ***polysaccharides*** (Fig. 1-3A), the amino acids which form the ***polypeptides*** or ***proteins*** (Fig. 1-3B), and the nucleotides which form the ***polynucleotides*** or ***nucleic acids*** (Fig. 1-3C). In most cases, the sugar units of the polysaccharides are all identical, as in the cases of starch, glycogen, and cellulose, in which all of the monomers are glucose molecules. The only requirement for the synthesis of these polysaccharides is, therefore, a source of energy to promote the formation of the bonds linking the glucose monomers together.

The proteins and the nucleic acids, however, are composed of nonidentical monomers; the amino acids in polypeptide chains and the nucleotides in polynucleotide chains are joined together in unique sequences that are characteristic of each macromolecule. Accordingly, their biosynthesis requires not only the energy to activate bond formation between the monomers, but also an organizing principle that will ensure that the correct sequential order of the polymers is faithfully reproduced. The elucidation of the mechanisms for the assembly of these macromolecules has been the most exciting and fruitful single advance in modern biology and has led to a revolutionary concept of the genetics and chemistry of living organisms at a molecular level. The following brief account is intended to set the stage for further discussions of the genetic control of metabolism and the nature of inherited metabolic diseases.

BIOLOGIC INFORMATION TRANSFER

Just as linguistic information is conveyed by the sequence of printed letters on a page, the transfer of biologic information is conveyed in terms of linear sequences of monomeric units in macromolecules. We can exclude the polysaccharides as carriers of information; their repetitive sequences of monomers makes for dull reading, with no possible variations

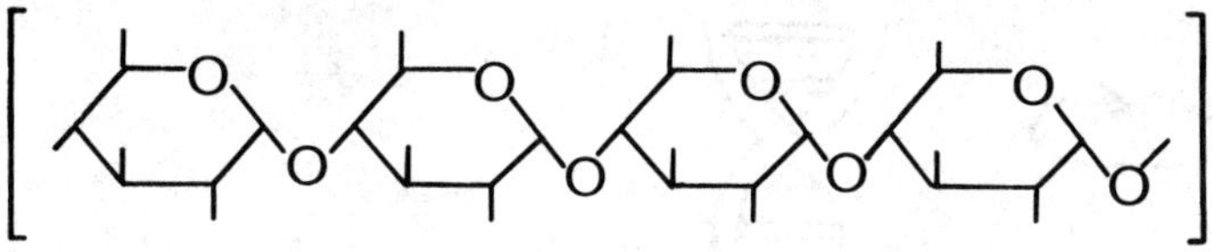

A. Polysaccharide

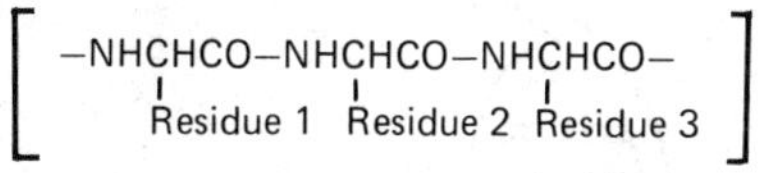

B. Polypeptide or protein

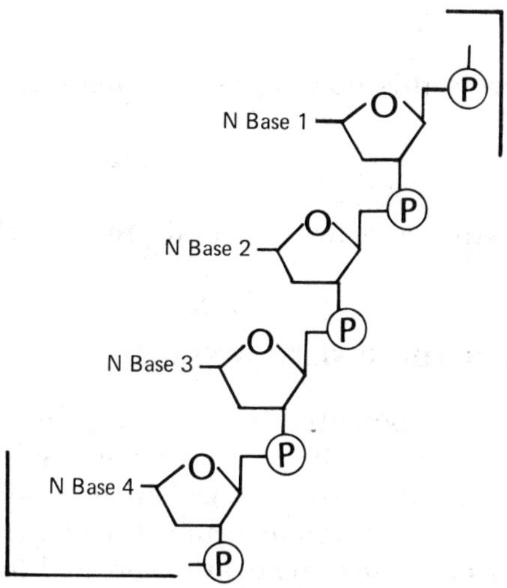

C. Polynucleotide or nucleic acid

FIG. 1-3. Macromolecules.

in the "message." The nucleic acids and the proteins, however, may serve as discrete memory banks of information storage in the cell, each linear sequence corresponding to a definite set of instructions about how that cell will be structured and programmed to perform its functions.

The impetus for these instructions derives from the ancestry of each individual, in the genetic messages he has had handed down through both of his progenitors. Modern biology has established that the substance of these genetic messages is contained within those particular nucleic acids of the ovum and the sperm, designated as the ***deoxyribonucleic acids*** or DNAs. It is also clear that the ultimate expression of these instructions from our genetic origins lies in the amino-acid sequences of the structural and functional proteins of our cells. Between the origin and the embodiment of these instructions for the living process lie the ***ribonucleic acids*** or RNAs, acting as both vehicles and translators in the information trans-

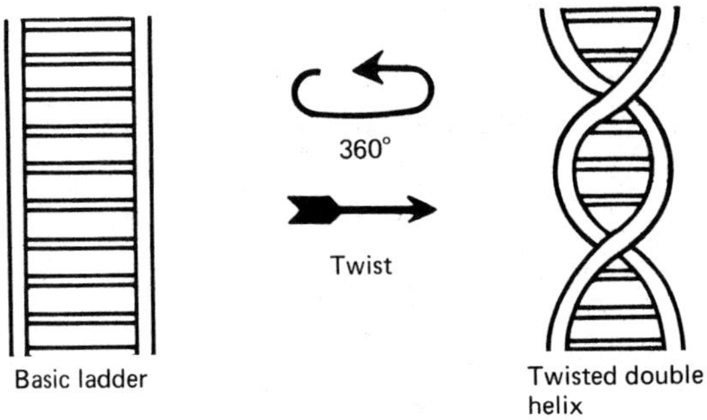

FIG. 1-4. DNA double helix.

fer. It is customary to express this flow of biologic information by the shorthand notation:

DNA → RNA → Protein

This simple statement constitutes a fundamental credo of the modern molecular biologist.

DNA AND THE REPLICATION OF THE GENETIC MESSAGE

The fertilized ovum contains the potential for the development of all the features and functions of a mature individual. This potential, referred to as the ***genotype,*** has its molecular basis in the unit of inheritance, the ***gene;*** the latter consists of a DNA macromolecule. The characteristics of the individual, referred to as the ***phenotype,*** are expressed finally in the functional and structural proteins of the cells; the DNA molecule of a given gene specifies the form of each polypeptide.

Thus, in order for the phenotypes of successive generations of individuals to possess the same basic characteristics, a mechanism must exist for the faithful transmission of the genetic message. Moreover, in order that successive generations of cells within the same growing individual may carry the instructions for differentiation, a similar precise copying of the genetic message must occur with each cell division. This process is known as DNA ***replication*** and the remarkable fidelity of the copying system is an intrinsic property of the DNA molecule. To understand how the message is transmitted, it is first necessary to examine its unique structural basis.

To visualize the DNA structure, picture a ladder with ten rigid rungs and with very flexible uprights (as in a nautical rope ladder); then, with the bottom rung anchored in fixed position, twist the top rung counterclockwise in a complete circle through 360 degrees (Fig. 1-4). This will provide a good facsimile of the helical twist in the two-stranded DNA structure, which is referred to as the *double helix.* In reality, each strand

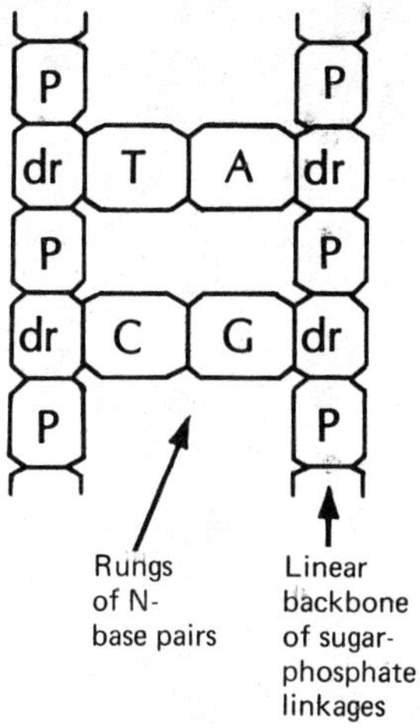

FIG. 1-5. Backbone repeating structure in DNA. (P, phosphate; dr, deoxyribose; A, adenine; T, thymine; G, guanine; C, cytosine.)

(represented by the rope of our model) is a chain of the deoxyribose components in DNA joined together by phosphate bridges to form the linear backbone of the polynucleotide (Fig. 1-5). Attached to each deoxyribose and projecting inward is one of four nitrogen-ring structures—*adenine* (A), *guanine* (G), *cytosine* (C), or *thymine* (T)—which complete the monomeric nucleotide units. Rungs joining the two strands of polynucleotides are formed in two specific ways: by the adenine of one strand linking to a thymine of the other (A–T) or by the guanine of one strand combining selectively with a cytosine of the other (G–C). Although these cross-linkages are extremely specific (*i.e.*, A or T never links to C or G), they are individually quite weak in comparison with the strong phosphate bridges along the strands. The cohesive attraction between the strands of the double helix is due to the summation of all these weak forces, as the two halves of a zipper are held by the multitude of weak interactions of individual teeth. Consequently, although the integrity of the individual DNA strands is seldom broken in life, the two strands may separate on occasion by "unzippering" the A–T and G–C linkages. The latter transverse combinations joining the two strands together are referred to as ***complementary base pairs*** (Fig. 1-6). The restrictions placed upon these faithfully exclusive matings not only serve to maintain the double-stranded structure but, more importantly, also ensure that copies of the nucleotide sequence in DNA are precisely reproduced when the genetic message is to be transmitted.

The cells of our bodies are *diploid,* that is to say, with the exceptions of spermatozoa and unfertilized ova, each cell contains two genes for each characteristic, one from each parent. The amount of DNA in each cell will remain constant except at the time when the cell divides. Just prior to cell division, each DNA molecule will precisely replicate itself to provide two identical genetic messages, one for each daughter cell. The mecha-

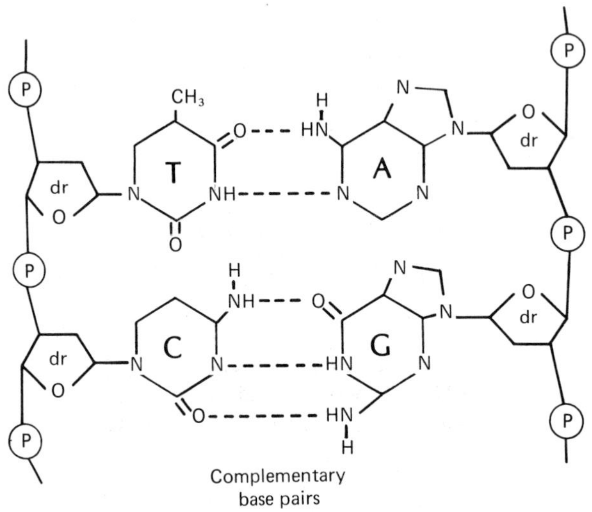

FIG. 1-6. Complementary base pairs in DNA.

nism for DNA replication involves the operation of a particular enzyme, *DNA polymerase,* which acts to unzip the two strands and insert new incoming nucleotides in place of those originally present in the opposite strand (Fig. 1-7). Since the same rules for combining new base pairs will apply as for forging the links between the two original strands, the new strands will be exact counterparts of those they displaced. An external source of energy input is required, as noted above for polysaccharide formation, to join the deoxyribose sugars together by forming the phosphate bridges along the new strand, but the information necessary to specify the sequence of monomers in the new strands is embodied in the original molecule, which acts as the template for its own renewal.

The concept of the template—the complementarity of fit between opposing strands—is central to our understanding of both gene duplication and gene function. The interlocking halves of the double helix serve as the positive and negative faces of a mold: when pulled apart, new material entering the positive contour forms an exact replica of the negative form, and vice versa. The opposite strands are seen to be negative images or complements of one another, and once the sequence of bases in one strand is established, the other strand must follow the pattern set by its opposite number. Thus the template principle allows for an exact copy of the genetic message to be transmitted between generations of cells, and it also provides the means whereby that message may be read within the daughter cells, as will be detailed in the next two sections.

RNA AND THE TRANSCRIPTION OF THE GENETIC MESSAGE

Although the DNA molecules serve ideally as repositories of genetic information due to their stability, complexity, and immense size, these

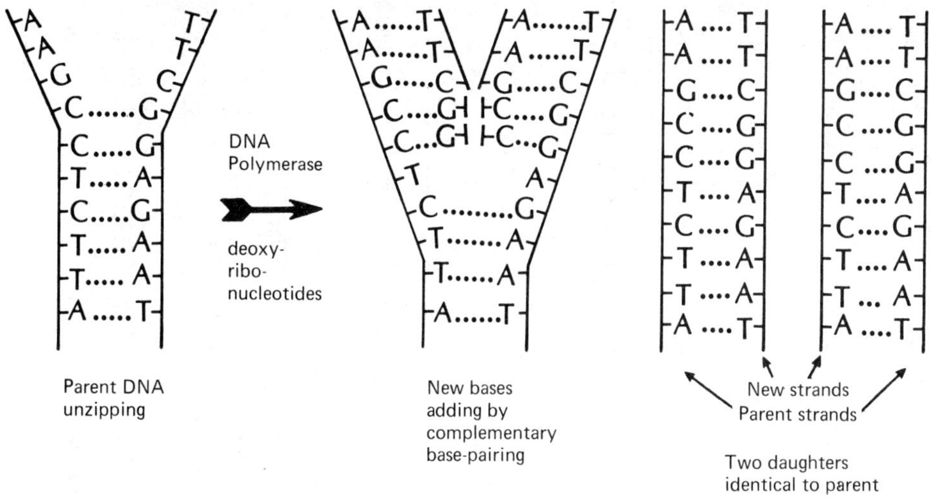

FIG. 1-7. Replication of a DNA molecule to form two identical double-stranded molecules.

very properties make them unsuited to act directly in the transmission of information within the cells. The bulky DNA molecules are complexed together with basic proteins in compactly coiled structures known as ***chromosomes,*** and these are confined within the membrane-enveloped nucleus of the cell. Since the events involved in reading the genetic message occur in the cytoplasm outside the nuclear envelope, it is apparent that intermediate agents must be employed to transport the information embodied in the DNA molecules. These agents are relatively short-lived molecules of RNA, which, because of their function, are termed ***messenger RNA.***

The formation of messenger RNA is simply a means of converting the DNA template into a mobile form. The process involves transcribing the information into a more readable format though in the same language, namely, the order of nucleotide bases in a chain. In the ***transcription*** of the gene, there are parallels to replication, but several key differences are found as well (Fig. 1-8). First, the two strands of DNA separate but only one serves as a template for production of the RNA; thus the resulting messenger is only single-stranded. Second, the reaction is catalyzed by a different enzyme, DNA-dependent RNA polymerase, which uses only those nucleotides that contain ribose instead of deoxyribose. Third, the rules of base pairing are followed in inserting the new nucleotides but with uracil (U), the RNA equivalent of thymine, taking the place of T in the sequence. Finally, the timing and the control of the transcription process are fundamentally different from those for replication, which occurs solely at the point of cell divisions. Messenger RNA will be synthesized throughout the life of a cell, not in relation to the generation of new cells, but in response

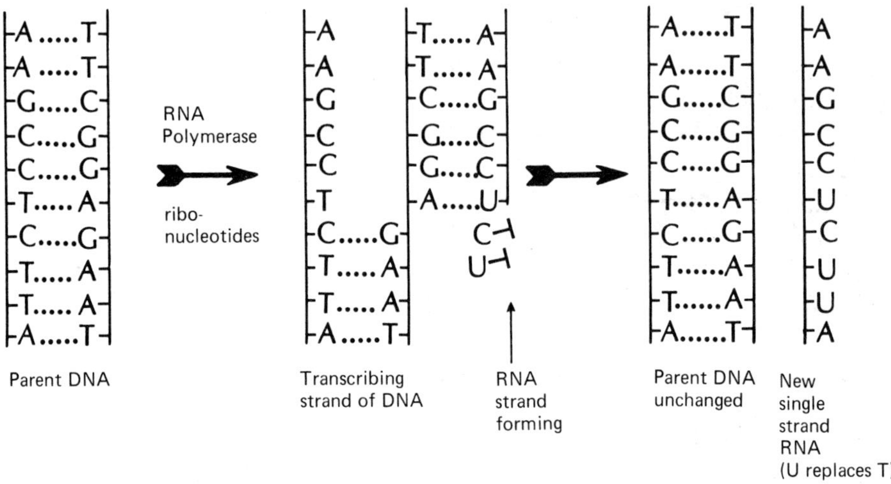

FIG. 1-8. Transcription of one strand of the DNA molecule to form single-stranded RNA that is identical in base sequence to one strand of the DNA, but with U (uracil) replacing T (thymine).

to the needs of existing cells for new proteins. Moreover, in contrast with DNA replication which involves total duplication of all the genes simultaneously, transcription will produce a copy of a single gene, often independently of all the other genetic information, to result in a discrete messenger RNA. In terms of the lifetime of the cell, the latter will have but a fleeting existence, unlike the enduring DNA molecules locked inside the nucleus. Upon passage through the pores in the nuclear membrane into the cytoplasm to its site of action in protein biosynthesis, the messenger RNA "struts and frets his hour upon the stage," and then is broken down to its component nucleotides.

PROTEIN SYNTHESIS AND THE TRANSLATION OF THE GENETIC MESSAGE

If the events in DNA replication and transcription are the tale of the nucleus, then the process of translation is the saga of the ribosome. This tiny spherical particle, visible in electron micrographs of the cytoplasm of all cells from the most primitive bacterium to the most complex nerve cell, is in reality a gigantic conglomerate in molecular terms. In the structural hierarchy of the subcellular components, the ribosome ranks between the macromolecule and the organelle; it contains several dozen macromolecules (proteins and nucleic acids) assembled together in organized fashion, yet it lacks the membrane envelope that is characteristic of organelles such as the nucleus or the mitochondrion.

The ***ribosome,*** so designated because of its high content of RNA, consists of two different components or subunits corresponding to the two

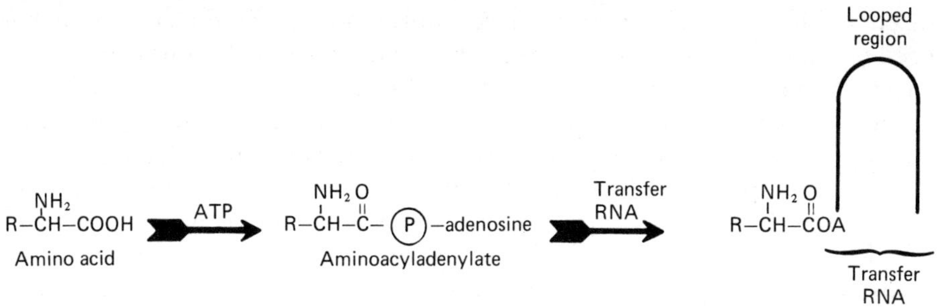

FIG. 1-9. Activation of an amino acid by ATP, and its transfer to the specific transfer RNA.

functional tasks it performs: the smaller subunit participates in *information transfer* in the decoding of the genetic message from the nucleus; the larger subunit acts to organize the *energy transfer* required for joining the amino acids together to form the polypeptide chain. The term ***translation*** is appropriate to this overall process, since the information of the messenger RNA, framed in a polynucleotide sequence, is expressed by the ribosome in a different language, the sequence of amino acids in the protein being synthesized. Both the recognition of amino acids for information transfer and their activation for energy transfer require the cooperation of soluble participants from the cytoplasm that act in concert with the ribosome and the messenger RNA.

Amino acids, which are the building blocks for protein synthesis, are found in free solution in the cytoplasm of the cell. This reserve of construction material is termed the *free amino-acid pool* of the cell. Since amino acids serve in many additional functions of the body besides the formation of proteins, the synthesis of the latter will be dependent upon the overall supply and demand for the amino acids. This entails that those amino acids not synthesized by the body *(essential amino acids)* be provided in adequate amounts by the diet, and that a total amount of protein sufficient to provide for the synthesis of other amino acids *(nonessential amino acids)* be ingested as well. In addition, since the biosynthetic process requires energy input and since amino acids themselves will be tapped off to provide energy when other sources fail, it is essential that a sufficient dietary caloric intake be maintained in the form of carbohydrates and fats to promote the maximal utilization of amino acids for biosynthesis of proteins.

Provided that these necessary conditions are met, the individual amino acids from the cytoplasmic pool may be activated and transferred to the ribosome for assembly (Fig. 1-9). The activation occurs in the cytoplasm and requires the presence of catalytic amounts of a selective enzymatic protein as well as stoichiometric amounts of an energy source, which is provided by the nucleotide, ***adenosine triphosphate*** (ATP). As discussed further in Chapter 3, energy stored in the ATP molecule may be used for many purposes in the body. In the activation of amino acids, the energy is conserved by a very reactive intermediate formed through the carboxyl group, *aminoacyl adenylate.* This highly unstable compound exists only momentarily and is bound to the enzymatic protein. The latter catalyzes the passage of the active aminoacyl moiety to a small, soluble molecule of RNA, which also occurs free in the cytoplasm. Since it is in this form that the amino acid, still energized at its carboxyl group, is transferred to the ribosome, the carrier molecule is known as ***transfer RNA.*** It is also in this form that the amino acid is recognized and lined up at the appropriate position in response to the genetic message during polypeptide assembly.

Transfer RNA possesses several properties that equip it ideally for its carrier and recognition functions: first, its molecular size is small, consisting of only about 80 nucleotides versus several hundred for messenger RNA, and it therefore is very mobile; second, it has such a high degree of selectivity that a given transfer RNA can accept one and only one activated amino acid (as a corollary, it follows that for each amino acid that may be incorporated into a protein molecule, there is a distinct transfer RNA molecule); third, despite the uniqueness of each of the transfer RNAs, they all have a common feature at the one end where the energized amino acids become linked by an activated carboxyl group to the terminal adenosine; finally, and most importantly, at a bend in the middle of the transfer RNA, remote from the amino acid, there is a looped region possessing a particular sequence of nucleotides that provides the unique recognition site for the interaction of the appropriate amino acid and its designated position along the genetic message.

CODONS AND ANTICODONS

Before proceeding any further with the mechanics of polypeptide chain formation, it is necessary to consider the cryptography of the genetic message, the so-called *genetic code,* and its unit "words" or codons. The language of the gene is in reality a very simple one. Each word that corresponds to a discrete bit of information is comprised of only three letters, with each letter designating one of the four nucleotide bases in a DNA strand (*i.e.,* A, adenine; C, cytosine; G, guanine; T, thymine). When the DNA strand is transcribed, it produces messenger RNA that is its mirror image with "complementary" RNA bases (*i.e.,* U, uracil; G, guanine; C, cytosine; A, adenine, respectively). The three-letter word in the messenger RNA sequence specifies how a particular amino acid will be lined up at the ribosome assembly point, and hence this triplet of bases is named the ***codon.*** The lining up is accomplished through the mediation of the transfer RNA for that particular amino acid. At its recognition site in the exposed loop of its midriff, the transfer RNA also has three bases that are complementary to those of the appropriate codon, and hence this opposite triplet of bases, which may link selectively with the codon, is termed the ***anticodon.*** An anticodon will therefore be a replica of the triplet sequence in the DNA strand that is responsible for transcribing the messenger RNA.

The following provides an example of some codons and corresponding anticodons:

Nontranscribing strand of DNA	— TCT	— GCT	— GTA	— TCC	—
Transcribing strand of DNA	— AGA	— CGA	— CAT	— AGG	—
Messenger RNA codons	— UCU	— GCU	— GUA	— UCC	—
Transfer RNA anticodons	— AGI	— CGI	— CAI	— AGI	—
Amino acids	$serine_1$	alanine	valine	$serine_2$	

Several features of the coding system are illustrated. Note first that a different letter (I, inosine) creeps into the last place of many anticodon words. This provides some flexibility or "wobble" in the system due to the fact that inosine may pair up with several bases. Thus, two slightly different codons—UCU and UCC—may line up the same transfer RNA that contains serine, as shown. In fact there are 64 possible codons arising from all the combinations of four nucleotides in triplets. Allowing for "wobble," some of the 20 amino acids incorporated into proteins may use two or more triplets, thus accounting for 61 different codons; the remaining three code for no amino acids, but act as signals either to initiate or to stop reading a sequence.

The second major point to note is the consequence of changing a letter in one of the codons. This could arise from some chemical or physical modification such that one of the bases in DNA is replaced by another. A change of this type is termed a ***mutation*** of the genetic material. If, for instance, it resulted in altering a triplet in the transcribing strand from *A*GA to *C*GA, the amino-acid sequence of our example would contain two alanines instead of two serines. The effects of this single amino-acid substitution among the several hundred in the finished polypeptide chain may be trivial if that part of the protein is not crucial to its function. If the serine in our example were essential, as it is in the catalytic action of many enzyme proteins, or if it were necessary to provide the proper structure and shape of the protein, then consequences of the mutation would be severe, possibly even fatal, for the afflicted person, as will be described later in detail.

PUTTING THE SEQUENCE TOGETHER

The mechanism of reading the genetic code has some of the playback features of a tape-recorder system, with messenger RNA acting as the tape and the ribosome as the head of the recorder. In the absence of a taped message, the ribosome will obviously be "silent," but when the cell nucleus is stimulated to transcribe a particular segment of its DNA, the resulting strand of messenger RNA will pass out into the cytoplasm and commence to feed into the translation mechanism of the ribosome in the smaller of its two subunits. This process is unidirectional and thereby ensures that the tape cannot be read backward to destroy the sense of the message. The taped message proper is preceded by an *initiator codon,* and other factors from the cytoplasm are also needed to get the messenger RNA lined up properly to feed into the playback head. When lined up, the large subunit of the ribosome can join with the small subunit still attached to the messenger tape to form a functional whole. In stages, the tape may then feed through the ribosome, which will read its message one codon—*i.e.,* three nucleotide bases—at a time and thereby add one amino acid at a time to a growing polypeptide chain (Fig. 1-10).

The events may be visualized in terms of the lining up of appropriate transfer RNAs along the messenger tape that bind through their selective anticodon-codon linkages as diagrammed earlier. Joining the adjacent amino acids by peptide bonds will then produce the linear sequence corresponding to the coded message. In practice, however, the ribosome and messenger RNA only join up with two transfer RNAs at any given time. The amino acid attached to the first transfer RNA ($serine_1$ in our example) then leaves and links through its carboxyl end to the free amino end of the amino acid attached to the second transfer RNA (alanine in the diagram). The first transfer RNA is now free to slip off the messenger tape, and the latter will be fed along through the ribosome, shifting the second transfer RNA with seryl-alanine still attached into the first position of the ribosome. This leaves the adjacent codon available for combination with the anticodon of the transfer RNA for the next amino acid of the sequence (*i.e.,* valine). Again the aminoacyl residue occupying the first position of the ribosome (seryl-alanyl) will join to the free NH_2 of the adjacent amino acid (valine) of the newly bound transfer RNA. This is the point in time frozen in Figure 1-10.

Thus the tape slides through the ribosome in discontinuous steps of one codon of the messenger RNA at a time, while the peptide chain, still anchored through the last transfer RNA, will grow in length by one amino acid at a time. When it reaches the end of the message, which is signaled by one of the *terminator codons,* the polypeptide chain slips off and the ribosome may separate into its two constituent subunits preparatory to receiving another message. In electron micrographs, the ribosomes at work in the cell may be recognized by their being combined to the strand of messenger RNA. In practice, one tape may be read sequentially and simultaneously by many ribosomes. Such a string of ribosomes spaced at regular intervals along a single messenger RNA strand is termed a ***polyribosome*** or ***polysome,*** and it is particularly abundant in growing cells.

The synthesis of proteins requires a considerable amount of cellular energy. As noted previously, the energy for the activation of their constituent amino acids comes from a high-energy phosphate compound (ATP)

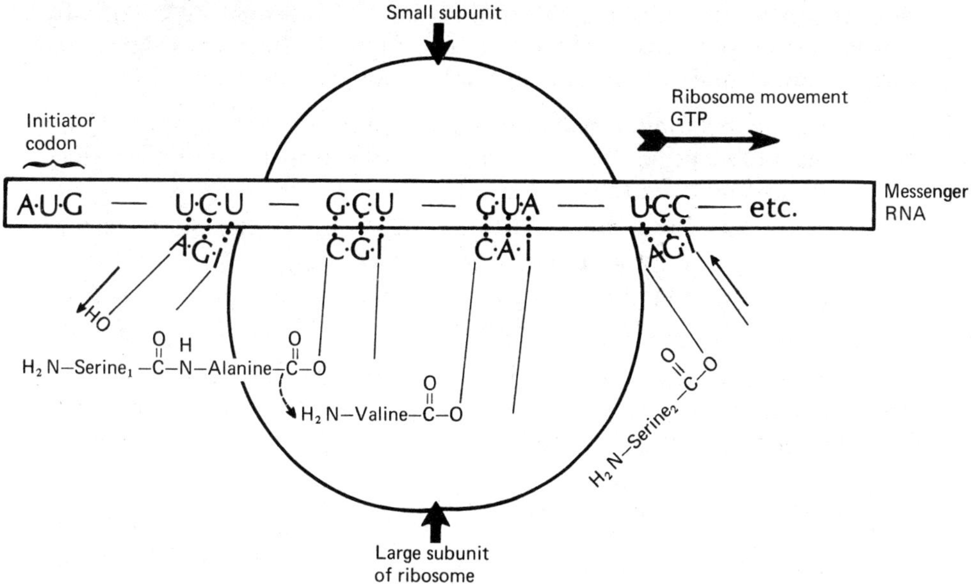

FIG. 1-10. Translation of messenger RNA by the ribosome.

and is conserved in the reactive acyl link to transfer RNA. This is the energy source for the formation of the actual peptide bond in proteins; additional energy input is needed for the mechanical events involved in ribosomal function, such as the binding of incoming transfer RNA and movement along the messenger RNA. The intermediary that drives these processes is another high-energy compound, *guanosine triphosphate,* (GTP). As will be noted in a later section, however, the ultimate energy source to replenish GTP used in this way is ATP, which is the primary, common driving force for virtually all synthetic reactions in the body.

RELATION BETWEEN DNA AND PROTEIN SEQUENCE

The foregoing is a simplified outline of the means whereby the sequence in DNA molecules of our genes predestines the protein sequence. The concept of correspondence in the linear sequences of genes and proteins is a valid, central principle of molecular genetics. However, the colinearity of polynucleotide and polypeptide is far from being exact. If such were the case we might expect to find a strict correspondence between a series of codons in the DNA and the 20 amino acids for which there are matching anticodons in the protein. In fact there are many more amino acid derivatives in the protein end-product, which usually differs in length and composition from expectations, based upon its DNA template. There is obviously considerable editing of messages during transfer of genetic information so that the classical language of DNA may be eventually translated into the vernacular of functional protein structures. These editorial changes may occur at three levels:

1. post-translational modifications made by either clipping or joining polypeptide chains, or by altering substituents on residues projecting from the chains;
2. post-transcriptional modifications made by retailoring messenger RNA molecules; and, less commonly,
3. recombination of DNA segments within the gene prior to its transcription into discrete messages.

Each of these processes requires highly selective enzyme catalysts and is subject to controlling influences in keeping with the physiologic needs or the differentiated state of the cell.

PROTEIN MODIFICATIONS

As noted in the next chapter, the action of enzymes may be regulated by covalent modifications of their polypeptide chains. In several instances proteins are synthesized by cells in inactive precursor forms that may be converted to functional molecules at the appropriate time and place in the body by a limited, selective cleavage, which removes a blocking segment. Such proteolytic modifications are important in the activation of digestive enzymes in the gastrointestinal tract, of clotting factors in the bloodstream, and of the precursor forms of polypeptide hormones (*e.g.*, insulin). Moreover, many secretory proteins are first formed with leading "signal" segments that allow them to penetrate membranes and are then clipped off once the protein has crossed the membrane barrier. Important also in the secretory process are the additions of sugar residues to produce the characteristic recognition groups of glycoproteins, and the addition of hydroxyl groups on proline and lysine residues during the synthesis of collagen by fibroblasts. Many other special functions of proteins are mediated by coupling particular factors to their amino acid residues, as in the iodination of tyrosine rings during synthesis of thyroid hormones, the phosphorylation of serine, threonine, or tyrosine-hydroxyl groups of regulatory enzymes, and the addition of vitamin coenzymes (biotin, pyridoxine, retinal) to lysine ϵ-amino groups. As well as the addition reactions that may alter the charge ratios, conformations, and catalytic actions of the polypeptide chain, there are a number of modifications that will bring about linkages within (intrachain) or between (interchain) polypeptides. Of greatest significance as cross-linking agents are the cysteine residues, which are capable of undergoing reversible oxidation to the diaminoacid, cystine, to form a disulfide bridge:

$$\text{Chain 1} - \text{cySH} + \text{HScy} - \text{Chain 2} \rightleftarrows \text{Chain 1} - \text{cyS} - \text{Scy} - \text{Chain 2} + 2\text{H}$$

NUCLEIC ACID MODIFICATIONS

Most mammalian genes that have been sequenced have intervening spacer DNA stretches ***(introns)*** that divide them into several pieces. Only the expressed DNA stretches ***(exons)*** are copied in the final messenger RNA. Thus, although the RNA polymerase reads the whole sequence of the gene during transcription, there is a subsequent cleavage to remove the introns, and an end-to-end linkage of the exons. Other post-transcriptional events leading to mature messenger RNA include modifications of the leading sequence ahead of the initiator codon (facilitating its recognition by the ribosome), and addition of a tail on its trailing end (stabilizing the molecule with multiple A residues). Thus there is a very extensive degree of processing involved in the synthesis of messenger RNA, and similar cutting, splicing and other modifications are required for the production of mature species of other cytoplasmic RNAs, such as ribosomal RNA or transfer RNA.

Changes in the DNA by the above reactions would clearly violate its role as a stable carrier of the genetic information from cell generation to generation. However, there are experimental techniques available whereby sections of DNA may be cut out of one genome and spliced back into another by the process known as *DNA recombination.* So far such genetic engineering has been most fruitful in transferring human ***recombinant DNA*** into bacteria. This transfer may be effected by insertion of the gene coding for a particular human protein into the DNA of a bacterial virus or other infectious agent. The bacteria are then infected with the agent, which carries the human gene along as a passenger. Repeated divisions of the infected bacteria in culture will thereby replicate multiple identical copies or clones of the same human gene. Provided that suitable conditions are met for the expression of the inserted DNA, the gene cloning technique may yield a potentially unlimited source for production of the particular human protein. The pharmaceutical industry is currently involved in the development of synthetic or appropriately redesigned DNA sequences that could be cloned, transcribed and translated by readily cultivated microorganisms, producing such medicinally important proteins as human interferon, growth hormone and insulin.

It has also been possible, for example by coupling DNA to animal viruses or by cell-fusion techniques, to transfer genes or whole chromosomes from one human cell to another. Such technology eventually may provide the means for a type of gene transplantation that could correct for inherited protein defects or enzyme deficiencies in the inborn errors of metabolism. An extremely important application of this sort of transfer of human genetic information is the development of monoclonal antibodies. In this technique an antibody-producing lymphocyte, which can generate a single immunoglobulin that is specific for one antigen, is fused with a cancerous lymphoma cell that has high capacity to grow and divide but not to produce antibodies. The resultant "hybridoma" cells retain properties of the parents, namely, to synthesize and secrete large quantities of the particular species of antibody and also to reproduce indefinitely in the culture medium.

Recombination of DNA can occur naturally during meiotic divisions of germ cells, but happens rarely in the somatic cells of the body. One exception is the somatic rearrangement of DNA in the lymphocytes that secrete antibodies. Our immune cells must be able to respond to millions of foreign antigenic substances in order to produce the corresponding neutralizing antibodies. However, the germ cells that develop into the antibody-producing cells contain none of the mature genes for the antibodies, but rather a flexible, do-it-yourself kit of gene fragments. As the lymphocyte differentiates to meet the challenge from exposure to some particular antigen, it must select the appropriate DNA components from the kit in order to assemble the gene for an appropriate antibody protein specially designed to bind the antigen. Thus the enormous diversity of antibody proteins arises from recombination of DNA fragments chosen from a limited pool of variable antibody genes present in the germ–line. The extent to which this type of rearrangement of genes may participate in the irreversible determination of differentiated cells outside the immune system is not known at present. However, it is becoming clear that the diversity of expression of the total genome, which is present in embryonic cells during somatic cell differentiation, can involve the reshuffling of the DNA as well as its RNA transcripts.

PROTEINS AND THEIR FUNCTIONS

In the preceding section, it was noted that the molecular architecture of living cells is the main feature that distinguishes them from inanimate matter. Of all the macromolecules contributing to this unique biologic superstructure, the proteins play the most significant part. Not only are they the most bountiful organic component of the human body, constituting over half of the solid content of most cells, but they also possess the widest range of variation with respect to size (ranging from several thousand to millions in molecular weight), shape (from long rods to compact spheres), and physical properties (from water-soluble to fat-soluble to totally insoluble). They also differ widely in their physiologic roles, acting as catalysts, mobile carriers, or structural components. It is biologically meaningful to classify the proteins in terms of the categories of their functions in the body only, and that is the rationale followed below.

It requires only a cursory survey of the major functional groups of proteins and a few examples to be impressed by the importance of proteins in human health and disease. First and foremost are the vast numbers of ***enzymes,*** protein molecules that have been designed to combine selectively with other molecules in cells and thereby catalyze reactions to provide new products and the energy to sustain life. In the final analysis, all physiologic processes in our bodies require enzymes to promote chemical changes that would not otherwise occur at temperatures compatible with survival. Digestive enzymes carry out the breakdown of foodstuffs in the intestinal tract to smaller absorbable parts (failure to produce such enzymes, *e.g.*, in cystic fibrosis, will result in serious malnutrition due to defective digestion). Blood coagulation is dependent upon certain enzymes to initiate the formation of the clot in response to injury (absence of one of these clotting factors can result in a bleeding disorder known as hemophilia). Tissue energy production is governed by certain respiratory enzymes (the inhalation of toxic gases such as hydrogen cyanide that interfere with such enzymes can prove fatal because of a deficiency of cellular energy). Other enzymes maintain the excitability and contractility of neuromuscular tissues (poisons or pharmacologic agents that act upon such enzymes may therefore produce anesthesia, paralysis, or convulsive disorders). Finally, and of relevance to clinical diagnosis, many enzymes may be released into the blood stream following damage to a particular tissue or organ; the detection in a blood sample of large amounts of an enzyme that is normally confined to heart muscle can, for example, be used to monitor cardiac disease.

Second are the regulatory proteins that control many body functions, the *polypeptide hormones.* In several instances these hormones control the synthesis or the activity of enzymes, thus providing a finely tuned mechanism whereby the body may respond to its circumstances. Some polypeptide hormones, such as those secreted into the blood from the pancreas, serve as chemical messengers, inducing other tissues to maintain a balance between the use and the production of blood sugar. The absence of one such hormone, insulin, results in diabetes mellitus, a disease in which blood sugar levels rise uncontrollably following a carbohydrate meal.

Third, there are the proteins that play a *mechanical or structural role.* These include the contractile proteins of the muscles and the strong fibrous proteins of the elastic and bony frameworks in the body known collectively as the connective tissue. A defect, for example, in the production and maintenance of one such connective tissue protein, collagen, will result from a deficiency of vitamin C in the body and will produce the clinical symptoms of weakened blood vessel walls that are observed in scurvy.

A fourth function is performed by a diverse group of proteins that bind with other substances and act as *carriers* in the bloodstream. Cardinal among these is the ***hemoglobin*** of red blood cells, which acts not only to carry oxygen from the lungs to tissues but also to carry carbon dioxide in the opposite direction. A deficiency of the iron necessary for the formation of hemoglobin will result in anemia and a reduced ability of the blood to supply oxygen to the tissues for energy generation. Carrier proteins of many diverse types exist to transport insoluble lipids, steroid hormones, and metals between tissues through the bloodstream and even within the cells in the aqueous cytoplasm. The detection of abnormal concentrations of carriers such as the lipoproteins in blood may provide valuable information in the diagnosis of diseases of lipid metabolism, for example.

Fifth and last is the category of the ***immunoproteins***, which form the first line in the defense of the body against invaders from without. Foreign proteins or organisms that reach the bloodstream will stimulate the lymphoid tissues to generate specific ***antibodies***, which are proteins that have the unique capacity to bind and inactivate the foreign invader substance or ***antigen.*** In bacterial infections, for example, the amounts of immunoproteins present in the γ-globulin fraction of the circulating proteins will be grossly increased to meet the threat of the invading organism.

LEVELS OF ORGANIZATION OF PROTEINS

Understandably, the student may interject impatiently at this point, "Enough! I am convinced that proteins are important compounds. But please spare me the agonizing details of all their three-dimensional tortuosities. Get to the point of what they do and why. Let me out of these peptide chains, release me from hydrogen bondage!" I could take the coward's way out and tell him to skip the next section. Instead, I have chosen to give a bare-bones account of the main features of the structures of human proteins, for just as it is essential to know something of the main features of the body's structure to comprehend its physiology and pathology, so must we appreciate the basic anatomy of the proteins before considering the mechanisms and implications of their functions. There are four key levels of organization within proteins, and these are quite logically referred to as their *primary, secondary, tertiary,* or *quaternary* structures.

Primary Structure

The *primary structure* refers to the linear sequence of amino acids in the polypeptide chain. We can tell something about the characteristics of different proteins in terms of the total amounts of each of the amino-acid components, but apart from dietary criteria, this type of analysis is not too helpful. It is known, for example, that other factors governing protein shape and function depend upon precisely where certain key amino acids appear along the chain. It is not enough to state that a particular catalytic function requires a serine molecule in the chain; the crucial serine must be exactly 195th from the end of the polypeptide chain, for example. Moreover, the way in which proteins coil and fold upon themselves will depend on the precise arrangement of amino acids in neighboring chains. In fact, once the linear sequence of the protein has been set, it is fair to say that theoretically, all the other aspects of the form and performance of that molecule may automatically fall into place.

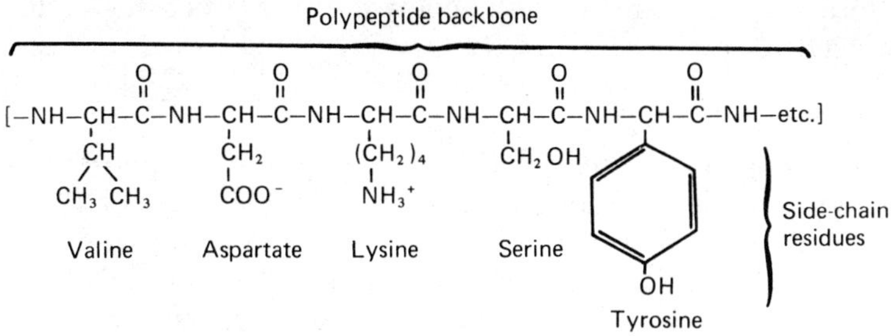

FIG. 1-11. Primary structure of proteins.

Just as all the nucleic acids have a common backbone held together by sugar-phosphate linkages, so too the proteins all have an identical peptide backbone. The differences among proteins hinge upon the differences in the distinct amino-acid residues that project sideways from the linear polypeptide chain (Fig. 1-11). It is for this reason that these residues (R) are termed the *side-chains* of the peptide. When a majority of these side-chains in a protein are aliphatic residues (*e.g.*, alanine, valine, and leucine), the overall properties of the protein will be hydrophobic; that is, it will tend to associate with the lipids in cells to form a fixed membrane component or possibly with a circulating carrier lipoprotein. On the other hand, when the protein possesses an abundance of side-chains with free ionic charges, it will be more hydrophilic and enter into salt linkages with other, oppositely charged ions.

If those amino acids with free negative charges (*e.g.*, aspartate and glutamate) predominate, the whole protein will be acidic at the neutral *p*H of body tissues; if there is a preponderance of positively charged side-chains (*e.g.*, lysine, arginine, and histidine), a basic protein, such as the histones that associate with DNA, will result. In practice, the proteins may be readily differentiated on the basis of slight differences in their net charge. A process known as *electrophoresis* permits their separation according to their varying rates of migration in an electrical field at various *p*Hs.

Alteration of the charge on particular amino acids in proteins may cause profound alterations of their biologic activity in the body. The most dramatic evidence of this phenomenon is the effect of adding a phosphate group: in the nucleus, for example, the association of DNA with chromatin histones may be disrupted to unleash the production of messenger RNA and thereby stimulate the synthesis of proteins. In the case of hemoglobin, the replacement of a single, negatively charged residue by a neutral amino acid can produce such a drastic structural alteration of the protein molecule—and hence of the red blood cell—that the person afflicted by this inherited disorder, known as sickle-cell anemia, may die. In other instances, the catalytic function of certain regulatory enzymes may be either totally prevented or fully activated by phosphorylation or dephosphorylation. Such alterations are central to the control of life processes in adaptation to the needs of the body.

Secondary Structure

As in the case of nucleic acids, the polypeptides tend to coil upon themselves in a corkscrew or helical form. Unlike the double helix of DNA, however, the polypeptide coil is formed within a single chain by interactions at every fourth peptide linkage. This tight coil is termed the α *helix* of proteins (Fig. 1-12A). It occurs to varying degrees in different proteins, and it causes the polypeptide chains to become more rigid than they are when extended in the less-ordered form known as the *random coil.* The presence of certain animo acids whose side-chains interfere with one another—either due to the large size of the residues or to repulsion by like charges—will prevent the helix from forming and favor the random coil (Fig. 1-12B). The amino acid, proline, cannot participate with other amino acids to form an α helix, and where it occurs, the polypeptide chain will form a twist or loop that interrupts the helical form.

In another type of arrangement, called a β *structure,* the polypeptide chain is fully stretched with the side-chains projecting upward or downward. Such a structure can give rise to a regular packing of polypeptide chains in a pleated sheet (Fig. 1-12C). Different proteins will contain varying proportions of each of the various secondary levels of organization. It should be noted that the α-helical and β-pleated sheet conformations provide a regular ordering and stability that is most important in providing strength and rigidity for structural purposes, such as in the connective tissue proteins.

Tertiary Structure

The total protein structure in the body is determined not only by the strong covalent linkages of the peptide bonds (which determine the primary structure) and the weaker, repeating linkages involving nearby amino acids in the polypeptide chains (which establish the secondary configuration), but also by interactions between residues at a considerable distance, in terms of the linear sequence, from one another. The resultant bending and looping of the protein molecule back upon itself in the manner of a pretzel is referred to as the *tertiary structure* (Fig. 1-13). Several types of interactions maintain this configuration: first, there is the abhorrence of the hydrophobic residues for surrounding water, with the result that they tend to cluster together, usually in the interior of the protein molecule. Second, oppositely charged residues will exert various degrees of electrostatic attraction for one another at different places along the chain. Third, a very strong bond may form when the –SH groups of two cysteine molecules come together and are oxidized to produce the –S–S– disulfide bridge. These and other weaker attractions between particular amino-acid side-chains, often at a great distance from one another in the

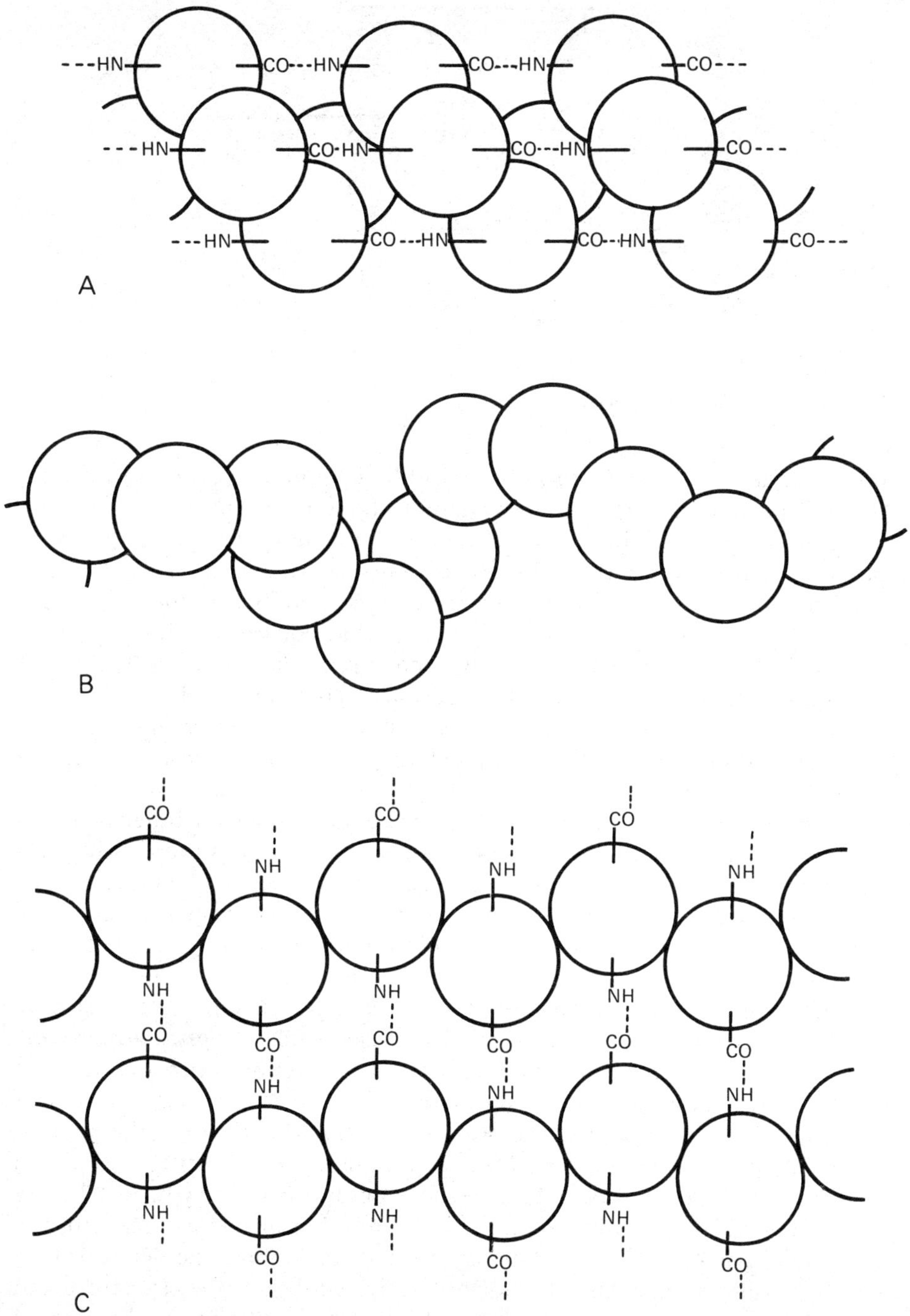

FIG. 1-12. Secondary structure of proteins. **A.** The α helix. **B.** The random coil. **C.** The β-pleated sheet.

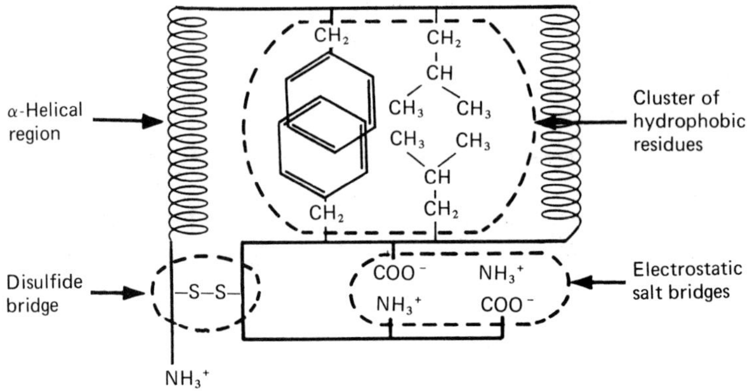

FIG. 1-13. Tertiary structure of proteins.

linear sequence, cause the protein molecule to adopt a unique, convoluted, three-dimensional shape. This particular shape is crucial for the normal function of that protein molecule, and, like the secondary structure, it is dictated ultimately by the primary sequence of the polypeptide chain.

The overall shape of a protein is related to its functional and structural role in the cell. Thus, those proteins that constitute the long fibers in mechanical elements of tissues will have relatively little in the way of loops and bends; rather, more interactions that tend to reinforce the ordered helical forms will occur, thus intensifying the longitudinal extension and rigidity of the molecule. These long, thin molecules are generally referred to as *fibrous proteins.*

A more compact, spheroidal shape is advantageous for other proteins to enable their organization in membranes or their interaction with small molecules. Such proteins may possess regions of ordered structure, either the α helix or the β structure, but kinks and bends braced by disulfide bridges and other cross-linkages will fold the molecules into tangled balls; these structures are referred to as *globular proteins.*

The importance and specificity of folding in protein molecules may be apparent from consideration of the consequences of ***protein denaturation.***. This is a term applied to the effects of high temperature or chemical agents that induce the unfolding of the native tertiary conformation of protein molecules. Usually, the normal functional activity of the protein is lost upon denaturation (see Chap. 2), and often its solubility in biologic fluids will be irreversibly lost, as in the coagulation of the white of an egg when cooked. In some instances, if the unfolding of tertiary structure is conducted gently under carefully controlled conditions, the denaturation can be reversed to restore the normal active configuration. Denaturation of bacterial proteins is the purpose of using heat for the sterilization of surgical intruments, the pasteurization of milk, and so forth. Strong acids or bases, oxidizing agents, heavy-metal compounds, urea, alcohols, and phenols are some of the more potent protein-denaturing agents.

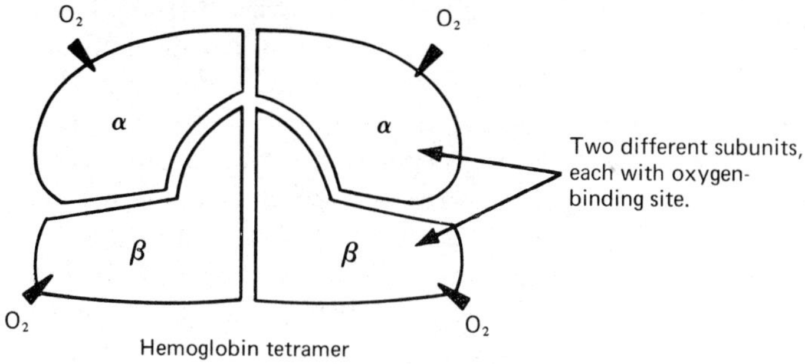

FIG. 1-14. Quarternary structure of proteins.

Quaternary Structure

The fourth and final level of protein organization is of great significance in the control of activity of certain enzymes and other proteins. The functional unit of such molecules consists not of one but of several polypeptides packed together (Fig. 1-14). Each polypeptide chain is referred to as a *subunit*. The subunits may differ considerably from one another, but when they interact in the whole, they will exert an influence upon one another, as will be discussed with regard to regulatory enzymes (Chap. 2).

Another example is the O_2-carrying protein of the blood, hemoglobin, which consists of four polypeptide chains, each capable of binding one molecule of O_2. Rather than acting independently, the subunits interact to bind O_2 cooperatively so that oxygenation of one binding site alters the conformation in adjacent polypeptide chains to facilitate O_2 binding by the neighboring subunit. Thus, hemoglobin has an enhanced affinity for O_2 in the lungs, where most of the binding sites are occupied, but it tends to unload O_2 to the peripheral tissues more readily as the binding sites become progressively vacant.

Not all proteins possess a distinct quaternary structure, but those that do will generally exhibit the cooperative interactions between the subunits that are so important for the regulation of metabolic and physiologic functions in the body.

MEMBRANES AND ORGANELLES

The complexity, diversity, and versatility of living forms derive from the interactions of molecules both large and small. In the preceding account, we have seen how the giant macromolecules can interact to duplicate living systems. It should be apparent that the interactions discussed so far are chiefly *linear* in nature: the monomers of DNA form a chain that corresponds to that of messenger RNA, which in turn corresponds to the chain of amino acids in the proteins of the cell. In describing these correspondences in the long dimensions of these macromolecules, it may be stated that DNA, messenger RNA, and polypeptide are *colinear.* Thus, while we must not oversimplify by implying that width, depth, or other conformational states are of lesser importance, our major preoccupation in the information transfers that determine the sequences of these polymers is basically *one dimensional.* On the other hand, the consideration of biologic membranes focuses on the interactions of smaller molecules that will confer a basic *two-dimensional* organizing principle in cells. The two-dimensional nature of biologic membranes isolates and protects cells from their environment while, at the same time, it provides an immense surface area for many important physiologic processes.

The structure and composition of membranes vary with respect to the tissue of origin and localization within the cell. Nonetheless, there are common features of all membranes. First, they all contain proteins and ***lipids*** in varying proportions; second, it is the lipids that impose the two-dimensional order upon their structure; third, the components have an asymmetric arrangement with respect to the two faces of the membrane; finally, the membranes have a considerable amount of flexibility and mobility in cells, allowing them to distend, to pinch off and divide, and, on occasion, to fuse together. In order to appreciate these features of the membranes, it is essential first to understand something of their component lipids and how they tend to associate together in layers.

The major membrane lipids contain both phosphorus and glycerol and are consequently termed ***phosphoglycerides.*** There are variations upon the basic pattern of their structures (as noted in Chap. 5), but for the sake of our present discussion, it suffices to state that in general, it is characterized by the fact that the small phosphorus–glycerol portion has two long hydrocarbon chains of fatty acids attached to it. The fatty acids are quite incompatible with and insoluble in water ***(hydrophobic),*** whereas the P-glycerol part exists quite happily in water solutions ***(hydrophilic).*** The result of this unusual arrangement of conflicting properties within the same molecule is that when the phosphoglycerides are put in an aqueous environment, such as the major fluids of the body, they will arrange themselves quite spontaneously into two thin sheets or ***lipid bilayers*** (Fig. 1-15). The development of this natural sandwich begins with the desire for the hydrophobic fatty-acid tails to escape from the water; the long, alkyl chains would prefer to associate with each other rather than with the surrounding water molecules, so they pack together lengthwise to form a ***monolayer*** one molecule thick. In the lung a monolayer of the phosphoglyceride, lecithin, acts as a natural surface active agent, coating the alveolus and preventing its collapse upon the expiration of air. In cell membranes, two monolayers come together with the greasy hydrocarbon groups as the filling in the middle of the sandwich and the hydrophilic P-glycerol groups forming outer surfaces that are exposed to an aqueous medium on both sides of the membrane. This is the fundamental bilayer structure that acts as the skeletal framework for all cell membranes to build upon. Because of its hydrophobic core, it retards the pas-

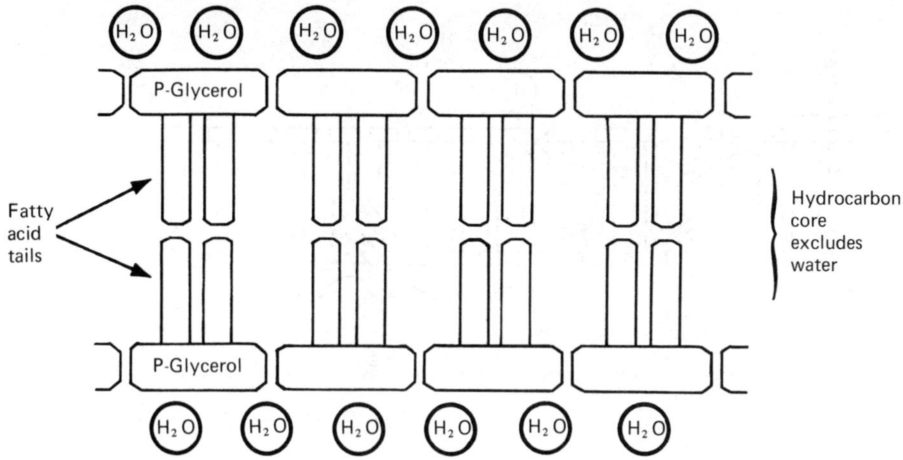

FIG. 1-15. Lipid bilayer structure of biologic membranes that is produced by phosphoglycerides.

sage of many water-soluble compounds, thereby allowing the retention of such essential substances within the cells.

The movement of substances across membranes is a crucial property of living systems. Superimposed upon the permeability restrictions of the lipid bilayer is a host of selective devices that meter the flow of nutrients—for example, sugars and amino acids—into the cells and waste substances—for example, CO_2 and NH_3—out of them. Analogous mechanisms exist to control the gates that permit small ionic species—for example, H^+, Na^+, or K^+—to migrate in or out of cells and thus sustain the electrical properties of tissues such as nerve and muscle. These specific transport properties are conferred upon the membranes by plugging in particular proteins that span the lipid bilayer and thus have access to both sides of the membrane surface (Fig. 1-16). Other protein molecules, depending upon whether they interact only with extracellular or intracellular components, may be located on only one side, bound to either the peripheral or the internal monolayers.

Other membrane proteins are specially designed and located to serve as receptors for hormones or other factors that modify the response of particular cells to their environment. Still others are active catalytically in the processes of metabolism, which are conducted within the highly organized membrane structures of subcellular organelles. There may be some variation in lipid composition (*e.g.*, in the proportions of different phosphoglycerides or in the occurrence of other lipids such as the sterols), but, in the main, the different properties that exist among tissue membranes as well as among subcellular membranes of the same tissue are due to the differences in composition and proportions of their characteristic proteins.

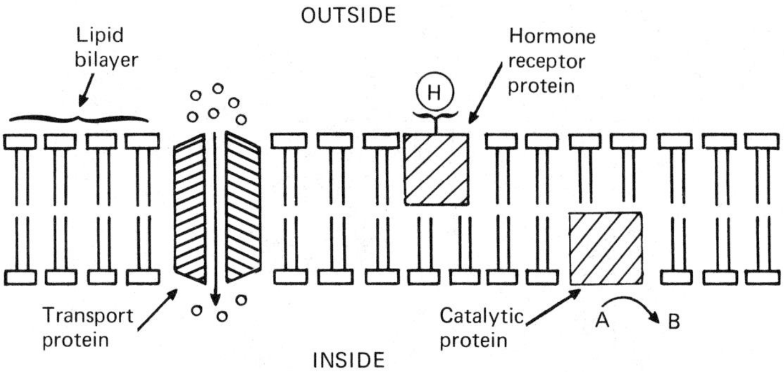

FIG. 1-16. Positions of protein molecules in the bilayer skeleton of cell membranes.

THE PLASMA MEMBRANE

This is the partition that ensures the individual existence of each cell, keeping it discrete from its neighbors and protected from the vagaries of its surroundings. In some cases, for example, that of the red blood cell, the plasma membrane will present a smooth, undifferentiated surface to the outside; in others, for example, in the secretory cells of the kidney or the absorptive cells of the intestine, it may contain a multitude of invaginations or evaginations to increase the surface area and hence facilitate the traffic of substances in or out of the cell (Fig. 1-17). In other situations, such as where two nerve cells touch one another, visible specialized structures may develop at the points of contact between the cell membranes. Beyond the level of detection by microscopy, there are specific molecular adaptations of the outer surface of cell membranes that enable the interactions of particular hormones with particular target cells. Like the bikini, the plasma membrane must reconcile and optimize two opposing functions: maximal exposure to the world together with adequate protection.

To be described later are the energy requirements for the movement of ions (Chap. 3) and the uptake of glucose (Chap. 4) and amino acids (Chap. 6). Each of these transport processes appears to depend upon a reaction of the substance being carried across the plasma membrane with a selective protein *carrier* molecule. Combination of the substance carried with its specific carrier resembles in many respects the reaction of a substrate with its specific enzyme (Chap. 2). It is possible that the energy needed for active transport of substances against concentration gradients is involved with the rotation of the protein carriers from an exterior orientation to the opposite side facing the cellular interior, or with smaller conformational changes of the carriers to alter the binding properties on either side of the membrane. In other instances, where substances that

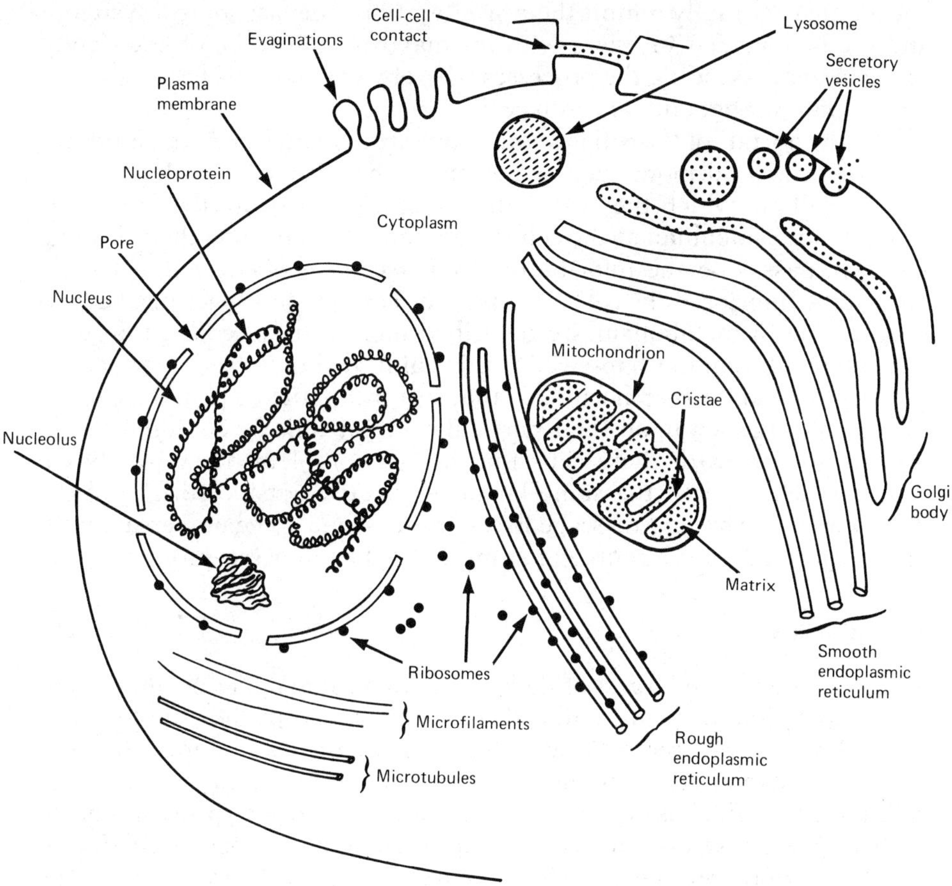

FIG. 1-17. Membranes and organelles of a typical cell.

are not normally soluble in the hydrocarbon interior of the membrane move passively in either direction, the carrier protein may simply provide a transmembrane pore, that is, a hydrophilic channel through the hydrophobic barrier surrounding the cell. Whether substances are passed by such a process will depend on their size and shape in relation to the pores.

Recognition markers on the outer surfaces of plasma membranes frequently involve sugar residues attached to the peripheral proteins ***(glycoproteins)*** or lipids ***(glycolipids).*** Invariably these sugar residues are localized solely on the cell exterior. The recognition points are of crucial importance in the way that the immune system detects foreign cells in the body, in the way that individual cells identify one another in associating together to form a tissue, and in the way that cells in contact with one another mutually inhibit their proliferation. The last control system—and the factors that interfere with it—may turn out to be of key significance in understanding the processes whereby normal cells become transformed into malignant cancerous cells.

At the portals of the cell is another important watchman, the ***hormone receptor***. Many hormones act as messengers between cells without entering the cells to which they carry their messages. Instead, they relay the signal to the watchman at the cell surface, and he in turn, without leaving his post, passes on the information by a ***second messenger*** that travels only in the interior of the cell. Details of the second messenger mechanism and its role in metabolism are described more fully in connection with enzyme regulation in Chapter 2. At this point, it should be noted that the polypeptides and other hormones that are incapable of penetrating the cell membranes act upon their particular target cells by combining very selectively only with those cells that bear a complementary watchman receptor, who speaks the same language, on the outside surface of the membrane. The second message in the inside seems to be universal for all cells, regardless of the source of stimulus at the outer receptor.

THE NUCLEUS

This is of course the largest of the organelles, visible under ordinary light microscopy, and present in all cells of the body except the red blood cell or erythrocyte. The red cell lacks the other organelles (the mitochondria and endoplasmic reticulum) as well. Nuclei contain the major part of the cellular DNA, which is organized in close association with proteins known as ***histones***. These proteins contain an abundance of basic amino acids (*e.g.*, lysine and arginine) and therefore have a positive overall charge. The positively charged histones readily form salt linkages with the negatively charged phosphate groups along the backbone of DNA. The ***nucleoprotein*** complex thus formed has two important properties: first, it may regulate the activity of DNA in the cell, as noted in Chapter 2, and second, it will facilitate the tight folding of long DNA molecules into the compact structures termed chromosomes. The latter storage problem will be appreciated when one considers that the nucleus of an average human cell must contain roughly five billion nucleotide pairs equivalent of DNA, an amount which, if fully extended, would approximate two meters in length. In addition to the proteins, chromosomes contain considerable amounts of RNA, which resembles that found in cytoplasmic ribosomes. Most of this RNA is found on a specialized region of the chromosomes in a dense body termed the ***nucleolus;*** the latter is the factory for the synthesis of *ribosomal* RNA.

Surrounding the nucleus is a characteristic double-membrane system. The inner membrane belongs to the nucleus proper and occasionally has fibers of DNA attached to it; the outer membrane is dotted with ribosomes and seems to be part of and continuous with the endoplasmic reticulum system. At intervals, broaching the double-membrane barriers are openings or *nuclear pores*, which are large enough to permit the traffic of messenger RNA outward to the cytoplasm and of the proteins inward to the nuclear space. This interrelationship between the nucleus and the cytoplasm is obviously of great significance in controlling macromolecular synthesis.

THE MITOCHONDRION

Like the nuclei, the mitochondria are virtually ubiquitous in body cells, but there are generally many (up to several hundred) mitochondria versus only one nucleus per cell. The mitochondrion is just discernible at the limit of resolution of light microscopy; under the electron microscope, it is seen to vary considerably in size and shape from tissue to tissue. In secretory cells, such as those of the kidney, it is long and filamentous, and disposed near the secretory membrane. In muscle tissue, it is tightly packed in among the fibers of the contractile system. In other tissues, such as the liver where there are fewer constraints, the mitochondria are more nearly spherical and move freely and rapidly throughout the cytoplasm.

The mitochondrion, like the nucleus, is surrounded by a double-membrane system; however, there are no pores connecting the interior space or *matrix* with the cytoplasm. All substances passing into the mitochondrial compartment must first traverse the smooth *outer membrane*, which is relatively porous, and then the *inner membrane*, which is the actual permeability barrier. The latter is also the true functional membrane of the mitochondrion, and it contains a highly organized array of proteins that are involved in the oxidative energy-yielding processes in the cells. In order to provide the maximum surface area for these important reactions, the inner membrane is folded inward in multiple sheets termed the *cristae*. The amount of surface area generated by this infolding of the mitochondrial inner membrane packed into the average human liver would exceed that of a field 100 meters on each side (equivalent to about two acres). Both the numbers of the mitochondria and the extent of the inner membrane invaginations within mitochondria correlate with the respiratory capacity of different cell types.

Mitochondria contain small amounts of DNA as well as ribosomes that are distinct from their cytoplasmic counterparts. They have, however, a limited ability to synthesize only a few of their membrane proteins and lipids, the bulk being imported from sites of synthesis in the cytoplasm.

THE ENDOPLASMIC RETICULUM

This is a network of membrane tubules that runs throughout the cytoplasm and shows continuities in places with the nucleus, the plasma membrane, and the ***Golgi apparatus***, which is a series of stacked tubules containing secretory proteins. In many places the endoplasmic reticulum is dotted with bound ribosomes. There is evidence that these bound ribosomes may play a special role in the synthesis of proteins for export from the cell, whereas the proteins synthesized on ribosomes that are free in the cytoplasm are for home consumption. The proteins for secretion from the cell or for the formation of new plasma membrane probably enter the lumen of the endoplasmic reticulum for transfer to the exterior; it is while passing through the channel of the Golgi system that the sugar portions of the glycoproteins are added on.

The smooth endoplasmic reticulum, devoid of ribosomes, contains the oxidative processes required to detoxify certain drugs, such as the barbiturates. When such compounds are ingested over prolonged periods, an adaptive increase in the smooth endoplasmic reticulum is noted in the liver due to the induced synthesis of new membranes to promote an increased level of detoxification.

LYSOSOMES

These are the scavengers of the cell and are particularly active in *phagocytosis*, which involves the digestion of foreign substances taken up by cells, and in *autolysis*, which involves the digestion of intracellular components as a result of degeneration or cell death. The lysosome consists of a single membrane surrounding a packet of digestive enzymes for the breakdown of cellular proteins, nucleic acids, lipids, and so forth. Lysosomes are probably involved during the normal life of the cell in the removal of worn-out or damaged organelles; they are frequently seen, for example, in association with partly degraded mitochondrial structures in liver cells. Under pathologic circumstances, lysosomes may proliferate to engulf and break down abnormal lipids, as occurs in the nervous tissues of patients with inherited lipidoses, and consequently these diseases are sometimes referred to as lysosomal diseases. Maximal uncontrolled lysosomal digestion occurs during the tissue breakdown of postmortem autolysis.

OTHER CELLULAR INCLUSIONS

The foregoing comprise the major membranes and organelles common to most body cells. Other membrane structures specialized for a variety of different functions are found to varying degrees in different tissues. These include secretory bodies such as the ***synaptic vesicles*** of the nerve cells, which contain neurotransmitters, the ***chromaffin granules*** of the adrenal medulla, which contain catecholamines, and the ***zymogen granules*** of the pancreas, which contain digestive enzymes. In each case, the membrane of the vesicle encases potentially reactive substances, allowing them to be stored within the cytoplasm in a safe form until needed.

In addition to the inclusions bounded by membranes, many cells contain filaments and tubules composed only of proteins. An example is the ***microtubule***, which acts to maintain the internal rigidity of cells, to direct the flow of cytoplasmic components, and to participate in chromosome movement during cell division. ***Microfilaments***, which are smaller than the microtubules in cross section, also play a role in the movements of many types of cells in a simple analog of the contraction of muscle. Such microfilaments appear to involve protein types that are similar to those found in muscle.

CONNECTIVE TISSUE

We have become conditioned by the cell theory to think of living organisms as the expression of the sum of the properties of their constituent cells; however, we must recall that a very large component of our bodies is outside the cells. This extracellular component includes the fluids bathing the tissues (blood plasma, lymph, cerebrospinal fluid) and the connective elements between and around cells, or fluid compartments forming important support structures (skin, skeleton, blood vessels). Tensile strength and adhesive properties of the extracellular ground substances are contributed by long, fibrous proteins (keratin, collagen, elastin), polysaccharides (glycosaminoglycans), and combinations of the two (proteoglycans). In the bones and teeth, connective-tissue proteins of the organic matrix act as reinforcing rods running through the inorganic hydroxyapatite ($Ca(OH)_2 \cdot 3Ca_3(PO_4)_2$) cement.

KERATINS

Beauty may be more than skin deep but we gain much of our initial assessment of an individual from his integument. So too with protein structure, our basic ideas about organization and form arose from studies of the epidermis and its appendages. Protein components responsible for providing the horny protective outer coatings of the body are organized in long rigid fibers, as found in the keratins of skin, nails and hair. The keratins are highly insoluble fibrous proteins that have a considerable content of the amino acid, cysteine. As a result of oxidative cross-linking between chains by cystine –S–S–bridges, the keratins acquire high tensile strength and low water-solubility and permeability. It was through x-ray diffraction studies of the α-keratins of hair that the repeating pattern attributed to the α-helical coil was revealed. Multiples of from two to seven strands may be intertwined in fibrous α-keratins to produce stable, rope-like coiled coils. A second type of diffraction pattern was seen with β-keratins in certain fibers such as silk, which possessed longer, regular spacings consistent with the so-called β-pleated sheet. The extended β-structure could be seen when α-keratins were stretched in steam, a process that is exploited to set the fibrous structures in clothes-pressing or hair-dressing.

In the skin, keratin synthesis proceeds at a rapid rate in specialized cells, *keratinocytes*, found in the lower stratum–spinosum layer of the epidermis. As the cells migrate outward into the stratum–granulosum layer they accumulate dense granules termed *keratohyalin*. Eventually the cells become completely filled with keratin filaments, and the nuclei and other cytoplasmic organelles disappear when they have reached the outermost stratum–corneum layer of the epidermis. Keratin differs from the other connective-tissue structural proteins by remaining an intracellular component of the skin.

COLLAGEN

This important component of cartilage, bone and skin accounts for 30% to 40% of total protein found in the body. Collagen fibers exhibit periodic striations in electron micrographs that are produced by regular, but staggered, arrays of monomers known as tropocollagen. It is in the gaps between tropocollagen monomers that the hydroxyapatite crystals in growing bone are first deposited. Each tropocollagen unit contains three polypeptide chains of about 1000 amino acids, twisted together tightly in a triple helix that confers great rigidity upon the molecule.

The tight helical packing of tropocollagen strands is determined by their primary structure, which has several unique features. First of all is glycine, the major amino acid representing one-third of the total residues, and appearing regularly along the chain. Absence of bulky side-chains on these positions allows the three chains to pack closely together. Second, a very large number of residues, up to one fifth of the total, are α-iminoacids such as proline. These prevent formation of the α-helical secondary structure but favor packing in the collagen triple helix. Third, up to one half of the proline residues and several of the lysines are hydroxylated. The hydroxyprolines favor the triple-helix conformation, while the hydroxylysines effect cross-links between tropocollagens, thereby stabilizing the structure of collagen. Thus the tropocollagen primary sequence is made up of the repeating triplets $(\text{–Gly–X–Y–})_n$, where either X or Y can be proline, the hydroxylated derivatives are found only in position Y, and n exceeds 300. The amino- and carboxy-terminal sequences of 15 to 20 residues are less regular and may play a role in associations of tropocollagen rods into collagen fibrils.

Collagen is produced and secreted by characteristic cell types known as fibroblasts, or their counterparts, the chondroblasts of cartilage, and the osteoblasts of bone tissue. The biosynthesis and processing of collagen exemplifies many of the features of the translation and modification mechanisms noted earlier. Messenger RNAs have been identified and isolated for each of the polypeptide chains. The latter are synthesized on ribosomes that are attached to endoplasmic reticulum membranes. The amino-terminal ends of newly synthesized chains comprise the "signal" sequences that pass through the membrane to enter the cisternae of the endoplasmic reticulum, where they are subject to several modifications. The signal segments are pruned off, and then hydroxyl groups are grafted to some of the prolines and lysines. The hydroxylation process requires molecular oxygen, ferrous iron, vitamin C, and enzyme proteins termed *hydroxylases* as selective catalysts. The correct proportion of hydroxylated residues then allows the individual chains to coil together in the triple-helical structure. Several of the hydroxylysines have sugar residues added to them simultaneously, but once the triple helix has formed, the amino acid residues are no longer exposed enough to allow further glycosylation or hydroxylation. Another process that promotes the triple-helix formation is joining of the chains by –S–S– cross-linkages at their carboxyl ends. These ends, like the amino terminals, extend somewhat beyond the triple-helical region. It is in this precursor form, termed procollagen, that the molecules are passed into the Golgi apparatus, and thence become excreted through the fibroblast–plasma membrane into extracellular spaces.

Procollagen is processed into mature collagen fibrils outside the cells. The extension peptides are clipped off either end and the helical tropocollagen molecules assemble spontaneously into fibrils. However, stable mature collagen is not produced until covalent cross-linking bonds are produced. The cross-links result from a copper-requiring oxidase enzyme that removes ϵ-amino groups from lysine or hydroxylysine residues to yield reactive aldehydes. These aldehydes join with amino groups of adjacent tropocollagens to make a strong bond between molecules.

Several genetic and acquired diseases of connective tissues in humans underline the importance of the key steps of collagen processing. In scurvy, a once common disease among seafarers who lacked fresh supplies of vegetables or fruit, dietary deficiency of vitamin C prevents the normal hydroxylation from taking place. The consequence is fragility of blood vessels and other connective-tissue components because of the inability of the fibroblasts to secrete a stable triple-helical form of collagen. Consequently there is resulting hemorrhage and defective healing of wounds. Interference with normal copper absorption from the diet also has been shown to lead to blood vessel disruption by lowering lysine oxidase and contributing to the failure to form stable cross-linkers in extracellular collagen. Some rare inherited defects of collagen production are also known to cause pathologic changes in the connective tissues. In the cases of two sisters, both afflicted with recurring joint dislocations, curvature of the spine (scoliosis) and abnormally stretchy skin, the collagen of their tissues was found to lack hydroxylysine. Their skin cells were markedly deficient in the enzyme, lysine hydroxylase, whereas cells from a parent had one half of the normal activity. The findings in their case are consistent with expectations for the transmission of a mutant gene for the lysine-hydroxylase enzyme by an autosomal-recessive mode. Thus failure to inherit the normal gene that synthesizes the enzyme for lysine hydroxylation will result in marked structural derangements throughout the body, owing to decreased collagen-fiber stability.

ELASTIN

As its name implies, this protein is a key component of elastic connective tissues. It is particularly important in maintaining the resiliency of major blood vessels such as the large arteries. Elastin differs from collagen in amino acid composition and structure. It contains a preponderance of hydrophobic residues (alanine, valine, leucine) as well as glycine, proline and lysine. The pulling apart of hydrophobic residues stores elastic energy, and the molecule resumes its original shape with little net-energy exchange upon relaxation. The lysine residues act as cross-linkers among large numbers of chains of precursor tropoelastin monomers to form a distensible network in the mature elastin fibers. The biologic importance of elastin in humans is indicated by the involvement of elastic connective tissue in degenerative diseases of the blood vessels (arteriosclerosis), and of the lung (emphysema).

SUGGESTED READINGS

Anderson WF, Diacumakos EG: Genetic engineering in mammalian cells. Sci Am 245 (1):106–121, 1981

Baxter JD: Recombinant DNA and medical progress. Hosp Pract 15, No. 2:57–67, 1980

Capaldi R: A dynamic model of cell membranes. Sci Am 230, No. 3:26–33, 1974

Frieden E: The chemical elements of life. Sci Am 227, No. 1:52–64, 1972

Chambon P: Split genes. Sci Am 244, No. 5:60–71, 1981

Davidson JN: The Biochemistry of Nucleic Acids, 9th edition. New York, Academic Press, 1981

Finean JB, Coleman R, Michell RH: Membranes and their Cellular Functions, 2nd edition. New York, John Wiley & Sons, 1978

Garrett RA: The structure, assembly, and function of ribosomes. In Offord RE (ed): Chemistry of Macromolecules IIB, Macromolecular Complexes, pp 121–177. Baltimore, University Park Press, 1979

Jukes TH: The amino acid code. Adv Enzymol 47:375–432, 1978

Klobutcher LA, Ruddle FH: Chromosome mediated gene transfer. Annu Rev Biochem 50:533–554, 1981

Kolter R, Helinski DR: Regulation of initiation of DNA replication. Annu Rev Genet 13:355–391, 1979

Kornberg A: DNA Replication. San Francisco, Freeman, 1980

Kreil G: Transfer of proteins across membranes. Annu Rev Biochem 50:317–348, 1981

Lake JA: The ribosome. Sci Am 245, No. 2:84–97, 1981

Leder P: The genetics of antibody diversity. Sci Am 246, No. 5:102–115, 1982

Miller A: Structure and function of fibrous proteins. In Offord RE (ed): Chemistry of Macromolecules IIA, Simple Macromolecules, pp 171–210. Baltimore, University Park Press, 1979

Muirhead H: Three-dimensional structure of globular proteins. In Offord RD (ed): Chemistry of Macromolecules IIA, Simple Macromolecules, pp 37–75. Baltimore, University Park Press, 1979

Novikoff AB, Holtzman E: Cells and organelles, 2nd ed. New York, Holt, Rinehart & Winston, 1976

Prockop DJ, Kivirikko KI, Tuderman L, Guzman NA: The biosynthesis of collagen and its disorders. N Engl J Med 301:13–23, 77–85, 1979

Quinn PJ: The Molecular Biology of Cell Membranes. Baltimore, University Park Press, 1976

Rosenberg M, Court D: Regulatory sequences involved in the promotion and termination of RNA transcription. Annu Rev Genet 13:319–353, 1979

Rossman MG, Argos P: Protein folding. Annu Rev Biochem 50:497–532, 1981

Sandberg LB, Soskel NT, Leslie JG: Elastin structure, biosynthesis, and relation to disease states. N Engl J Med 304:566–579, 1981

Schulz GE, Schirmer RH: Principles of Protein Structure. New York, Springer–Verlag, 1979

Tanford C: The Hydrophobic Effect, 2nd edition. New York, Wiley–Interscience, 1979

Tonegawa S: Somatic recombination and mosaic structure of immunoglobulin genes. Harvey Lect 75:61–83, 1981

Weissman SM: Gene structure and function. In Bondy PK, Rosenberg LE (eds): Metabolic Control and Disease, 8th edition, pp 1–26. Philadelphia, WB Saunders, 1980

Wold F: *In vivo* chemical modification of proteins (post-translational modification). Annu Rev Biochem 50:783–814, 1981

Woodhead-Galloway J: Collagen: The Anatomy of a Protein. London, Edward Arnold, 1980

Yelton DE, Scharff MD: Monoclonal antibodies: A powerful new tool in biology and medicine. Annu Rev Biochem 50:657–680, 1981

Chapter Two Enzymes

Metabolism is the process whereby substances either become built up into other substances (usually more complex in structure) that are required for specific functions in the body, or become broken down to waste substances (usually simpler in structure) to produce energy. The building process is known as ***anabolism,*** whereas the breaking down is referred to as ***catabolism*** (Fig. 2-1). The whole chemical economy of the body is dependent upon a precise balance between the anabolic and the catabolic processes. Thus, the healthy adult organism exists in a ***steady-state*** with respect to its tissues; it continually breaks down some tissue components for energy and replenishes these components by anabolism from the foods it takes in. In disease, fever, starvation, or with inadequate diets, the catabolic processes may predominate, leading to tissue wasting and death. In recovery from illness, wound healing, pregnancy, lactation, or growth, anabolism will exceed the catabolic reactions; carried to the pathologic extreme, excessive anabolism may lead to gigantism or obesity. Aberrant metabolism is most commonly due to a failure of the normal mechanisms that serve to maintain balance by means of regulating the rates of anabolic and catabolic processes in the various tissues of the body.

The rates of metabolic processes are governed by the actions of a large group of biologic catalysts, the *enzymes.* Essentially, the study of metabolism is the study of enzyme functions and the factors that affect those functions (Fig. 2-2). Enzymology provides the basis for an understanding of most aspects of human physiology and pathology at the molecular level. The digestion of foodstuffs and their utilization for energy production, the formation of the structural and functional components of tissues, the contraction of the muscles, the transmission of electrical signals in the nerves, the detection of light in the eye, the coagulation of the blood—each of these physiologic mechanisms relies upon the catalytic action of particular enzymes. Numerous pathologic conditions have been recognized to

METABOLISM

ANABOLISM	CATABOLISM
Biosynthesis	Degradation
Small → Large molecules	Large → Small molecules
Energy required $\Delta F+$	Energy released $\Delta F-$
Disorder decreased $\Delta S-$	Disorder increased $\Delta S+$
Often reductive	Often oxidative

EXAMPLES

Gluconeogenesis	Glycolysis
Fat synthesis	Fat oxidation
Protein synthesis	Proteolysis

FIG. 2-1. Comparison of anabolism and catabolism.

result directly from abnormalities of enzyme catalysis, and the measurement of enzyme activities in the blood and other human tissues has provided an increasingly valuable tool for the physician involved with the detection of disease. Obviously, a knowledge of some of the major characteristics of the enzymes and their reactions is a requisite to a rational study of human diseases and their diagnosis.

NATURE OF ENZYME CATALYSIS

The mechanism of action of the enzymes is complex and as yet incompletely understood. The essential features of the process may be summarized by the following sequence of steps (Fig. 2-3):

1. The enzyme (E) combines with the compound upon which it reacts, known as the ***substrate*** (S):

   $$E + S \rightleftarrows E\text{–}S$$

   As indicated by the arrows, the reaction is reversible.
2. The result of this union is the formation of E–S, the ***enzyme–substrate complex.***
3. When combined with the enzyme, the substrate becomes activated so that its constituent electrons and atoms rearrange readily to form the ***product*** of the reaction (P):

   $$E\text{–}S \rightarrow E\text{–}P$$

4. The complex then dissociates, liberating the product and the free enzyme:

   $$E\text{–}P \rightarrow E + P$$

It is noteworthy that only the substrate is modified by the reaction; there

ATTRIBUTES OF ENZYMES

AS PROTEINS	AS CATALYSTS
Composed of amino acids (+ prosthetic groups)	Active in small quantities
High molecular weight	Not altered by reaction
Sensitive to pH	Stimulate rate only
Unstable to heating	Do not add energy
Inactivated by denaturants (heavy metals; heavy anions)	Do not alter equilibrium
	Readily poisoned

FIG. 2-2. Summary of the properties of enzymes.

is no net change in the enzyme, which has thus functioned simply as a catalyst. The unmodified enzyme may therefore return to react again and again with more substrate. Because these reactions occur very rapidly, a small quantity of enzyme will catalyze the conversion of a large amount of substrate to the product in a short period of time.

In principle, all enzymatically catalyzed reactions may be reversed to some extent; that is, the same enzyme may react with the product of the reaction to form the substrate. In the body, the degree to which the reverse reaction will proceed depends upon the concentrations of the substances involved. A large accumulation of the product may slow down the forward reaction and may eventually cause the back reaction to predominate. If the energy content of the product is much lower than that of the substrate, however, the reverse reaction will be quite improbable, because this would be equivalent to moving the components uphill against an energy gradient. Such reactions are termed *irreversible.* In these cases, the conversion of the product back to the substrate requires a different enzyme to catalyze it.

Although the substrates and products may vary greatly in chemical composition and size, the enzymes are invariably proteins with molecular weights ranging from several thousand to several million. Obviously, a small substrate molecule such as a sugar could conceivably combine with such a long protein molecule at many points along the chain. In fact, only a particular region of the enzyme, the ***active center,*** is involved in the complex formation with the substrate. The active center may include several amino acids in a sequence along the chain, or it may involve the interaction of amino acids that seem to be located at a distance if measured linearly along the chain. In the latter situation, the enzyme in its active configuration is coiled back upon itself to bring the reactive amino acids of the active center into proximity.

In addition to the amino-acid residues, the enzyme's active center may contain a nonprotein ***prosthetic group.*** The latter is sometimes an organic molecule, sometimes an inorganic atom, commonly a metal. The prosthetic group plays an essential role in the enzyme's action, either facilitating its combination with the substrate or acting to transfer elec-

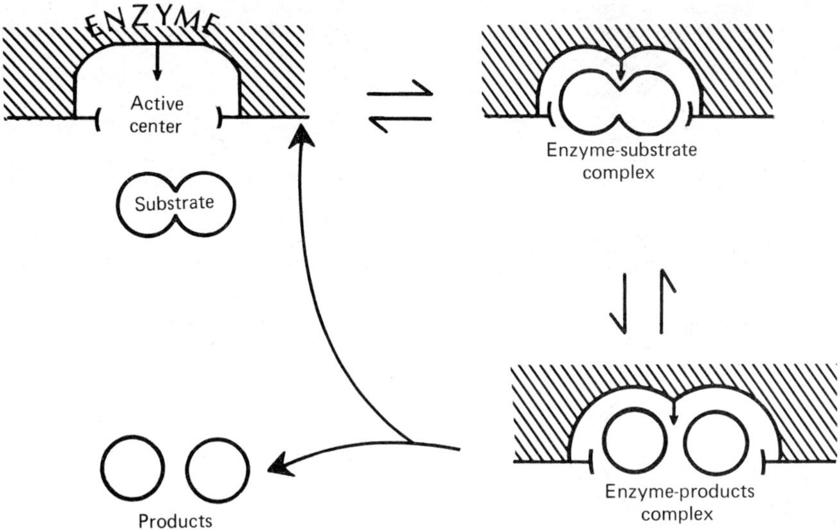

FIG. 2-3. Reaction of an enzyme with its substrate to produce the enzyme–substrate complex.

trons, atoms, or ions between the substrate and the product. In its natural state in the body, the prosthetic group is tightly attached to the enzyme. When the latter is extracted and isolated from the tissues, the prosthetic group remains bound in place. If the prosthetic group is removed by drastic means that dissociate the whole enzyme, or ***holoenzyme,*** the protein remaining, termed the ***apoenzyme,*** will lose its catalytic activity. Reconstitution of the structure and function of the holoenzyme may be achieved by recombination, under proper conditions, of the apoenzyme with its prosthetic group; this demonstrates the essential role of the latter.

By contrast with the firm linkage between enzymes and their prosthetic groups, the complex formed with the substrate at the active center is a loose and tenuous one, with but a brief existence in time. In most cases it is not possible to isolate the enzyme–substrate complex, so labile and transitory is the union; the existence of such complexes must be detected by rapid physical techniques, such as stopped-flow spectrophotometry, which do not disturb the bashful chemical copulation.

A second important feature of the enzyme–substrate complex is the highly specific nature of the combination. In a chemical sense, the substrate must fit exactly into the contours of the enzyme's active center as a hand into a glove (Fig. 2-3). The glove may be somewhat flexible so that the enzyme can react with another substance that is structurally analogous to the proper substrate, for example, a closely related sugar molecule. If this difference is minimal, the analog may fit into the active center and

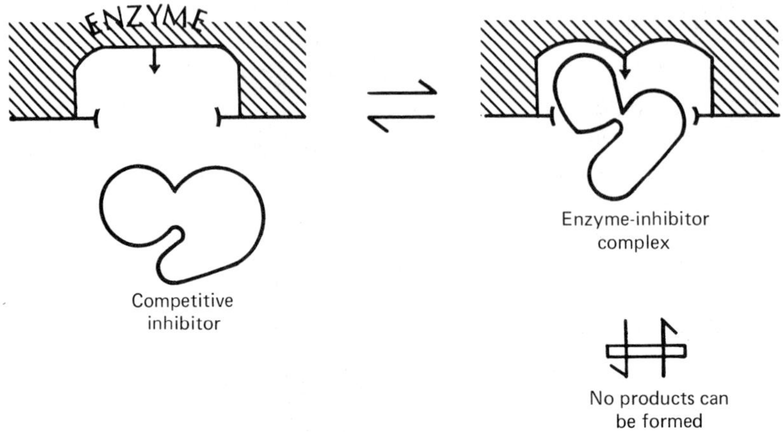

FIG. 2-4. Reaction of an enzyme with a structural analog of the substrate, which acts as a competitive inhibitor.

be converted to products, though probably less efficiently than the natural substrate would be. If the structural difference is somewhat greater, the analog may still react with the enzyme to form a complex, but the fit at the active center may figuratively leave one finger out in the cold, thus preventing the catalytic action. In the latter case, the analog will not only fail to react itself but will also take up some of the active centers unprofitably, thereby preventing the natural substrate from reacting. Thus, the analog will act as an ***inhibitor*** of the normal enzymatically catalyzed reaction (Fig. 2-4). The addition of a large excess of the normal substrate would displace the inhibitor from the active center and restore the enzyme's catalytic activity. Since the analog competes with the substrate in reacting with the enzyme, its effect is known as ***competitive inhibition.*** This principle forms the basis of the phenomenon of ***metabolic antagonism,*** whereby the structural analog of an essential component in the metabolism of an organism may act as a selective poison for a particular type of cell. Examples of the practical application of this important phenomenon of enzymology in clinical medicine include the use of many of the antibiotics, which act exclusively upon bacterial metabolism, and the employment of ***antimetabolites*** in cancer chemotherapy, which selectively block the growth of rapidly dividing cells.

The reaction of the enzyme with the normal substrate may be so selective that no other substance can react. As an example, urease acts upon urea alone and thus exhibits *absolute specificity*. In many instances, an enzyme will fail to react with an optical isomer of the normal substrate; D-amino-acid oxidase fails to oxidize the L isomers, for example. The hand-in-glove analogy is particularly apt in such a case; although providing a perfect fit for one of the hands, a glove could not be expected to conform to the mirror-image topography of the opposite member unless a mechanism existed to turn it inside out.

EFFECT OF SUBSTRATE CONCENTRATION

Many factors combine to modify the rate at which an enzyme acts upon its substrate. Of great importance in the situation found in the cells of the body is the influence of the concentration of the substrate molecules. At any instant of time, the enzymes present in the cell will be limited and fixed in concentration, since such large protein molecules are not free to move from cell to cell. Substrates, on the other hand, are generally small enough to diffuse freely through cell membranes from the blood and other extracellular fluids. As a consequence, the concentrations of substrates in the cell may fluctuate considerably, depending upon the immediate nutritional state of the body, the demands of the various organs and tissues in response to physical activity, and so forth. If little substrate is available, the chances that it will collide with the appropriate enzyme molecule will be quite low, and there will be only a slow formation of product. Increasing the substrate concentration will increase the probability of collision with the fixed numbers of enzyme molecules; in fact, at this stage, doubling the substrate concentration will result in doubling the rate of product formation (Fig. 2-5A). Under these conditions in the cell, the process described previously in steps (1) and (2), which lead to the union of enzyme with substrate, will be the rate-limiting factors. Thus, a graph of the enzyme reaction rate or velocity (v) against the substrate concentration [S] will yield a linearly increasing plot described by the equation:

$$v = C \cdot [S]$$

where C is a constant. This straight-line relation, however, cannot continue indefinitely.

Assume, for example, that the person is basking in the postprandial glow of a heavy meal. While he relaxes in spirit, foodstuffs from the intestines are being absorbed and are transported busily through the blood to his tissues. In the cells, the influx of these substrates causes steadily rising rates of metabolism, but the increases in enzyme reaction velocity gradually taper off until they reach a steady or maximum velocity; beyond this point, further increases in the substrate concentrations can cause no further increase in the rates. The reason for this phenomenon is quite simple: the number of enzyme molecules in the cell is fixed, and when the cell becomes swamped with substrate at any time, all the active centers of the enzyme will be occupied by substrate molecules. In short, the union of enzyme with substrate may not be promoted further by any conditions favoring collision of the two, because the enzyme has now been saturated to its full capacity with the substrate (Fig. 2-5B). At this point, the rate-limiting processes will be those described previously in steps (3) and (4),

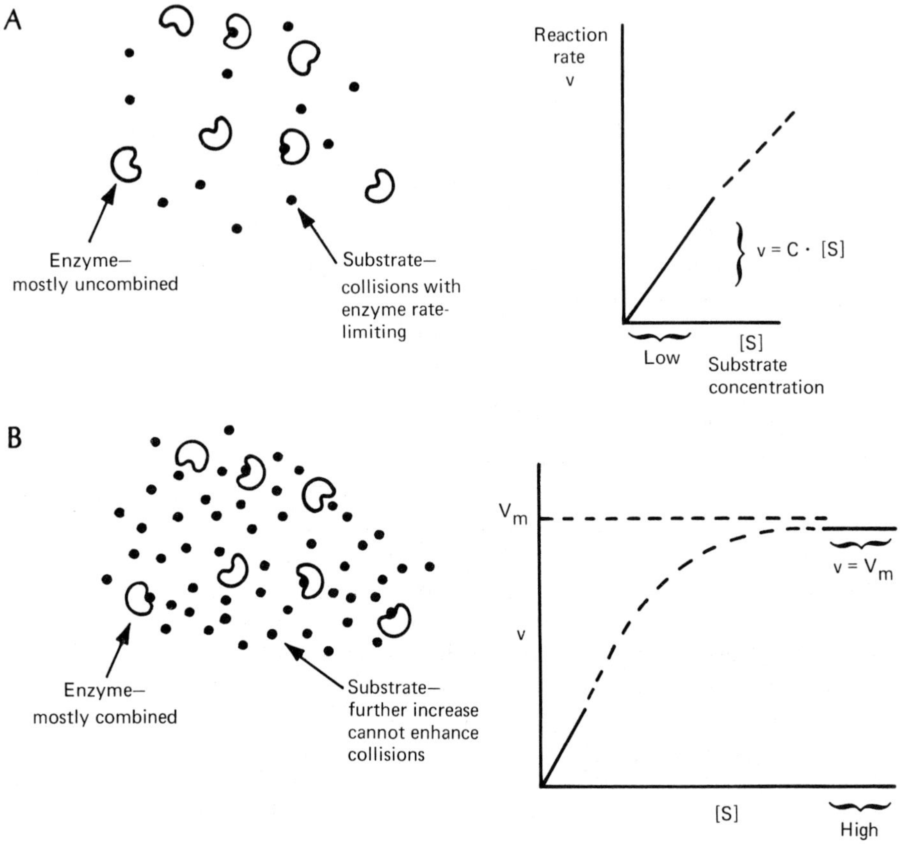

FIG. 2-5. Effect of substrate concentration upon the rate of an enzyme reaction where substrate availability is rate-limiting **(A)** and where enzyme is saturated with substrate **(B)**.

whereby the enzyme–substrate complex dissociates to form the products. The rate will then be described as:

$$v = V_m \text{ or } V_{max}$$

where V_m is the constant maximum velocity.

A mathematic relationship describing the dependence of velocity upon substrate concentration in the presence of a fixed amount of enzyme was derived by Michaelis and Menten around the turn of the century, based upon the above account of enzyme–substrate interaction:

$$v = \frac{V_m\,[\mathrm{S}]}{(K_m + [\mathrm{S}])}$$

where K_m is a combined rate constant known as the ***Michaelis constant.*** This equation follows the change of velocity for all values of substrate concentration and yields a curve in the shape of a rectangular hyperbola

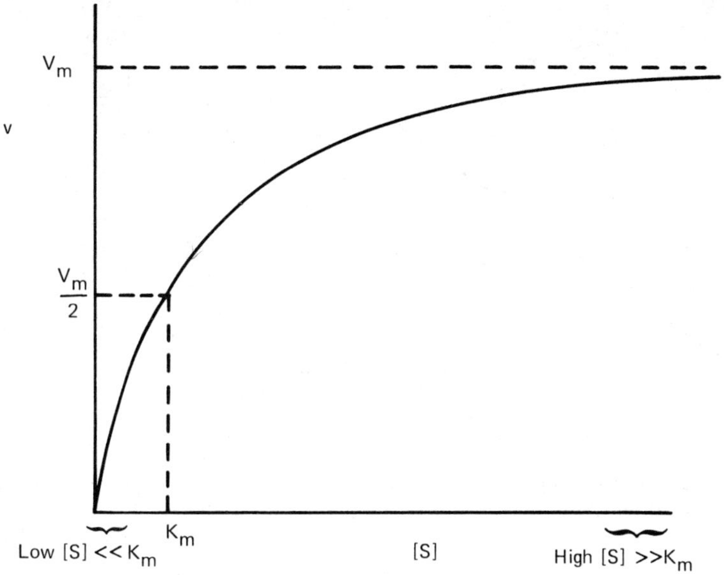

$$v = \frac{V_m\,[S]}{K_m + [S]}$$

FIG. 2-6. Michaelis–Menten relation between the rate of an enzyme reaction *(v)* and substrate concentration ([S]) where V_m is the maximum velocity at saturating levels of substrate and K_m is the concentration of substrate that results in a rate of one-half V_m.

when v is plotted on a graph against [S] (Fig. 2-6). It may be seen that the two linear portions of the curve described earlier derive readily from this equation. In the first case, where the substrate concentration is very low, [S] is much less than K_m. In this limiting condition, the denominator, K_m + [S], will approach the value of K_m since [S]will be so much smaller relative to K_m. The equation then reduces to $v = V_m\,[S]/K_m$, which is equivalent to the relation expressed earlier, $v = C \cdot [S]$, since both V_m and K_m are constants. In the second case, where the substrate concentration is very high, [S] is much greater than K_m. At the other limit, then, the value for the denominator, K_m + [S], will be approximated by [S], K_m being so much smaller that it may be ignored. The equation reduces to $v = V_m[S]/[S] = V_m$, as noted previously.

The Michaelis constant, K_m, is a useful quantitative parameter that aids in establishing the way that enzymes will react under the conditions of the living cell. The value for K_m may be determined graphically from the curve relating v to [S]: Letting K_m = [S] in the Michaelis–Menten equation,

$$v = \frac{V_m[S]}{([S] + [S])} = \frac{V_m \cdot [S]}{2[S]} = \tfrac{1}{2}\,V_m$$

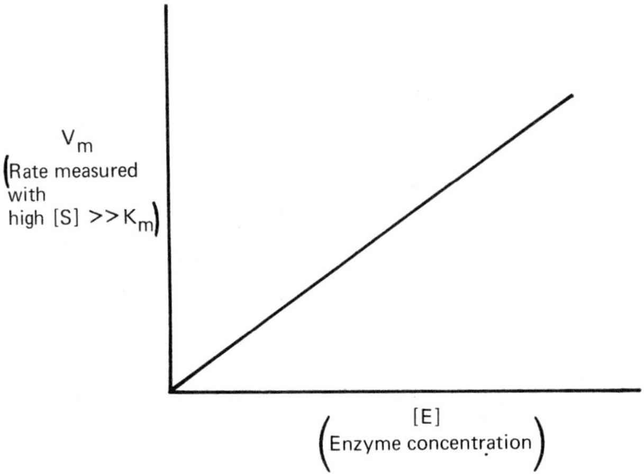

FIG. 2-7. Effect of enzyme concentration upon the rate (V_m) of an enzyme reaction in which enzyme is saturated by substrate.

In other words, K_m represents the concentration of substrate that will be found when the reaction is at half its maximum velocity. If an enzyme has a high value for K_m, this means that it will take a lot of substrate to raise the rate of the reaction to half the maximum; a low K_m implies that little substrate is needed to saturate the enzyme. A corollary of this statement is that enzyme–substrate pairs with low K_m values have a high affinity for one another, and *vice versa*.

The K_m value is useful for determining which of a group of substrates will combine most effectively with a given enzyme. If the enzyme has a low specificity under the usual conditions of limited concentrations of reactants in the cell, it will favor the reaction with the substrate that gives the lowest K_m value. Knowing the concentrations of different substrates in the cell and the K_m values for the enzymes allows one to determine the efficiency of the individual reactions of metabolism in the body as well as which of these reactions may be rate-limiting. This knowledge is imperative to an understanding of the effects of physiologic regulators, such as hormones, upon metabolic rates.

Although the value for K_m is a characteristic fixed constant for any enzyme–substrate pair, V_m will vary directly with the amount of enzyme present (Fig. 2-7). In fact, it may be shown algebraically that $V_m = k[E]$, where [E] is enzyme concentration and k is the rate constant for the reactions described previously under steps (3) and (4). A measurement of V_m—the velocity in the presence of an excess of substrate—is a measurement of the number of molecules of enzyme present in the cell. Although it is not always possible to express [E] in moles per liter because the value for k may not be known, it may be assumed that a measure of the activity

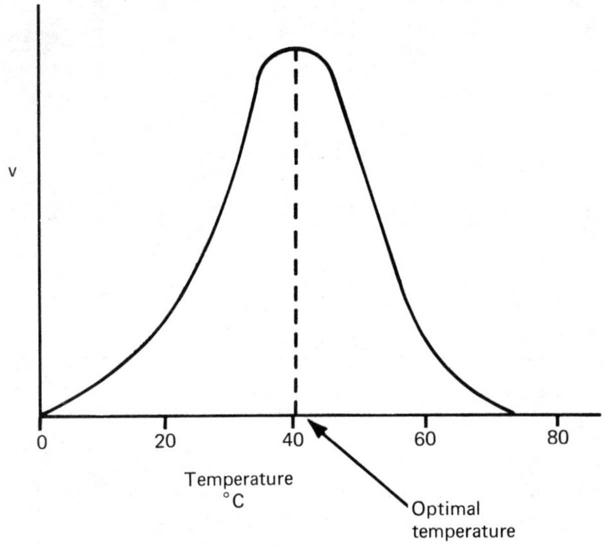

FIG. 2-8. Effect of varying the temperature upon an enzyme reaction.

of an enzyme under optimal conditions will yield a valid, relative measure of the amount of enzyme.

In addition to providing adequate substrate to saturate the enzyme, the optimal conditions for enzyme activity will include the appropriate physical conditions (temperature, *p*H, and ionic state) as well as the presence of any additional factors that may cooperate in the catalytic process.

TEMPERATURE EFFECTS

Just as in the case of chemical reactions occurring outside the body, enzymatically catalyzed reactions increase in rate as the temperature is raised. Unlike ordinary chemical reactions, however, the velocity of an enzymatic reaction does not increase indefinitely with temperature. After a certain point, further increases in temperature result in decreased enzyme action (Fig. 2-8). This point, where the enzyme reaction rate is at a maximum, is the *optimal temperature* for that enzyme. The explanation for this phenomenon tells us a good deal about the nature of enzymes and how they act.

The enhancement of velocity obtained by increasing the temperature in the range below the optimal point is related to the energy requirement of the reaction. The energy conditions to be satisfied are twofold: first, the energy content of the substrate for a reaction must be greater than that of the products (this may be restated by saying that in the energy sense, the reaction will go of itself only if it is running downhill); second, some

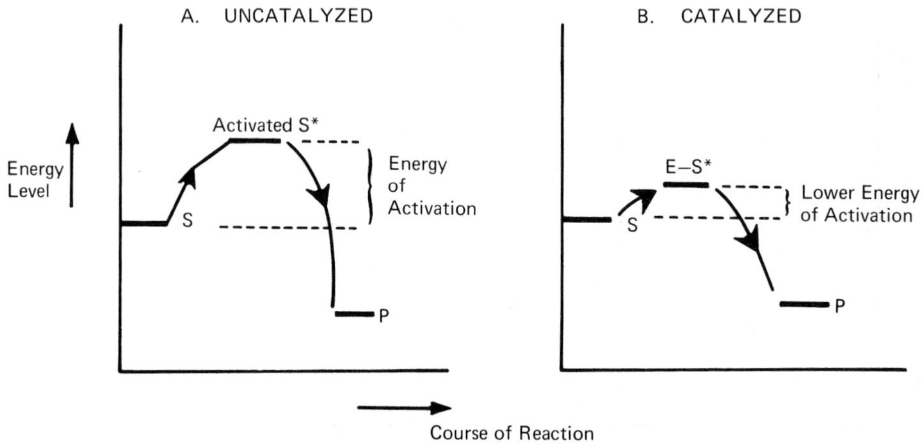

FIG. 2-9. Energy changes during conversion of a substrate, S, into product, P, in the absence (**A,** uncatalyzed) and presence (**B,** catalyzed) of an enzyme. The overall energy loss in going from S to P is the same in both cases, but the formation of enzyme–substrate complex (E-S*) results in a lowering of the energy of activation in the catalyzed reaction.

energy must be absorbed by the substrate to raise it to the higher energy level of an activated state before its electrons and atoms will rearrange to give those of the product (this is analogous to saying that there is a small energy barrier, or inertia, which prevents the downhill tumble from happening spontaneously).

The latter energy input is known as the *energy of activation* of the reaction. One source of this energy input is heat energy. In uncatalyzed reactions, the amount of heat energy required to start the reaction may be considerable, and therefore little conversion to products can occur unless the temperature is elevated considerably (Fig. 2-9A). In most cases, each 10°C rise in temperature will result in a doubling of the rate of reaction. This will also be true for most enzymatic reactions, but a significant reaction rate may still be obtained with much lower temperatures. This is because the enzyme *lowers* the energy of activation for the reaction (Fig. 2-9B). How this happens is not clear, but it is somehow related to the fact that when it is complexed to the active center of the enzyme, the substrate will already exist in a somewhat activated state. The result is that even though the ordinary chemical reaction would require several hundred degrees of temperature increase for a significant rate to proceed, with the enzyme present the same reaction can occur at measurable velocities in the ambient range (0°C–40°C).

Above the optimal temperature, a counteracting factor begins to interfere with the enzymatically catalyzed reaction. As noted previously, the enzyme molecule is a long chain of amino acids joined together. In its active state, this linear chain is coiled back upon itself in a very specific

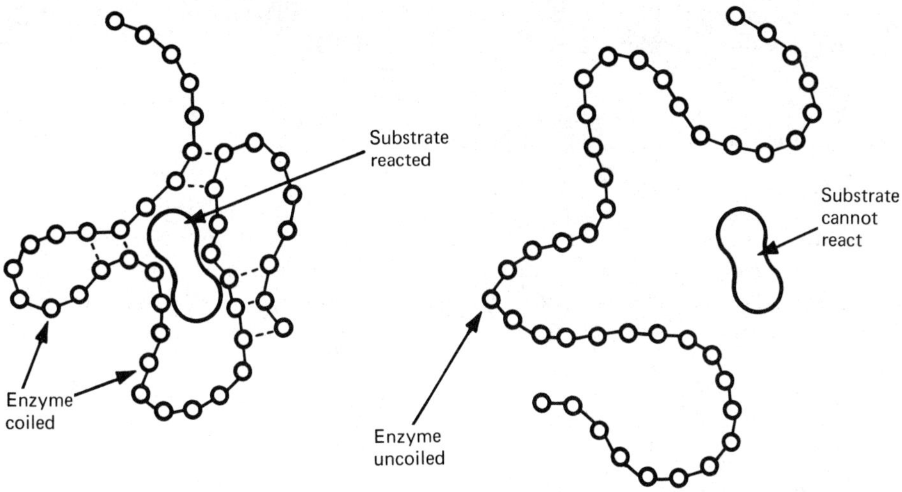

FIG. 2-10. Effect of temperature denaturation upon an enzyme reaction. Excessive heating causes unfolding of the enzyme, producing a denatured state in which the substrate is prevented from reacting at the active center.

way, so that the amino acids of the active center may be brought together to react with the substrate (Fig. 2-10A). Some of the chemical forces that hold the protein molecule in this active configuration are quite weak and are easily disrupted by heat (see Chap. 1). Hence, a sufficient rise in temperature will result in uncoiling of the proteins. This process is known as *denaturation* (Fig. 2-10B). The disruptive effect of excess heat on the structure of the active center of the enzyme explains why enzymes are inactivated at elevated temperatures; they rarely regain their activity if they are heated and subsequently cooled.

The optimal temperature for the reactions of most enzymes is close to the normal temperature of the body, 37°C. In humans, who are homeothermic organisms, the regulation of temperature is quite rigidly held about this optimal point. Fever, a defense mechanism against infection, may cause the inactivation of the enzymes of invading bacteria, but it will eventually produce deleterious effects upon the host as well. Refrigeration of isolated organs has been used to reduce their metabolism during surgical procedures, and the storage of tissues and body fluids at freezing temperature or below is a standard method of preventing autocatalytic decomposition.

*p*H EFFECTS

Enzymes, like all protein molecules, bear many charged groups, both positive and negative. The net charge of the enzyme molecule will depend upon the balance between these oppositely charged groups. The major factor that alters the balance is the presence of hydrogen ions (H^+) in the cell, which neutralize negatively charged groups:

$$—COO^- + H^+ \rightarrow —COOH \text{ (zero charge)}$$

and produce positively charged groups:

$$—NH_2 + H^+ \rightarrow —NH_3^+$$

Thus, an increased positive charge (or decreased negative charge) on the enzyme will be produced by an increase in hydrogen ion concentration (*i.e.*, a decrease in *p*H). Conversely, a decrease in hydrogen ion concentration (an increase in *p*H) will result in the dissociation of these protonated forms:

$$—COOH \rightarrow —COO^- + H^+$$

$$—NH_3^+ \rightarrow —NH_2 + H^+$$

In this case, there will be a consequent increase in the negative charge on the enzyme molecule. At some point there will be a hydrogen ion concentration for which the positive and negative charges will be exactly equal, so the enzyme will carry no net charge. This is known as the ***isoelectric point.***

Most enzymes show their greatest activity and stability at or near their isoelectric points. Like elevated temperatures, extremes of *p*H—either high acidity or high basicity—hasten the uncoiling of the molecule, leading to denaturation and irreversible inactivation. Frequently, charged groups on the enzyme in the vicinity of the active center will participate either in the binding of the substrate or in the catalytic action of the enzyme. For this reason, enzymes usually function best over a rather limited range of hydrogen ion concentration. A plot of enzymatic activity against *p*H will produce a bell-shaped curve in most instances (Fig. 2-11). The point of maximal enzymatic activity is referred to as the *optimal* p*H*, and it will be characteristic for a given enzyme reacting with a given substrate.

Intracellular enzymes usually have *p*H optimums near neutrality (*p*H 7.0), which is close to the normal *p*H of body fluids. Extracellular enzymes, such as pepsin, which functions in the highly acidic milieu of the stomach contents, may have *p*H optimums that are widely removed from neutrality. It is apparent, however, that just as for the body temperature, the regulation of *p*H in the human body is an essential requirement for maintaining optimal enzyme action. Imbalances of hydrogen ion concentration—either excesses *(acidosis)* or deficiencies *(alkalosis)*—will produce drastic alterations of metabolism.

IONIC EFFECTS

Other ions in the cell in addition to hydrogen ions may play vital roles in enzymatic reactions. The metabolic activity, especially of excitable tissues

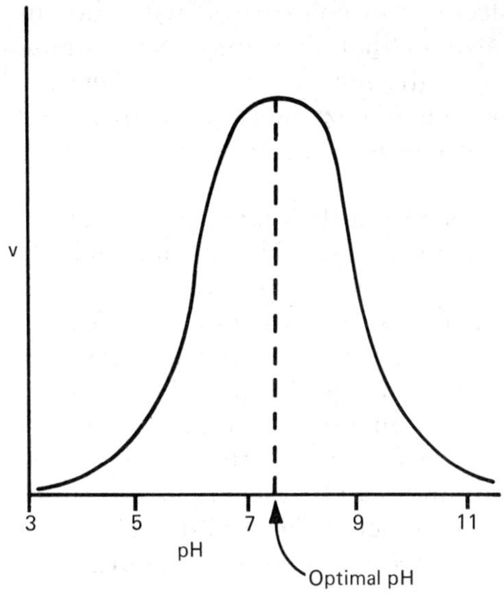

FIG. 2-11. Effect of varying the *p*H upon an enzyme reaction.

such as nerve and muscle, is greatly affected by calcium (Ca^{2+}) and potassium (K^{+}) cations. The propagation of nerve impulses and the contraction of the muscle protein, myosin, are delicately controlled by the effects of variations in the concentrations of ions, and so are the enzymes that release the energy required for these physiologic processes.

In many cases, cations (+) or anions (−) may play a direct role in an enzymatically catalyzed reaction. The role of metallic ions joined to enzymes as prosthetic groups has already been mentioned. In other instances, ions may combine loosely or reversibly with the enzyme's active centers, much in the way that substrates do, to activate the catalytic process. Magnesium (Mg^{2+}) cations frequently participate in metabolic reactions in this manner. Sometimes the ions may play an integral part in the reaction by serving as actual substrates. The involvement of bicarbonate (HCO_3^-) anions in carboxylation reactions, of phosphate (HPO_4^{2-}) anions in phosphorylation reactions, or of ammonium (NH_4^+) cations in amination and amidation reactions provides examples of the direct participation of inorganic ions as substrates in enzyme reactions. Excesses or deficiencies of these ionic species may exert drastic pathologic consequences through their effects upon the enzymatically catalyzed reactions of metabolism.

COFACTORS AND COENZYMES

The generic term for substances that collaborate in promoting the actions of enzymes and substrates is ***cofactor.*** The ions that serve as activators or prosthetic groups may thus be considered among the cofactors that facilitate enzyme catalysis. Another important group of substances in this category is the ***coenzymes.*** These substances, in contrast with the inorganic cofactors, are complex organic molecules; in comparison with the enzymes, they are much smaller in size so that they may usually pass through membranes, and they are generally not inactivated by heating. The coenzymes serve an intermediary role in enzyme catalysis by acting as the carriers of electrons, atoms, or groups of atoms from the substrates acted upon.

Often the coenzymes will contain as part of their structure one of the ***vitamins,*** organic substances that are not synthesized by the human body and that must be provided in the diet. The vitamin portion of the coenzyme molecule is the active component that acts to combine with the group being transferred. The remaining part of the coenzyme serves to bind it by a highly specific linkage and in a carefully aligned orientation at the active center of the enzyme. The binding may be loose and reversible, like that of the substrate or some of the ionic activators discussed previously. In other instances, the binding may be exceptionally firm or irreversible, in which case the coenzyme functions as an integral part of the enzyme's active center, that is, as a prosthetic group. The dramatic pathologic effects of nutritional vitamin deficiencies upon the human body may be directly related to the disturbances of metabolism incurred by insufficient concentrations of coenzymes for specific enzymatic reactions.

ENZYME INHIBITORS

Reference was made previously to the antagonism of enzyme catalysis that may be exerted by certain structural analogs of the substrate. Those substances were termed *competitive inhibitors.* This is but one type of enzyme inhibition, a phenomenon of considerable significance in understanding the control of metabolism as well as the action of many drugs and poisons that affect the body. As noted earlier, a number of physical agents—including high temperatures, extremes of *p*H, and so forth—can lead to enzyme inactivation by uncoiling the protein structure. Although the susceptibility of different enzymes may vary somewhat, physical denaturation is a general and nonselective type of enzyme inhibition. There are, however, chemical agents that will act much more specifically as inhibitors of particular enzymes and not others. The competitive inhibitors are typical of this group, as are the ***noncompetitive inhibitors.*** The distinction between these two types of inhibitors is based upon the observed ability of increased substrate concentration to overcome the competitive inhibition.

Thus, plots of the enzymatic reaction rate v against [S] in the presence of a fixed amount of competitive inhibitor will yield the same value for V_m as obtained without the inhibitor, since the large excess of substrate will compete effectively with the inhibitor for binding sites at the enzyme's active center and will eventually displace it. The value for K_m in

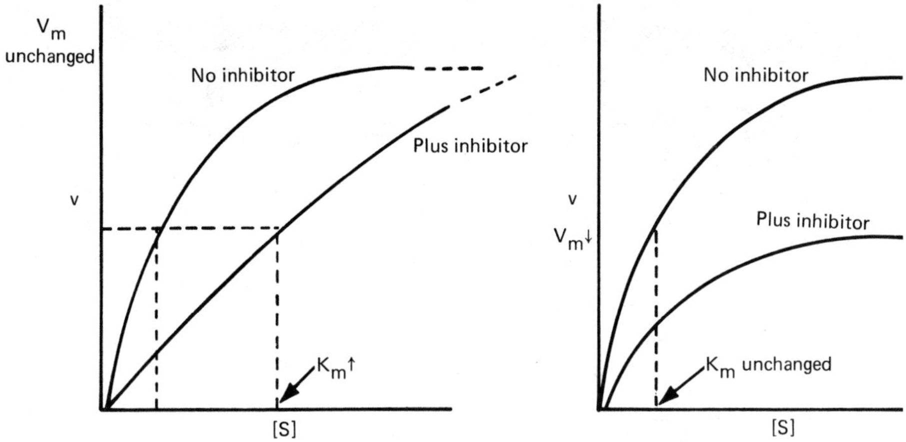

FIG. 2-12. Comparison of the effects of competitive and noncompetitive inhibitors upon enzyme reaction rates. **A**. Competitive inhibition: V_m unchanged, K_m increased. **B**. Noncompetitive inhibition: V_m decreased, K_m unchanged.

this case, however, will be increased, since it will require more substrate to produce half the maximal reaction velocity in the presence of inhibitor (Fig. 2-12A). This implies that the competitive inhibitor has interfered with the formation of the enzyme–substrate complex and hence has decreased the apparent affinity of the enzyme for its substrate; the decomposition of the enzyme–substrate complex to the products of the reaction is not affected (Fig. 2-13A).

When the same type of plot of v against [S] is determined for a noncompetitive inhibitor, it is found that the value for V_m will be decreased (Fig. 2-12B). In the presence of a fixed amount of inhibitor, no amount of substrate can displace it from the enzyme. This suggests that the substrate and the noncompetitive inhibitor must combine with the active center of the enzyme in different ways. The inhibitor interferes only with the catalytic action of the enzyme in decomposing the E–S complex to the products; it does not affect the binding of substrate to the enzyme to form the E–S complex (Fig. 2-13B). As a corollary of the last statement, it follows that the value for K_m in the presence of a noncompetitive inhibitor is unchanged.

Both the competitive and the noncompetitive types of inhibitors combine reversibly with the enzyme. A third type of agent is the *irreversible inhibitor,* which forms such a strong union with the enzyme that it may not be removed readily. This type of inhibitor may form covalent bonds with one of the amino-acid residues at the enzyme's active center. This effectively prevents any catalytic action of the part of the enzyme.

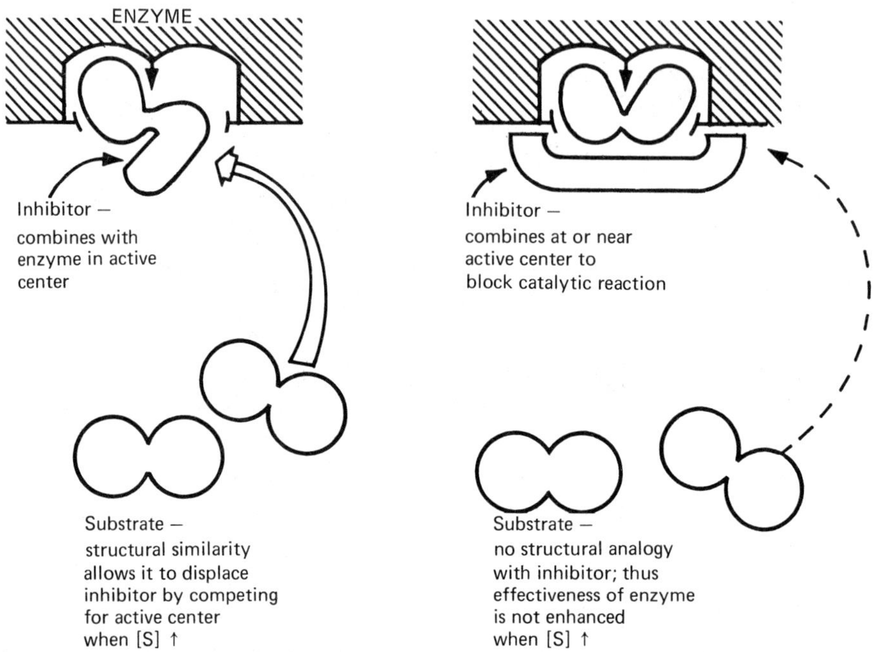

FIG. 2-13. Comparison of the reaction of competitive and noncompetitive inhibitors at enzyme active centers.

REGULATION OF ENZYMES

ALLOSTERIC MODIFIERS

Certain key enzymes in metabolism exhibit a phenomenon known as ***allosteric modification.*** The enzymes that display this particular effect are important in the overall regulation of a metabolic pathway and are consequently termed ***regulatory enzymes.*** The allosteric modifiers may be either inhibitory or stimulatory to the enzymatic reaction rate. When a plot of v versus [S] is drawn for a regulatory enzyme, the curve obtained differs from that predicted by the Michaelis–Menten equation. Instead of a rectangular hyperbola, the shape of the curve is S-shaped or sigmoid, with very low reaction rates being found at low substrate concentration (Fig. 2-14). As [S] is increased, the rate accelerates more sharply in a nonlinear fashion and eventually tapers off, reaching saturation at a maximal reaction velocity with high concentrations of substrate, as is observed with a normal enzyme.

The explanation for the sigmoid shape of this curve is that the regulatory enzyme consists, not of a single chain of amino acids, but of several proteins, or subunits, joined together. Each of these subunits is capable

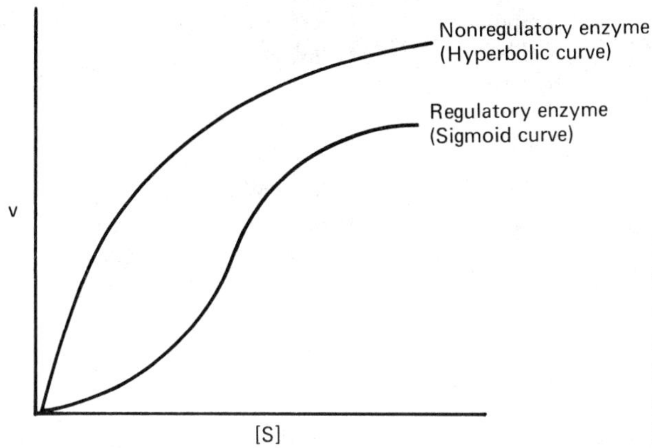

FIG. 2-14. Comparison of the relationship between enzyme reaction rate *(v)* and substrate concentration ([S]) for a regulatory enzyme with the normal Michaelis–Menten hyperbolic curve.

of binding a substrate molecule to form catalytically active enzyme–substrate complexes. When one of the subunits has already combined with substrate, it facilitates the union of additional substrate molecules with the remaining subunits of the regulatory enzyme. Thus, the subunits cooperate with one another to enhance the formation of E–S complexes when the concentration of substrate rises, as described for the cooperative binding of oxygen to hemoglobin subunits (Chap. 1). Eventually, the active centers of all the subunits become occupied to maximal capacity by substrate, and the limiting maximal value of V_m is then reached, as in the standard case.

In addition to the active centers that bind substrate and are responsible for the catalytic activity, regulatory enzymes also bear different binding sites that interact with the allosteric modifiers of the enzyme. This is the root meaning of the term (*allosteric,* another shape), and it refers to the fact that the modifiers exhibit different binding parameters than those that apply to the interaction of the substrate with the enzyme at the active center. If the modifier acts as an inhibitor of the reaction, it is known as a *negative modifier* (Fig. 2-15). In the presence of a constant amount of negative modifier, the curve of v against [S] will remain sigmoid in shape, but it will be shifted toward the right. Thus, the effect of the negative modifier will be to raise the amount of substrate required to saturate the enzyme and produce the maximal velocity, V_m. More importantly, it may be seen that under the conditions of low substrate concentration to be found in the normal cell, the negative modifier will produce a profound inhibition of the enzyme activity. This makes the effect of a negative modifier an extremely sensitive controlling influence upon regulatory enzymes under physiologic conditions.

The effect of a *positive modifier,* which is a substance capable of stimulating or activating the activity of a regulatory enzyme, is just the op-

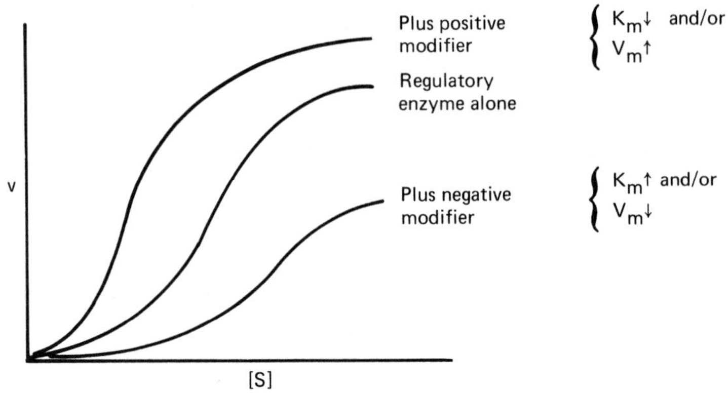

FIG. 2-15. Effects of allosteric modifiers upon a regulatory enzyme reaction. A negative modifier may cause a decrease in V_m or an increase in K_m whereas a positive modifier may increase V_m or decrease K_m.

posite. In this case, the cooperation between subunits is further enhanced by binding the modifier to the allosteric site. In the presence of a fixed amount of positive modifier, the sigmoid v-versus-[S] curve will be shifted to the left. The degree of the shift will depend upon the amount of positive modifier added, and, if this amount is large enough, the shape of the curve may resume that of the typical hyperbola (Fig. 2-15).

The significance of allosteric modification of the regulatory enzymes is demonstrated by the mechanisms used by the body to adapt to its metabolic requirements in response to changes of diet and physiologic activity. The regulatory enzymes usually occupy key positions in the pathways of metabolism, so that positive and negative modifiers may turn on and off entire reaction sequences. The compounds that act as the negative modifiers of these pacemaker enzymes are generally the end products of the total reaction sequence. Thus, if the regulatory enzyme is located at the beginning of the sequence, it is possible to prevent wasteful utilization of the substrates feeding into the pathway.

This type of control system, in which the products of a metabolic pathway inhibit an early step, is termed ***end-product inhibition,*** or ***negative feedback*** by analogy with electronic circuitry (Fig. 2-16). The above situation applies when there is an accumulation of the substances produced by metabolism. The opposite effect may often be obtained by the action of a substrate or precursor for the metabolic sequence that functions as a positive modifier of the same pacemaker regulatory enzyme. The latter constitutes a type of ***positive feedforward,*** which counteracts end-product inhibition and results in a delicately balanced control system that is capable of a wide response range to physiologic requirements from "on" at full capacity to "off" or total inactivity.

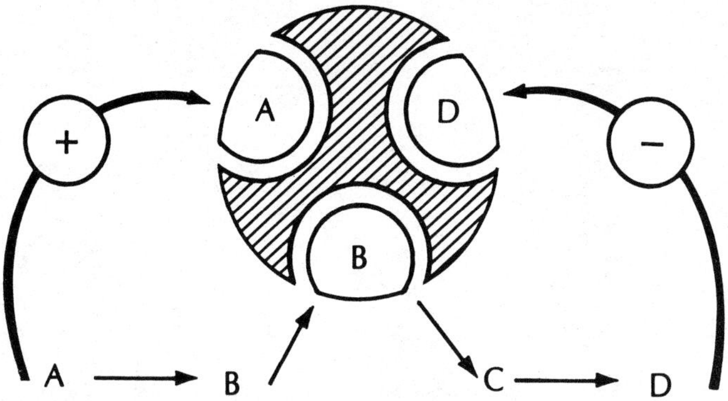

FIG. 2-16. Metabolite control by negative feedback or positive feedforward. The regulatory enzyme catalyzes the key step B → C in the metabolic pathway A → D. The end product, D, acts as a negative modifier to hinder step B → C and hence prevent accumulation of excessive amounts of D. The precursor, A, acts as a positive modifier to facilitate step B → C and hence assist utilization of excessive amounts of A.

ZYMOGEN ACTIVATION

Other means of modifying enzyme structure also provide "on–off" control systems for enzyme catalysis. Many of the enzymes that function in the digestion of foodstuffs are produced within the cells of the digestive glands in an inactive, or ***zymogen,*** form. It is only when they are released into the digestive tract that the zymogens become activated, since their activity would otherwise digest the cells producing them. In several enzyme systems studied, the zymogen is inactive because its catalytic action is masked by the attachment of an extra segment of the protein chain that hinders the accessibility of substrates to the enzyme's active center (Fig. 2-17). Activation involves the removal of the blocking group and leads to irreversible activation of the zymogen; it is not possible with these systems to return from the "on" to the "off" position with the same enzyme molecule.

REGULATION BY CHEMICAL MODIFICATION OF ENZYMES

There are also reversible "on–off" enzyme regulatory systems in which the chemical structure of the enzyme's active center may be altered. In some instances, an enzyme may exist in a cell normally in an inactive or weakly active state, but the addition of a phosphate group by a second phosphorylating enzyme, termed a ***kinase,*** to an amino acid near the active center can fully activate the catalytic capabilities of the enzyme (Fig. 2-18). Thus the action of enzyme 2, the kinase, has resulted in the acti-

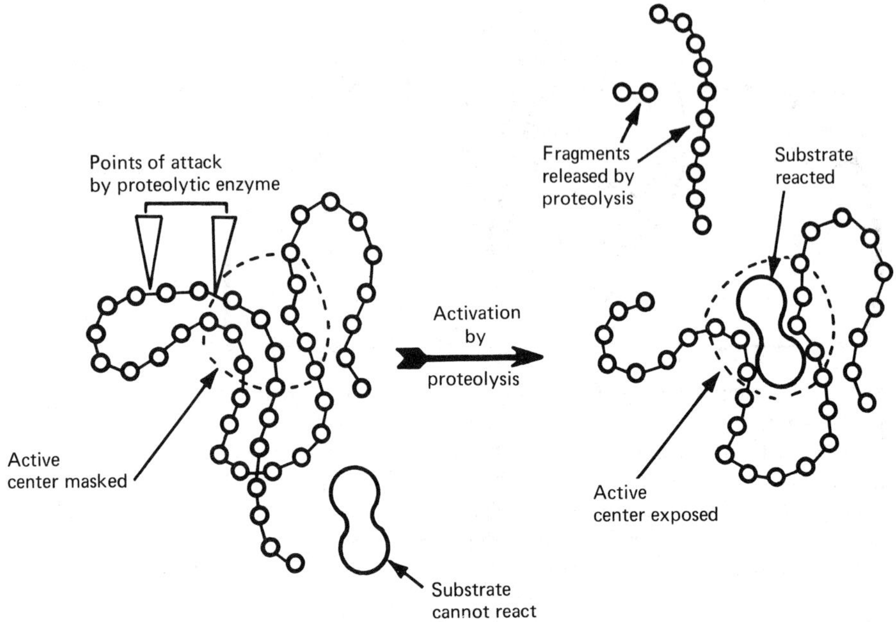

FIG. 2-17. Zymogen activation. Removal of a portion of the polypeptide chain of the zymogen unmasks the active center and produces active enzyme.

vation, by chemical modification, of enzyme 1. Such a system can be returned to the "off" position by the action of a third enzyme, a ***phosphatase,*** which removes the phosphate group from the activated enzyme. In this type of control system, the activity of enzyme 1 will obviously depend upon the balance between the activity of the kinase, which adds phosphate groups, and that of the phosphatase, which takes them off.

Some of the twenty enzymes that have been reported to undergo reversible phosphorylation/dephosphorylation include glycogen phosphorylase, glycogen synthetase, pyruvate dehydrogenase, pyruvate kinase, phosphofructokinase, acetyl CoA carboxylase, β-hydroxy,β-methyl glutaryl CoA reductase, the intracellular hormone-sensitive lipase, and hydroxylases for the aromatic amino acids. Not every protein phosphorylation will result in activation of catalytic activity; in several instances the enzyme may be rendered less active, or less responsive to substrate. For example, in the metabolism of glycogen as discussed in Chapter 4, the enzyme that is responsible for breakdown (phosphorylase) is activated upon phosphorylation, while the opposing enzyme involved in buildup (synthetase) is inactivated. In this way an exquisitely sensitive control system operates, such that when the catabolic process is turned on anabolism is turned off, and *vice versa.* In a few cases (*e.g.,* phosphofructokinase) no discernible difference has been found upon phosphorylation/dephosphorylation. In addition to the above enzymes of carbohydrate, lipid,

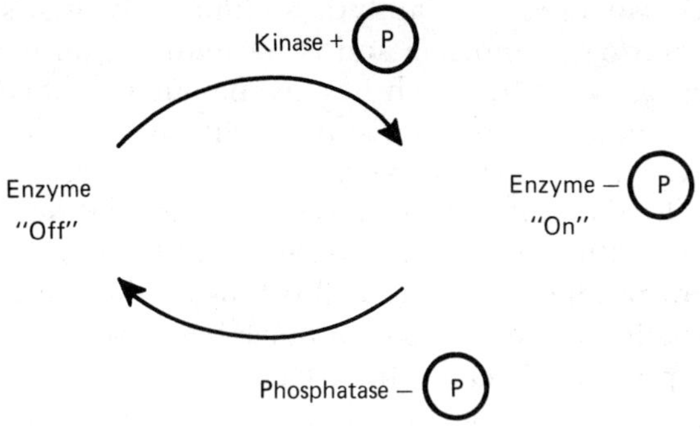

FIG. 2-18. Enzyme regulation by chemical modification. Phosphorylation produces active enzyme, whereas dephosphorylation inactivates the enzyme.

and amino-acid metabolism, ribosomal proteins that may regulate the rate of gene translation and membrane proteins that are involved in controlling cellular permeability and transport reactions are also subject to this type of covalent modification.

Levels of phosphorylation of regulatory enzymes may be controlled either by way of the ***protein kinase*** reactions or dephosphorylation by the opposing enzymes, the phosphatases. Current research suggests that both types of control are important and are effected by physiologic regulators such as the hormones. There are three principal factors involved in the modulation of protein phosphorylation: the nucleotide messengers, cyclic AMP and cyclic GMP, and the ionic messenger, Ca^{2+}. Cyclic nucleotides are generated inside cells by the following type of reaction:

$$\text{ATP} \rightarrow 3',5'\text{-cyclic AMP} + \text{P–P}_i$$

which is catalyzed by the membrane-bound enzyme complex termed ***adenylate cyclase.*** The two terminal phosphates of ATP are removed as inorganic pyrophosphate (P–P_i), and the energy from this cleavage is used for an internal esterification of the remaining phosphate upon its own ribose to form the cyclic ester. A similar enzyme may react with GTP to produce the corresponding 3′,5′-cyclic GMP. In addition to the cyclase enzymes that synthesize the cyclic nucleotides, cells contain counteracting enzymes known as *phosphodiesterases* that break them down. The balance of cyclase and phosphodiesterase activities will be the determining factor in regulating intracellular cyclic nucleotide levels. For example, the methyl-xanthine drugs (*e.g.*, caffeine) exert a potent inhibition of the phosphodiesterase, thereby raising the cellular concentration of cyclic AMP and stimulating key protein kinases to activate metabolic processes. Most of the protein kinases are activated by one or the other of the cyclic nucleotides, some by Ca^{2+}, while still others may respond to quite different intracellular messengers.

Overriding control of the intracellular protein kinases is generally exerted by extracellular primary messengers, or hormones, with cyclic nucleotides or Ca^{2+} acting as the intracellular second messengers. The adenylate-cyclase system consists of (1) its catalytic subunit, which sits on the inner surface of the plasma membrane and is normally in an inactive state, and (2) its receptor subunit, which lies on the outer surface of the membrane and is capable of binding selectively with a hormone that is specific for the particular cell. When the receptor subunit reacts with the hormone exterior to the cell, it undergoes a conformational change allowing it to interact with the catalytic subunit and turn on its cyclase activity. The resultant production of cyclic AMP inside the cell is followed by activation of protein kinases and hence phosphorylation of a number of enzymes or other proteins. Because the effect of hormone may be amplified considerably at each stage of this cascade of reactions, the consequence is usually a profound metabolic response to a miniscule triggering amount of hormonal agent.

Many cell types, especially those in excitable tissues such as brain and muscle, use calcium ions as intracellular modulators of metabolic control systems. For example, both the protein kinase that phosphorylates glycogen phosphorylase in skeletal muscle and the contractile system are activated simultaneously by Ca^{2+}. In this way the energy demands for muscle contraction may be furnished by coordinated breakdown of glycogen to glucose (see Chaps. 3 and 4). Moreover, the calcium ion levels in cells may exert indirect effects by alterations of cyclic nucleotide levels, besides their direct effects upon protein phosphorylation. The enzyme that degrades cyclic AMP, phosphodiesterase, is activated by Ca^{2+}, whereas the enzyme that produces cyclic AMP, adenylate cyclase, may be either stimulated or inhibited depending upon the intracellular concentration of Ca^{2+}. These intracellular effects of calcium ions are modulated by their binding to a cytoplasmic protein receptor known as *calmodulin*. Calmodulin, normally present in cells in an inactive state, is converted to its active form when Ca^{2+} flows into the cytoplasm, either from the extracellular fluid or from bound storage sites within the cell in the endoplasmic reticulum cisternae or the matrix of the mitochondria.

ADAPTATION BY ENZYME SYNTHESIS

The effects of the inhibitors and activators that control metabolism are exerted upon the enzyme molecules that preexist in the cell. In the previous discussion, we have considered the situation in which the amount of enzyme remains relatively constant, and the variable factors considered were the low-molecular-weight metabolites and other conditions that might change in the adult organism in a steady state. Another type of metabolic response to the changing needs of the body is the alteration in the enzyme content within the cell. During the growth of the fetus, for example, and in particular during the first hesitant days after birth when it must adapt from a sheltered, regulated life in the womb to the hostile environment outside, the enzyme contents of the various organs and tissues undergo dramatic changes to meet this challenge. Throughout the stages of development of the body, a carefully timed sequence of enzymatic differentiation accompanies the morphogenesis of the specialized cells and determines their ultimate physiologic roles. Such changes of enzyme composition may require weeks or even months; however, some alterations in the cellular enzyme content may occur more rapidly. Adaptations to starvation on the one hand or to overfeeding on the other are associated with changes in the amounts of the enzymes involved with carbohydrate and fat metabolism and may occur in a matter of several hours. Nonetheless, these adaptive changes in the enzyme patterns of the body are relatively slow in contrast to the feedback effects of inhibitors and activators upon existing enzyme molecules.

It is now apparent that enzyme molecules, like other proteins in the body, are continually synthesized and broken down. The rates of synthesis and breakdown are equivalent in the normal adult in a steady state, so that there is no net change with time. The rate of breakdown is characteristic of each enzyme. In most cases, total cessation of enzyme synthesis would lead to the disappearance of half the enzyme in several days, but some enzymes turn over much more rapidly. The rate of synthesis of enzymes may vary from zero to full capacity, whereas the rate of breakdown appears to remain constant. Thus, any substance that can affect the rate of enzyme synthesis can exert drastic effects upon the regulation of metabolism by changing the enzymatic profile of the body. Many effects of hormones upon human metabolism have been traced to just such a controlling influence upon the production of enzymatic proteins.

A substance that interferes selectively with the synthesis of a particular enzyme is termed a ***repressor.*** The phenomenon of repression is an additional mechanism by which the end products of metabolism may exert control over their own formation by means of a negative feedback loop. In several systems, it has been proved that the accumulation of the metabolites that are produced by a sequence of reactions will prevent the synthesis of one of the enzymes that functions at the beginning of the pathway (Fig. 2-19). The product in such a case acts as a specific repressor of the synthesis of that enzyme, preventing both the unnecessary consumption of the substrates feeding into the pathway and the wasteful utilization of the energy and amino acids required for formation of the enzymatic protein.

The opposite of repression is known as *enzyme induction* or ***derepression.*** In the typical situation, the substrate for a particular enzyme may trigger an induced synthesis of that enzyme, which in turn promotes utilization of the substrate. The inducer probably acts either directly or indirectly to counteract the action of a repressor upon the machinery of enzyme synthesis. The balance between the repressor (end product) and the inducer (substrate) will thus determine the amount of key pacemaker enzymes and ensure that the metabolic sequences may adapt to the

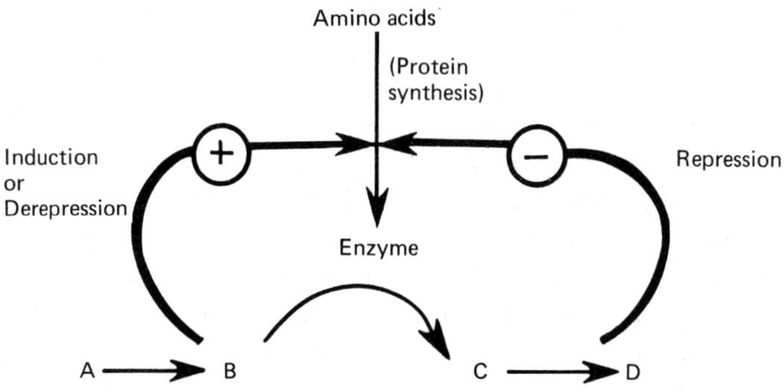

FIG. 2-19. Regulation of enzyme synthesis. The end product, D, of the metabolic pathway lowers the concentration of the enzyme responsible for step B → C by repressing its synthesis. The substrate, B, induces the synthesis of the enzyme that converts it to C by interfering with the repressor.

amounts of these metabolites coming into the cells of the body from the diet.

As in the case of regulatory enzymes, only a few key enzymes are capable of responding in this manner to changing physiologic requirements. These enzymes are referred to as ***inducible enzymes*** (or adaptive enzymes); enzymes that do not change in this way are termed ***constitutive enzymes,*** because they constitute a fixed content in the cell. In the human body, adaptive enzymes are influenced by endocrine factors probably to a more significant extent than they are by the intermediates of metabolism. Thus, the hormones of the adrenal cortex promote the synthesis of enzymes required for the formation of the blood sugar, glucose, whereas insulin counteracts these effects. The adrenal corticosteroids directly or indirectly play the role of enzyme inducers, whereas insulin enhances the process of repression. The balance of enzyme synthesis obtained by the offsetting influences of induction and repression explains the physiologic regulation of blood glucose levels by these two opposing endocrine systems.

GENETIC DETERMINATION OF ENZYMES

The final determinant of the enzyme complement of the body is the genetic inheritance of the individual. We may define normality in the human species within the ranges of physique, size, coloration, and so forth, but just as there are wide variations in the anatomy, features, and other physical attributes, so too are there differences in the metabolic activities of different people. These metabolic differences give rise to characteristic biochemical features in each individual, a type of *biochemical individuality.* The biochemical characteristics of the individual are transmitted from one generation to the next in the same manner as the inheritance of the color of the eyes or hair, that is, by way of either dominant or recessive genes in the Mendelian sense. The expression of the metabolic variations carried in the genotype occurs at the level of enzyme synthesis in the phenotype. In some cases the variations are trivial or benign; in others they may exert tragic effects.

We have noted in Chapter 1 the manner by which one gene (a DNA molecule) may specifically order the sequence of amino acids in one enzyme (a protein molecule). The specificity of base-pairing between the nucleic acids involved in the transcription and translation of the message from that particular gene will thus ensure that its enzyme product can fold into the required conformation for its catalytic activity. The gene, which is the unit of inheritance, is seen to govern the formation and function of the enzyme, which is the unit of metabolism. Consequently, any alteration of the gene by mutation of one of its nucleotide bases will cause a corresponding alteration of one of the amino acids in its enzyme product, and this may result in profound derangement of the enzyme's catalytic role in the body.

Interference with the genetic machinery for the biosynthesis of enzymatic proteins may produce several sequelae. First, it could alter the enzyme-synthesizing system so that it was no longer sensitive to the action of repressors or inducers. In this case, a normal enzyme would be produced, but its formation would not be subject to regulation, so that instead of being an adaptive enzyme, it would become a constitutive enzyme. Thus, the mutation would affect the control of metabolism, but not the course of metabolism. A second consequence could be that the enzyme-synthesizing system would, in scanning the three-letter groups of nucleotides, interpret the triplet bearing the alteration to signify an amino acid differing from that normally found in that position in the enzyme molecule (Fig. 2-20). If the substitution is a minimal deviation such that the properties of the new amino acid are quite similar to those of the usual one or if the substitution occurs in a region of the protein chain that does not affect the active center of the enzyme, then the metabolic effects of the mutation may be insignificant. On the other hand, upon substitution with a radically different amino acid either at the active center or in a position that determines the critical folding of the enzyme protein, the metabolic consequences may be extreme. A similar situation would arise if the mutation deleted one of the nucleotides near the beginning of the message; in this case, the three-letter sequence in the gene would be put out of register, so that the system for enzyme synthesis would read the genetic message as nonsense and produce no active enzyme.

Either of the last two situations will give rise to the pathologic condition described as an ***inborn error of metabolism.*** The conversion of substance A to end product Z in the body requires the operation of a series of metabolic reactions, each catalyzed in its turn by a separate, specific

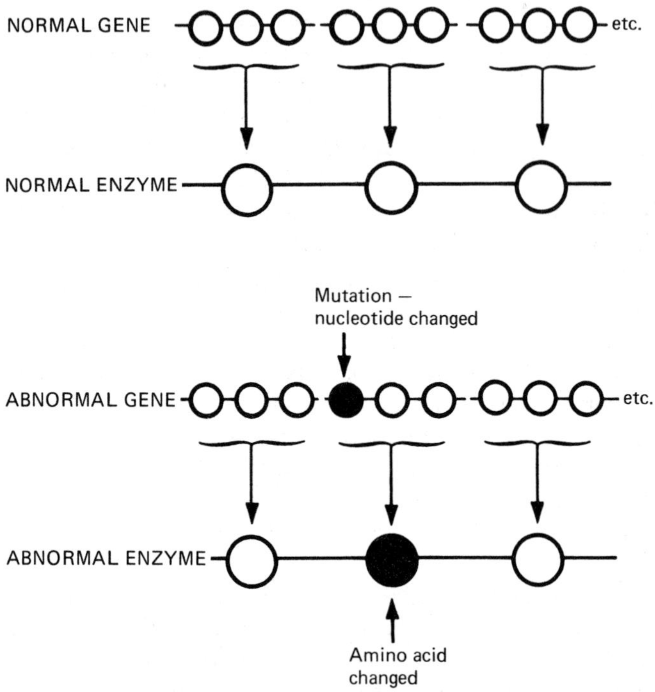

FIG. 2-20. Genetic determination of enzymes. Mutation, by alteration of one nucleotide in a triplet codon of the gene, will cause alteration of one amino acid in the enzyme, which may affect its catalytic functions.

enzyme, with each enzyme produced under the instructions from a particular gene (Fig. 2-21A). When those instructions are sufficiently garbled by a mutation to produce either no enzyme or a catalytically inactive enzyme, a blockage will occur in the normal sequence of metabolism. Resulting from such a failure at one of the steps of metabolism will be an accumulation of each of the metabolites occurring above the block and a deficiency of all those below the block (Fig. 2-21B). The pathologic consequences of this metabolic error may reult in the serious impairment of normal physiologic activities of the affected person, possibly threatening his ability to survive.

The range of defects may be seen in the following group of disorders (Fig. 2-22), which all affect the metabolism of the aromatic amino acid, phenylalanine, and are described in greater detail in Chapter 6. In *albinism,* there is the absence of an enzyme required for the conversion of phenylalanine to the normal body pigment, melanin. The albino consequently has white hair, pink irises in the eyes, and pale pigmentation of the skin. With the exception of an increased sensitivity to exposure to sunlight, the individual is asymptomatic and may lead a normal life. In

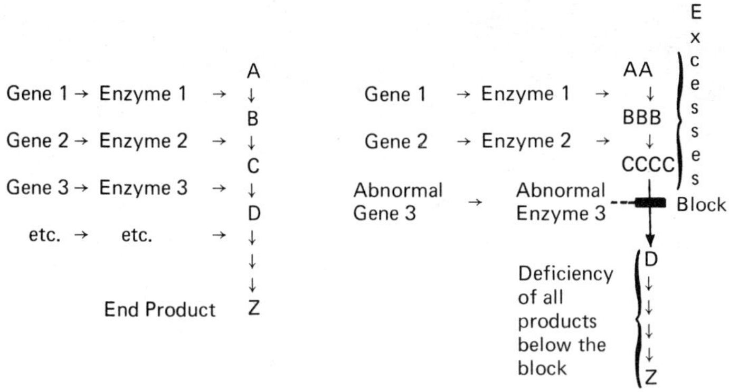

FIG. 2-21. Consequences of an inborn error of metabolism. Defective gene resulting in defective enzyme produces accumulations of metabolites above and shortages of those below the block in the pathway.

alkaptonuria, a step in the catabolism of the aromatic ring is absent, and an abnormal accumulation of a normal metabolite, homogentisic acid, occurs. The latter substance is rapidly removed from the blood by the kidney and hence produces few toxic manifestations. The condition will be undetected until the person notices that his urine darkens if it is allowed to stand, or until he obtains a falsely positive test for sugar in the urine due to the reducing action of homogentisic acid. In later life, an accumulation of abnormal aromatic metabolites will produce pigmentation and arthritis of the joints.

In *phenylketonuria,* the first step in phenylalanine metabolism in the liver is absent. Phenylalanine accumulates in body tissues, producing significant impairment of brain development in the infant and eventual mental retardation unless the dietary intake of phenylalanine is controlled. In *goitrous cretinism,* an enzyme in the thyroid gland necessary for the conversion of phenylalanine to thyroxine is missing. The consequent absence of thyroid hormone prevents normal growth and development and leads to dwarfism and mental retardation unless the infant is supplied with replacement thyroxine. Thus, the pathologic results of inborn errors of phenylalanine metabolism may range from benign (albinism), to moderate or ones of late onset (alkaptonuria), to severe due to the toxicity of accumulated metabolities (phenylketonuria) or to the lack of necessary biosynthetic products (cretinism).

CLINICAL ENZYMOLOGY

Researchers have found enzymes to have a variety of applications in industry and medicine. As reagents enzymes have many advantages because

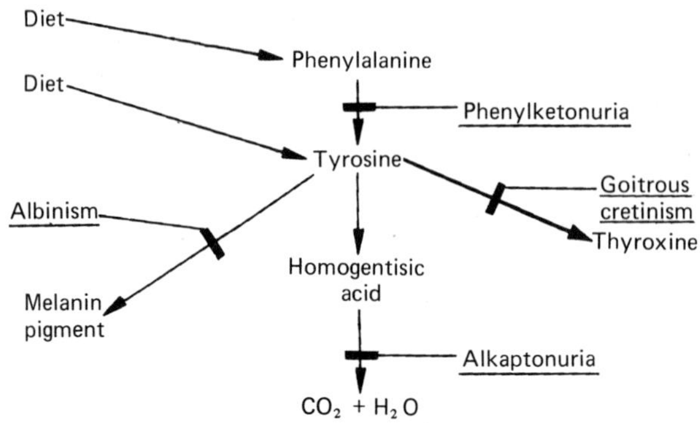

FIG. 2-22. Genetic disorders of phenylalanine metabolism.

of the high specificities of their catalytic actions, the absence of side-reactions or by-products, and their mild conditions of action at moderate temperatures. The use of particular enzymes can add or remove portions of drug or hormone molecules in the production of pharmaceuticals. Enzymes are also used as direct therapeutic agents, for example, in the replacement of deficient digestive enzymes in cystic fibrosis patients. However, the most common application of enzymes in medicine is in the diagnosis of pathologic conditions from the assay of enzyme activity in the tissues from patients. Obviously the sampling of living cells for enzyme determinations can present certain difficulties for doctor and patient; therefore, most routine enzyme measurements for diagnostic purposes are carried out on blood samples. Activities in blood cells may be useful in some metabolic disorders, particularly those involving hemolytic anemia. However, not all enzyme defects are reflected by activities in the blood cells, and biopsy of tissues may be indicated. If the enzyme of interest is present in tissue fibroblasts, these may be cultured *in vitro* from a small skin biopsy specimen in order to grow enough cellular material to characterize both the level of activity (V_{max}) and binding properties (K_m) of the enzyme.

The activity of an enzyme in blood serum reflects the difference between its release from the tissue of origin and its clearance from the circulation. Release of enzymes into the bloodstream is usually quite limited and occurs at rapid rates only when cell-membrane damage or cell death allows the escape of nondialysable components from the cytoplasm. Thus, increases of particular enzymes in serum present a profile that is characteristic of a particular tissue or organ and may give an indication of the extent of damage or necrosis at the site of injury. Clearance of different enzymes may occur at different rates and by different mechanisms. For example, some enzymes are excreted by the kidney and can be detected in their active states in urine.

Amylase is an enzyme of polysaccharide digestion, produced in the pancreas and salivary glands and normally confined to secretions in the gastrointestinal tract (see Chap. 4). Thus its elevation in serum suggests disease of the secretory organs, for example, inflammation of the parotid glands (as in mumps), pancreatitis, or neoplastic diseases. Amylase is a small protein and is cleared from the plasma by the kidney. Increased urinary amylase is found in the conditions noted above. In kidney failure the reduced ability to clear the enzyme causes an elevation of serum amylase with no increase of urinary amylase. The lipid hydrolyzing digestive enzyme, *lipase* (see Chap. 5), is produced by the pancreas but not by the salivary glands, and its assay in the serum is superior to amylase determinations as a diagnostic index for acute pancreatitis.

Phosphatases are tissue hydrolases that act on phosphate–ester linkages under specified conditions. They are present in a wide variety of cell types, but possess some distinctive properties that relate to their origins. *Acid phosphatase* is active only at low *p*Hs (around 5) and is quite unstable at *p*H values above neutrality. The activity in serum is elevated in men with metastatic cancer of the prostate, and successful chemotherapy of the condition may be followed by a corresponding decline of serum acid phosphatase. Because the enzyme is present in erythrocytes and blood platelets, an increase in serum activity of the enzyme also may be seen in both men and women with hemolytic or thromboembolic diseases. *Alkaline phosphatase* is a zinc-containing enzyme that is active at high *p*Hs (around 10). Various charged forms of the enzyme (isoenzymes), which are derived from various tissue sources, may be differentiated by electrophoresis or by studying the effects of inhibitors and denaturants. Major clinical applications of assays for alkaline phosphatase in serum are in skeletal disorders and hepatobiliary diseases, which are characterized by increases of the bone and liver isoenzymes respectively. Histochemical studies of bone localize the enzyme to the osteoblasts at sites of active mineralization. Raised serum alkaline phosphatase is associated with increased osteoblast activity during normal bone growth and calcium phosphate deposition in childhood and adolescence. Further pathologic elevations of the bone isoenzyme are noted in cases of rickets, tumors of the bone, or defects of bone formation, such as Paget's disease, in which osteoblast activity is stimulated. The liver isoenzyme increases in serum particularly in diseases in which there is obstruction of bile passages (cholestasis); the rise in alkaline phosphatase activity may often precede the rise in bile-pigment concentration in cases of obstructive jaundice. Other liver disorders (*e.g.*, alcoholic cirrhosis) may show normal serum alkaline phosphatase levels unless complicated by involvement of the biliary tract, where most of the liver isoenzyme seems to originate. As noted in Chapter 6, the combination of assays for alkaline phosphatase, the amino-transferases, and bile-pigment derivatives in blood serum are all invaluable adjuncts for diagnosis of different categories of liver disease.

Creatine phosphokinase is an important energy-balancing enzyme of neuromuscular tissues (see Chap. 3). Its elevation in the bloodstream is a clarion signal of trouble either in the skeletal or heart muscle cells. Because of the urgency in establishing a status report of damage in heart attacks, this enzyme is of great significance in clinical diagnosis. Creatine phosphokinase is a dimeric enzyme that may contain two type of subunits, M for muscle, B for brain. Three isoenzyme forms, which are separable by electrophoresis, may result from combinations of subunits: MM, which constitutes all of the enzyme in striated skeletal muscles; BB, which accounts for the total in nervous tissue; the hybrid form, MB, found only in heart muscle and contributing about one fifth of the total cardiac enzyme, with MM as the remainder. The isoenzyme MM will rise nonspecifically in serum upon trauma to either skeletal or cardiac muscle, for example, in cases of muscular dystrophy. However, selective determination of the MB isoenzyme provides a specific and sensitive indicator of injury to cardiac muscle. In cases of myocardial infarction, caused by interruption of the blood supply to the cardiac muscle, the ischemic damage to the heart tissue results in a precipitous rise of the MB form of creatine phosphokinase in the blood. Taken in conjunction with assays for other serum enzymes (aminotransferases—Chap. 6; lactate dehydrogenase isoenzymes—Chap. 4) the creatine phosphokinase isoenzyme measure provides considerable information about the extent of damage and the course of recovery in heart attack patients. In particular, creatine phosphokinase levels in the serum rise more rapidly than the other enzyme markers, reaching peak values within only a few hours of the onset of chest pain and thus facilitating a prompt confirmation of the diagnosis of ischemic heart injury.

SUGGESTED READINGS

Aurbach GD: Polypeptide and amine hormone regulation of adenylate cyclase. Annu Rev Physiol 44:653–666, 1982

Cheung WY: Calmodulin: An overview. Fed Proc 41:2253–2257, 1982

Chock PB, Rhee SG, Stadtman ER: Interconvertible enzyme cascades in cellular regulation. Annu Rev Biochem 49:813–843, 1980

Deswick RJ, Grabowski GA: Advances in the treatment of inherited metabolic diseases. Adv Hum Genet 11:281–370, 1981

Glass DB, Krebs EG: Protein phosphorylation catalysed by cyclic AMP-dependent and cyclic GMP-dependent protein kinases. Annu Rev Pharmacol Toxicol 20:363–388, 1980

Goldberg DM: Diagnostic enzymology. In Gornall AG (ed): Applied Biochemistry of Clinical Disorders, pp 26–40. Hagerstown, Harper & Row, 1980

Kirsch J: Mechanism of enzyme action. Annu Rev Biochem 42:205–234, 1973

Koshland Jr DE: Protein shape and biological control. Sci Am 229, No. 4:52–64, 1973

Krebs EG, Beavo JA: Phosphorylation–dephosphorylation of enzymes. Annu Rev Biochem 48:923–959, 1979

Mildvan AS: Mechanism of enzyme action. Annu Rev Biochem 43:357–399, 1974

Posen S, Doherty E: The measurement of serum alkaline phosphatase in clinical medicine. Adv Clin Chem 22:165–245, 1981

Roberts R, Sobel BE: CPK isoenzymes in evaluation of myocardial ischemic injury. Hosp Pract 11, No. 1:55–62, 1976

Rosenberg LE: Inborn errors of metabolism. In Bondy PK, Rosenberg LE (eds): Metabolic Control and Disease, 8th ed, pp 73–102. Philadelphia, WB Saunders, 1980

Sanwal BD: Regulation of enzyme activity. In Golberger RF (ed): Biological Regulation and Development, Vol 2, Molecular Organization and Cell Function, pp 1–42. New York, Plenum Press, 1980

Sols A: Multimodulation of enzyme activity. Curr Top Cell Regul 19:77–101, 1981

Stroud RM: A family of protein-cutting proteins. Sci Am 231, No. 1:74–88, 1974

Walker PR: The regulation of enzyme synthesis in animal cells. Essays Biochem 13:39–69, 1977

Walsh C: Enzymes: macromolecular biological catalysts of chemical change. In Offord RE (ed): Chemistry of Macromolecules IIA, Simple Macromolecules, pp 143–170. Baltimore, University Park Press, 1979

Wang JH, Waisman DM: Calmodulin and its role in the second-messenger system. Curr Top Cell Regul 15:47–107, 1979

Wilkinson JH: The Principles and Practice of Diagnostic Enzymology. Chicago, Year Book Medical Publishers, 1976

Wimmer MJ, Rose IA: Mechanisms of enzyme-catalysed group transfer reactions. Annu Rev Biochem 47:1031–1078, 1978

Chapter Three Energetics

METABOLIC SEQUENCES—ENERGY CONSERVATION AND REVERSIBILITY

In the foregoing account we have followed some of the important factors that influence the catalytic actions of individual enzymes. The processes of metabolism represent the sum total of these enzymatic reactions within the body. The conversions performed by the component enzymes of a metabolic pathway would each be expected to play a significant, purposeful role in the overall pathway, and yet, to the student uninitiated in the wisdom of the body's metabolic economy, all too often the paths of metabolism seem to wander meaninglessly through morasses of terminology and tangled thickets of chemical structures, with only an odd flash from a seemingly magical "high-energy" compound to light the way. Why, for example, should the chemically simple conversion

Glucose → Lactic acid

$C_6H_{12}O_6 \rightarrow 2C_3H_6O_3$

require the sequential action of 11 enzymes with five coenzymes or cofactors and produce along the way 10 ***intermediate*** organic compounds, nine of which are combined with one or two molecules of phosphate? At times the student must view this apparently unnecessary complexity either as a bad joke on the part of nature or as a diabolic scheme invented by a conspiracy of sadistic biochemists. Although teachers of biochemistry may be innocent of the latter charge, I suspect that many have been too busy—either in justifying the reality of the metabolic meanderings of nature with a plethora of experimental evidence or in describing the beauty spots along the way with loving, detailed care—to rationalize all the apparent detours and pitfalls on the metabolic highways and byways.

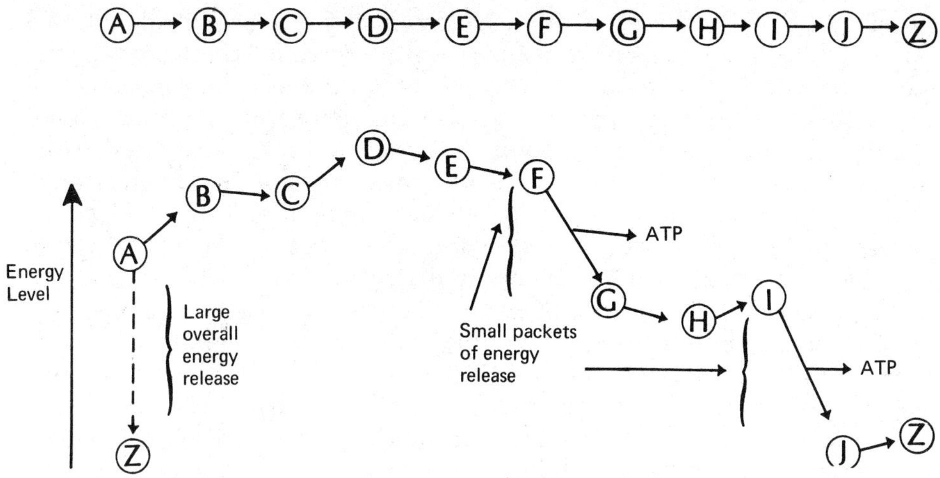

FIG. 3-1. Energy changes in a metabolic sequence.

A comprehension of the purposes of metabolic reactions must be based upon considerations of the energy changes involved as well as upon the ultimate products of the reactions. In short, it is not enough to consider the course of metabolism as a series of events in a horizontal continuum. It is essential to view the process in the vertical dimension as well, to recognize that the pathways traverse hills and valleys of energy content. Thus we should place ourselves not in the position of cartographers simply interested in charting metabolic routes on the flat surface of a map, but rather as the navigators and drivers faced with the problems in following those routes over the inclines and around the obstacles imposed by an undulating energy terrain between the starting point A and the finish line Z of a metabolic rally (Fig. 3-1).

If, in our metaphor, A is a molecule of glucose and Z is two molecules of lactic acid, then for most tissues of the body, the major rationale for the conversion is the capture of useful chemical energy needed for other physiologic functions. In the brain and nerves, the energy is needed to perform electrical work; in the skeletal muscles, for mechanical work; in the kidney and red blood cells, for osmotic work; and so forth. The total conversion of glucose to lactic acid, termed ***glycolysis,*** is indeed an energy-yielding process, and it therefore satisfies the requirement posed previously for individual enzyme reactions, namely, that the energy content of the substrate must be greater than that of the products. Thus, the overall conversion may proceed spontaneously in a downhill course from the standpoint of energetics. Quantitatively, the reaction sequence will involve the release of 56 Cal of energy for each mole of glucose used.

Were the energy to be released in a single, direct chemical conversion, little of that energy could be conserved for useful purposes. In fact, most would be released in the form of heat, which could be considered purposeful only if our sole aim in using foodstuffs were to keep warm against a temperature gradient. This, however, is but one of the minor goals of the body in relation to catabolic events; heat energy, for example, cannot be utilized by the body for growth or the replacement of essential tissue constituents by anabolism, nor for the physiologic work demands mentioned above. For these energy-requiring processes to be geared to the energy-yielding processes of catabolism, there must be an essential cog to transfer the useful chemical energy in discrete packages.

These packets or quanta of chemical energy are roughly 7.5 Cal each and correspond to the formation of the high-energy phosphate compound known as *adenosine triphosphate* (ATP). Once we accept this quantization of energy in packets, the stepwise nature of metabolism becomes clearer. It is necessary, then, to break down the 56-Cal glycolytic leap into shorter energy drops to prevent wasteful losses of energy as heat in order to conserve portions in the molecules of ATP that can serve as carriers of chemical energy for other energy-requiring reactions. The sequential reactions of glucose metabolism provide a staircase of small individual energy changes, instead of a precipitous energy drop.

Another advantage of the arrangement of an energy staircase relates to the problem of ascent, or the reversibility of the process. In the example chosen, catabolism of glucose to lactic acid leading to energy production predominates in most organs of the body, but in the liver, the reverse, anabolic process—the production of glucose from lactic acid—is more important physiologically. Other tissues remove glucose from the blood, utilize it for their energy needs, and pass the waste products, including lactic acid, back into the bloodstream. Vigorous activity on the part of these extrahepatic tissues may lead to excessive lactic-acid accumulation (causing a *p*H decrease or *metabolic acidosis*) as well as to deficient glucose concentrations in the blood (*hypoglycemia*). The liver acts to redress these imbalances by taking up lactic acid from the blood and resynthesizing glucose, which may be secreted into the circulation to meet the catabolic needs of extrahepatic tissues. This process of glucose synthesis from other carbon sources such as lactic acid is termed ***gluconeogenesis,*** and it occurs mainly by the reversal of the enzymatic reactions of glycolysis.

The reactions of glycolysis could not proceed readily in the reverse sense if they were accompanied by a single, large dissipation of energy. It is only when the energy differences between the substrate and the product are small that an enzyme may react reversibly to convert significant amounts of product back to substrate. In this situation, where the reaction is poised at or near a balanced equilibrium of substrate and product, the direction of the reaction will be determined chiefly by the rates at which either reactant is removed. In fact, the rapid removal of a reactant may force a reaction to proceed uphill against an energy gradient; this can occur only when the energy barrier is quite slight. If the barrier is large, an alternative enzyme reaction must be employed to circumvent the obstacle.

Gluconeogenesis employs both types of mechanism. In the case of three of the enzyme steps of glycolysis, energy considerations make the reverse reactions unfavorable, and entirely different enzymes must be employed to surmount these energy hurdles. The remaining eight enzymes employed in glycolysis, however, are poised near enough to the balance point to allow them to be used in the opposite direction for glucose synthesis. Thus the analogy with the staircase is complete. Not only does it provide a safer and more rational method to descend from energy heights, but the ability to climb back is facilitated by breaking down an insurmountable energy expenditure into easily negotiated steps.

Before passing on to an examination of human metabolism at closer range, one final, general point should be understood. This concerns the multitude of intermediate compounds involved in the sequences and cycles of metabolic reactions. In the majority of cases, the substances intervening between points A and Z are present in very low concentrations in the cells of the body. Metabolic pathways generally operate fairly efficiently to remove the intermediates as rapidly as they are formed. The necessity of such an efficient system is obvious from an economic standpoint: it would be as disadvantageous for the cell to accumulate partly modified molecules in significant concentrations as it would be for a manufacturing plant to allow a buildup of half-finished goods to occur on an assembly line. Apart from the wastefulness of such a blockage, the piling up of intermediate substances may produce considerable damage to the cell's metabolic machinery. In many instances, it has now been established that the effects of one inoperative enzyme in a chain of reactions may be extremely devastating, even fatal. The presence in the body of abnormally increased amounts of the normal intermediates of metabolism can thus be seen to have profound pathologic sequelae. When we set down the reactants for a particular segment of metabolism, we present essentially a stopped-action picture of events that are in constant flux. We should continually bear in mind that these events occur with great rapidity in the cell and that the substances appearing along the way exist for only a brief moment.

HIGH-ENERGY COMPOUNDS

The concept of the so-called ***high-energy compounds*** is of great importance in the comprehension of human metabolic events. A high-energy compound is defined as one in which a large amount of energy may be stored and is released upon *hydrolytic cleavage* of the compound, that is, by the addition of the elements of water. In a sense, then, the high-energy compounds are quite labile in aqueous solutions. A second consideration

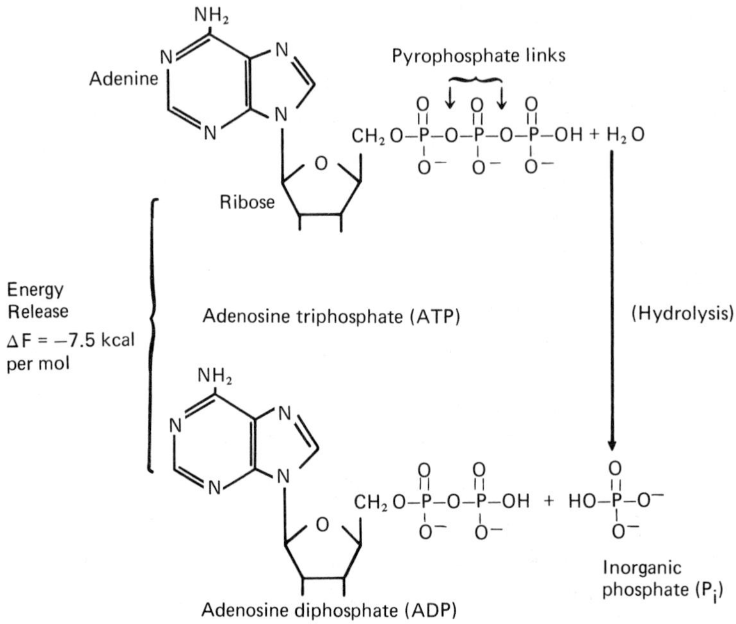

FIG. 3-2. The structure of adenosine triphosphate (ATP) and its hydrolysis products.

to remember is that it is the ***free energy,*** or useful chemical energy, that is obtained on hydrolysis, which confers upon the high-energy compounds their importance in metabolism. The free energy represents the component of energy that may be made available for chemical work in the body and is generally abbreviated as F or G; ΔF or ΔG, which is more commonly seen, is the *change of free energy* occurring during the reaction. Third, and most significant of all, is the fact that enzymes are capable of catalyzing the transfer of energy between pairs of reactants. This is the means whereby we are able to couple together the energy-yielding processes of catabolism with the energy-requiring processes of anabolism, muscle contraction, nerve conduction, and so forth.

The most important of the high-energy compounds is ATP, adenosine triphosphate (Fig. 3-2). It consists of a heterocyclic ring structure, adenine, a pentose sugar, ribose, and three phosphate groups attached together in a string by pyrophosphate linkages. It is the last that confer upon ATP its great ease of hydrolysis. Water will readily attack the pyrophosphate structure because the electrostatic repulsion between the negatively charged phosphate groups tends to force them apart. This strong tendency to split off the terminal phosphates of ATP also provides the driving force for the enzymatically catalyzed transfer that leads to the ***phosphorylation*** of other compounds in the body at the expense of ATP. This transfer is energetically feasible because it proceeds from a substrate with a higher energy content than that of the product.

A simple example to illustrate these principles is the phosphorylation of glucose by ATP, which takes place in the presence of the enzyme *hexokinase*. The products of the reaction are adenosine diphosphate (ADP, or ATP with the terminal phosphate group removed) and glucose-6-phosphate, a low-energy phosphate compound. The difference between the high-energy and low-energy phosphate compounds is exemplified in the reactions that describe their hydrolyses:

1. Adenosine—P~P~P + H_2O → Adenosine—P~P + H_3PO_4
(ATP) (ADP) $\Delta F = -7.5$ Cal

2. Glucose-6-P + H_2O → Glucose + H_3PO_4 $\Delta F = -3.2$ Cal

In reaction (2), as in the hydrolysis of any simple sugar–phosphate ester, there is a free-energy release of only 2 Cal to 3 Cal per mole of reactant. In contrast, the hydrolytic cleavage of one of the pyrophosphate linkages of ATP is accompanied by a much greater free-energy loss; the reactant is considered to be a high-energy compound—by definition—*when the decrease in free energy exceeds −5 Cal per mol of reactant.* Hexokinase couples these two energy transfers by utilizing the driving force from the large energy release in reaction (1) to promote reaction (2′) in the reverse direction, or uphill in an energy sense:

1. ATP + H_2O → ADP + H_3PO_4 ($\Delta F = -7.5$ Cal)

2′. Glucose + H_3PO_4 → Glucose-6-P + H_2O ($\Delta F = +3.2$ Cal)

3. ATP + Glucose → ADP + Glucose-6-P ($\Delta F = -4.3$ Cal)

Reaction (3) is the sum of reactions (1) and (2′), and the energy change of the total reactants will be the sum of the energy changes in (1) and (2′).

The following points should be noted in considering this example. First, in reversing reaction (2), it is necessary to reverse the sign of the free-energy change to indicate that energy must be put into the system to esterify the glucose molecule with phosphate (*i.e.,* +3.2 Cal per mol of glucose-6-P formed). Second, the reaction represented by the sum (3) would not proceed in the absence of the appropriate kinase enzyme, which brings the reactants together to catalyze the overall phosphate transfer. Third, the reactants H_2O and H_3PO_4 are shown as canceling out algebraically in the total reaction (3). In fact, the enzyme catalyzes the reaction by bringing ATP and glucose close together, each being bound at its active center, so that the phosphate group may be transferred directly without any production of H_3PO_4. Finally, the sum of the energy terms indicates that reaction (3) as written will cause a net release of −4.3 Cal per mol of reactants. The reaction will thus tend to occur in the direction of the phosphorylation of glucose, and the phosphorylation reaction will now become downhill in the energy sense as the result of coupling it to the dephosphorylation of ATP. Because there would be a considerable energy increase (+4.3 Cal per mol of reactants) required for the reverse of reaction (3), the enzyme hexokinase is unable to promote this process in the opposite or uphill direction to any significant extent.

Many different high-energy compounds are found in nature and they may play various roles in metabolism, depending upon the specialization of the organism and its tissues (Fig. 3-3). Several other ***nucleoside triphosphates*** are found in the cell, but ATP appears to play the major, preferred role as a donor of phosphate groups with the *kinase* enzymes. The other triphosphates have specialized functions in segments of metabolism: uridine triphosphate (UTP) acts to produce reactive derivatives of sugar phosphates (such as UDP-glucose) for anabolic processes in carbohydrate metabolism; cytidine triphosphate (CTP) plays an important role in generating activated intermediates (such as CDP-choline) for phospholipid synthesis; and guanosine triphosphate (GTP) has an important selective function in the promotion of peptide-bond formation during protein anabolism. Moreover, ATP, UTP, CTP, and GTP serve as the immediate precursors for the polynucleotide chain of ribonucleic acid (RNA); the corresponding deoxy forms of the sugar nucleotides—dATP, dCTP, dGTP, and dTTP (deoxythymidine triphosphate) replacing UTP—play the same role in deoxyribonucleic acid (DNA) biosynthesis (see Chap. 1).

Other phosphate compounds may yield sufficient energy upon hydrolysis to be considered in the high-energy class. These substances fall into two general divisions: (1) those that are generated during catabolism for the purposes of energy conversion and conservation, and (2) those that may act as emergency tissue energy stores. Included in the former group are the two carbohydrate derivatives 1,3-diphosphoglycerate (DPG) and phosphoenolpyruvate (PEP), both of which are produced as transient intermediates during anaerobic energy conversions in the glycolytic pathway. In DPG, the phosphate is esterified in position 1 by the very labile acyl-phosphate ester linkage; in PEP, the phosphate is attached with an unstable enol ester. In both cases, the energy lost upon hydrolysis is considerably more negative than even that for ATP hydrolysis ($\Delta F = -12$ Cal and -14 Cal per mol, respectively). Consequently, it is possible for these two substances to donate a phosphate group to ADP in order to synthesize ATP by a reversal of the usual kinase-type reaction.

The second group of substances listed consists of nitrogen-containing ***phosphagens,*** such as the phosphocreatine found in vertebrate tissues and the phosphoarginine in invertebrates. Phosphocreatine is particularly abundant in skeletal muscle, where it may provide energy for short bursts of rapid activity, such as in a sprint race. The hydrolysis of the N-phosphate linkage of the phosphagens releases quantities of energy similar to those of ATP hydrolysis, and thus the net energy loss upon the transfer of

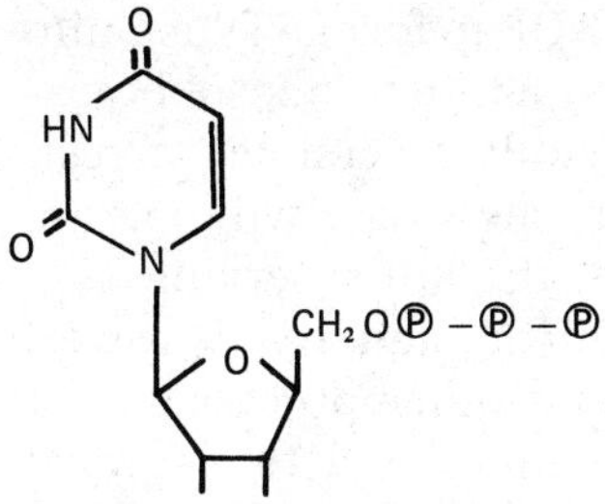

Uridine triphosphate (UTP)

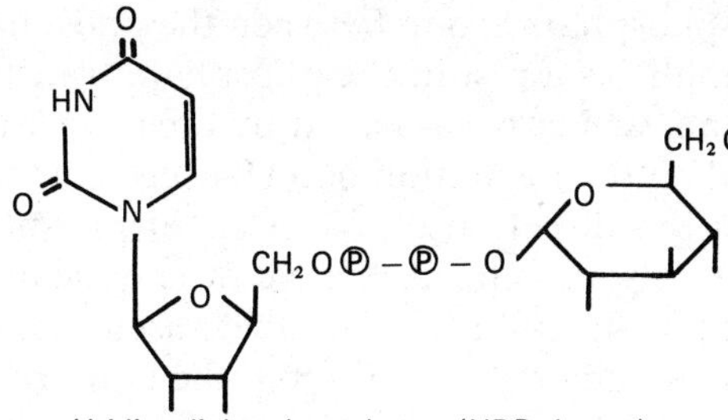

Uridine diphosphate glucose (UDP-glucose)

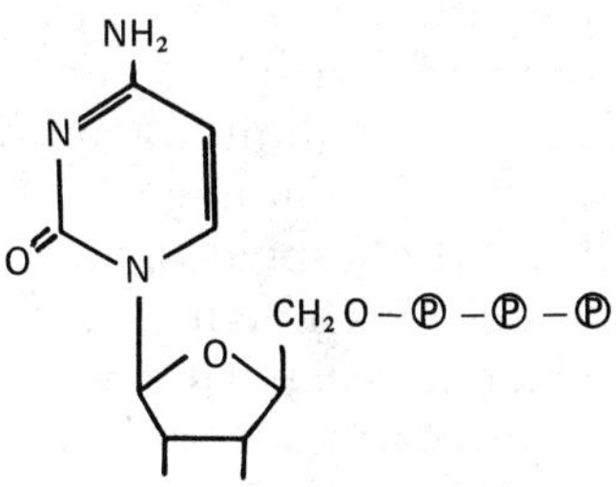

Cytidine triphosphate (CTP)

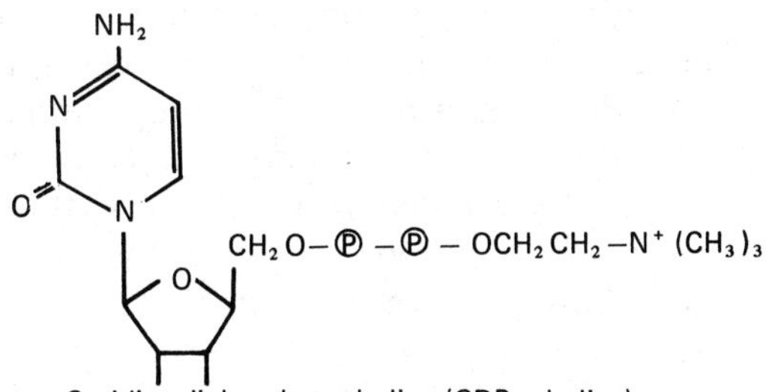

Cytidine diphosphate choline (CDP—choline)

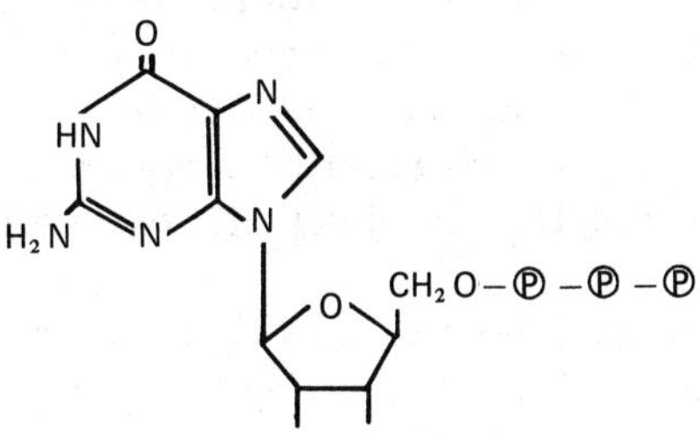

Guanosine triphosphate (GTP)

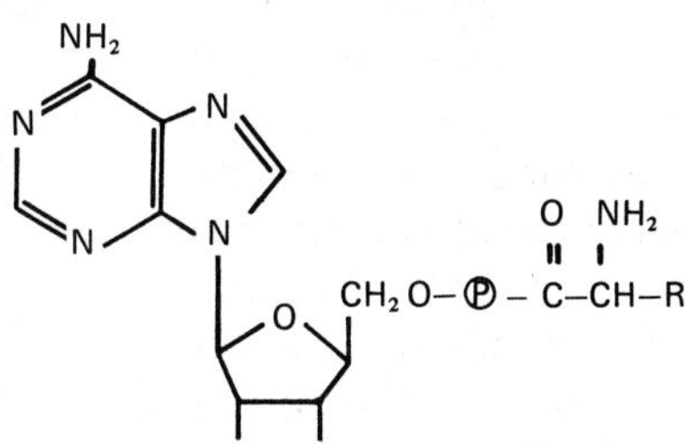

Aminoacyl adenylate (Aminoacyl AMP)

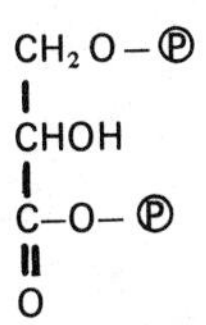

1,3-diphosphoglycerate (DPG)

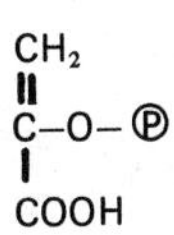

Phosphoenolpyruvate (PEP)

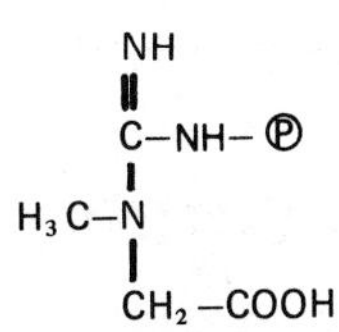

Phosphocreatine

Acetyl CoA

FIG. 3-3. Some important high-energy compounds. The circled *P*s represent phosphate groups.

a phosphate group between the phosphagen and ADP to form ATP is quite small. As a result, the phosphate transfer between ATP and phosphocreatine, which is catalyzed by creatine kinase, is readily reversible. At rest, when the formation of ATP exceeds its utilization, the kinase will operate in the direction of synthesis of phosphocreatine, the latter serving as a readily mobilized reserve of chemical energy. During activity, however, the utilization of ATP for muscle contraction and other processes that consume energy will force the equilibrium of the kinase reaction toward phosphocreatine breakdown in order to maintain a constant concentration of ATP to meet the tissue's energy requirements.

The formation of high-energy compounds is a common means whereby the cell may activate molecules for the transfer of chemical groups and other metabolic processes. The acyl groups of short-chain compounds (such as acetate) or long-chain fatty acids (such as palmitate) may be rendered highly reactive by conversion to the sulfur-containing (thio) esters of ***coenzyme A*** (CoA). These thio esters, like acyl phosphates, have large negative free energies of hydrolysis (-7 Cal/mol). Such high-energy acyl–CoA esters provide the reactive species for acylation reactions as well as for a large variety of intermediate steps in the anabolism and catabolism of lipids. Activation of a carboxyl group may also be brought about by the formation of the highly unstable acyl–adenylate compound. Such a reaction is important in the mechanism of production of acyl–CoA esters and in the activation of amino acids for peptide-bond formation. In either case, the carboxylic group forms an ester linkage to the adenosine monophosphate portion of ATP. The two terminal phosphates of ATP are released as inorganic pyrophosphate ($P\text{–}P_i$), and, since the latter may not be directly reutilized for ATP synthesis, the reaction in effect utilizes two high-energy equivalents.

Numerous other examples of high-energy intermediates that are activated in order to promote group transfer reactions are encountered throughout metabolic pathways. In all cases, their formation is dependent, either directly or indirectly, upon the energy available from ATP. Thus, in the energy economy of the cell, there may be several transferable forms of currency, each appropriate to its metabolic locale; the common currency integrating the energy interconversions, however, is the adenine–nucleotide system, with ATP as the universal gold standard.

OXIDATION–REDUCTION REACTIONS

The energy-yielding processes of the tissues are ultimately dependent upon enzymatically catalyzed oxidative reactions. Individual organs and cell types may vary greatly with respect to their oxidative capacities, but ultimately each will have an absolute requirement for molecular oxygen. For the student faced with the seemingly endless types of chains of oxidation processes, this simple essential principle may be too frequently

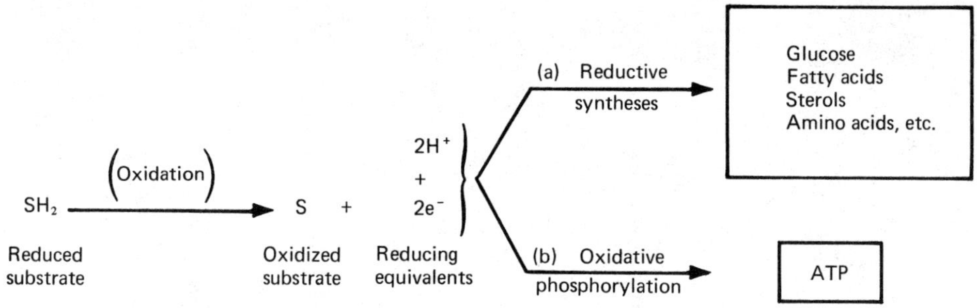

FIG. 3-4. Utilization of energy released by biologic oxidations.

forgotten. In all the welter of dehydrogenases and electron carriers, let us not lose sight of the central fact of biologic oxidations: that they exist primarily in order to provide the energy for cellular work, either *directly (a),* by furnishing mobile hydrogens for the reductive syntheses of glucose, fatty acids, sterols, and so forth, or *indirectly (b),* by combining these hydrogens with oxygen to form water, with the coupled production of ATP in the process of oxidative phosphorylation (Fig. 3-4). It is only by considering the oxidative functions of the cell in terms of these energy exchanges—either as *(a) reducing energy* or as *(b) phosphorylating energy*—that we can perceive their purpose and importance.

Just as ATP plays the pivotal role in cellular energy transfers by way of phosphorylation reactions, so the coenzymes NAD and NADP function centrally in oxidation–reduction reactions. ***Nicotinamide–adenine dinucleotide*** (NAD) contains an amide of the B vitamin, niacin or nicotinic acid, which is joined by a ribose molecule to the terminal phosphate of ADP (Fig. 3-5). The adenine–nucleotide end of these coenzymes indubitably participates chiefly in the binding reactions that determine their specificity of interaction with particular enzyme systems; this is apparent from the fact that ***nicotinamide–adenine dinucleotide phosphate*** (NADP) differs only by an extra phosphate attached to the adenine–ribose end, yet it plays a distinctly different role in metabolism from that of NAD. It is the nicotinamide part of both coenzymes, however, that confers upon them the unique property that allows them to occupy the center stage in the theater of oxidations and reductions. In the presence of the appropriate enzymes, the nicotinamide ring of the coenzymes is able to extract hydrogen from a wide variety of simple biologic molecules; this process is logically known as ***dehydrogenation,*** and the enzyme responsible, a *dehydrogenase* (Fig. 3-6). In a typical reaction, two hydrogen atoms (with their two electrons) are removed from the substrate to produce the oxidized form of the substrate: one hydrogen with two electrons ($H:^-$ or hydride ion) is transferred to the nicotinamide ring to produce reduced NAD or NADP (NADH or NADPH), and the other is released to the surroundings without electrons (H^+ or proton).

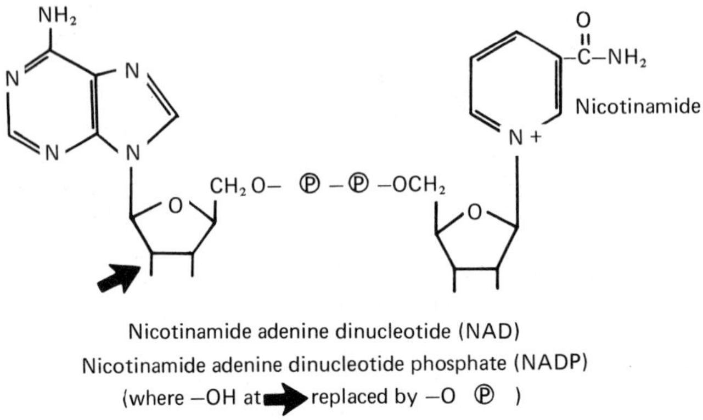

FIG. 3-5. The structure of nicotinamide-adenine dinucleotide (NAD) and its phosphate compound (NADP).

The chief features of this type of hydrogen transfer that make it so important in the economy of the cell are the following: (1) under most conditions, the energy change is small and therefore many of the dehydrogenase reactions may be reversed; (2) the coenzymes in their oxidized or reduced states are usually freely dissociable from their dehydrogenases; and (3) many different types of substrates—carbohydrates, lipids, amino acids, and so on—may participate in such reactions. As a result of these general properties, the nicotinamide–nucleotide oxidation–reduction system has a tremendous degree of flexibility. Its reversibility allows it to act as a means either to oxidize substrate or to reduce product, depending on the needs of the cell. Its mobility facilitates the shuttle of hydrogen atoms and electrons from one dehydrogenase system to another and from one part of the cell to another. Its wide range of substrate types provides a link whereby the reducing equivalents required for reductive anabolism of one category of biomolecules may be provided by the oxidative catabolism of another.

An important example of the NAD-coupled hydrogen transfers in human tissues is the reaction catalyzed by lactate dehydrogenase. As in the case of the hexokinase reaction discussed earlier, it is useful to break the process down into its constituent half-reactions to understand the driving force that makes the system work. With the oxidation–reduction reactions, however, the usual way to express this driving force is in terms of the ***reduction potentials*** (E_0') instead of the energy changes, but it is essentially another way of saying the same thing. The half-reactions are expressed as the tendency of each of the reactants to take on or give up electrons:

1. $NAD^+ + 2H^+ + 2e^- \rightarrow NADH + H^+$ $(E_0' = -0.32V)$

2. $\text{Pyruvate} + 2H^+ + 2e^- \rightarrow \text{Lactate}$ $(E_0' = -0.19V)$

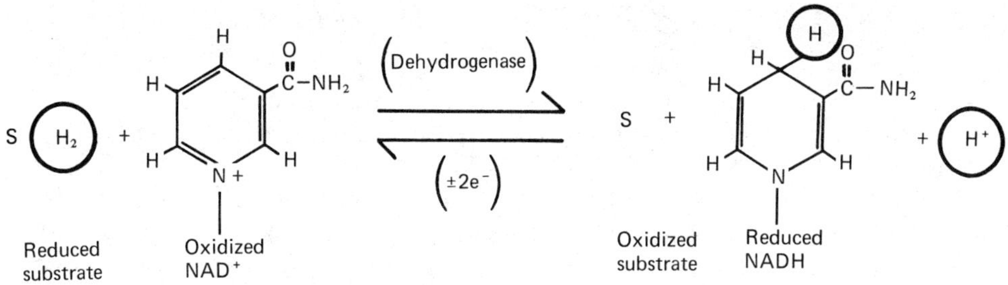

FIG. 3-6. Role of nicotinamide coenzymes in dehydrogenation.

The more positive the value for E_0' (in volts), the greater will be the tendency for the reaction to act as an oxidizing system by accepting electrons from another substance; the more negative the value for E_0' (in volts), the greater will be the tendency for the reaction to act as a reducing system by donating electrons to another substance. Thus, in our example, the NADH–NAD^+ couple of reaction (1) is more negative in terms of reduction potential and will tend to donate electrons (and hydrogen atoms) to reduce the more positive lactate–pyruvate couple of reaction (2). The overall reaction (3) is expressed as the sum of the half-reactions (1) and (2), canceling out the common reactants:

3. $NADH + H^+ + Pyruvate \rightarrow NAD^+ + Lactate$

The driving force for this reduction of pyruvate is provided by the strong reducing potential of NADH, whose E_0' value is low enough to allow it to act as chemical reductant for a host of important metabolites in the body. In this way, both NADH and NADPH (which has the same reduction potential) provide a form of *chemical reducing energy* in the biosynthesis of glucose by gluconeogenesis, in the production of the long alkyl chains of fatty acids, in the fixation of ammonia into glutamic acid, and in many other anabolic processes. Just as the ATP–ADP couple acts as the central energy source for chemical-group transfers, so the NADH–NAD and NADPH–NADP couples form the common energy currency for reductions involving hydrogen transfers.

The second form of energy exchange in biologic oxidation–reduction reactions is the utilization of the reducing energy, not for anabolic purposes, but for the generation of ATP. The latter is an integral aspect of the catabolic process of cellular respiration, and it serves as the principal source of ATP in virtually every cell of the body. Overall, the reactions may be written:

1. $NAD^+ + 2H^+ + 2e^- \rightarrow NADH + H^+ \quad (E_0' = -0.32V)$

4. $\frac{1}{2}O_2 + 2H^+ + 2e^- \rightarrow H_2O \quad (E_0' = +0.82V)$

Applying the criteria given previously, the coupled reaction would yield the sum:

5. $NADH + H^+ + \frac{1}{2}O_2 \rightarrow NAD^+ + H_2O \qquad (\Delta E_0' = +0.82 - (-0.32) = +1.14V)$

where $\Delta E_0'$ is the net difference in reduction potential between the two half-reactions. This large potential difference between the NADH–NAD couple and the H_2O–O_2 couple will result in the release of a large amount of energy, $\Delta F = -53$ Cal per mol of NADH oxidized by oxygen ($\Delta F = -46\Delta E_0'$). Just as in the case discussed previously for the glycolysis of glucose, if such a reaction occurred directly—as, for example, in the combustion of hydrogen by oxygen in a rocket engine—then the energy would be lost from the body as thermal and kinetic energy. To control the energy release, two organizational principles are utilized by the cell so that useful chemical energy may be trapped and conserved.

The first principle is strictly analogous to that propounded above for the glycolytic system, namely, a staircase of small energy drops is utilized rather than a single, suicidal leap. In the oxidation of NADH, this is accomplished by interposing a series of ***electron carriers*** with graduated increases of E_0' that eventually transfer the reducing equivalents to oxygen as the final electron acceptor. This stepwise cascade of electrons facilitates the quantization of the energy changes into several of the -7.5 Cal packets required to synthesize a mole of ATP (Fig. 3-7).

The second principle, which distinguishes this system from the mixture of soluble glycolytic enzymes of the cytoplasm, is based on the physical proximity of the electron carriers, which are bound together in rigid proportions in the membrane system of the mitochondrion. Thus, the order of reaction in the electron transfer is governed both by the progression of reduction potentials and by the sequential arrangement of the carriers in the semisolid state. The physical restraints of this highly ordered system contribute to the speed and efficiency of the process by eliminating the necessity for diffusible intermediates and by minimizing side reactions. Moreover, the orientation of the electron carriers with respect to the sides of the membrane is crucial for energy coupling and ion movements, as will be discussed later in detail.

THE RESPIRATORY CHAIN AND OXIDATIVE PHOSPHORYLATION

The electron-transport system of the mitochondrion contains four classes of carrier compounds, each capable of undergoing reversible oxidation and reduction by donating and accepting electrons. In some instances, hydrogen transfer may accompany the electrons; in others, hydrogen ions may be generated or utilized, producing H^+ concentration gradients.

In the first category of carriers are the ***flavoproteins,*** which contain the B vitamin, riboflavin, as their prosthetic group. The riboflavin plays an analogous role to that of nicotinamide in NAD by accepting hydrogens and electrons into its heterocyclic ring structures (Fig. 3-8). In its nucleo-

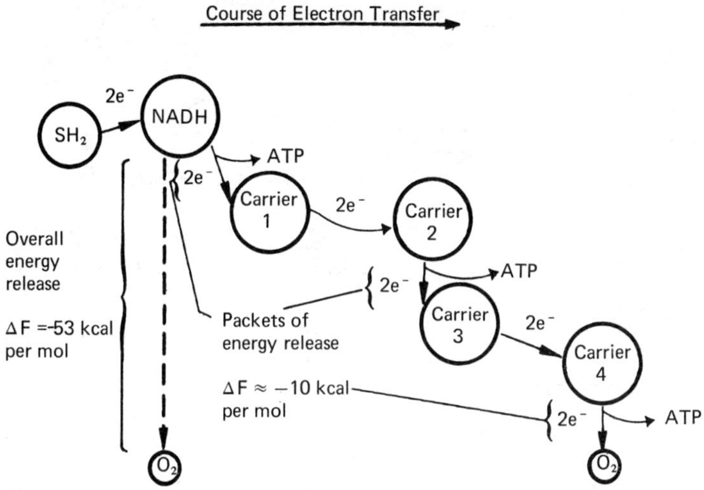

FIG. 3-7. Oxidation of NADH by oxygen through a series of electron carriers.

tide form as flavin–adenine dinucleotide (FAD) or as flavin mononucleotide (FMN; riboflavin phosphate), the flavin is rigidly bound to its dehydrogenase and, unlike NADH, is not free to carry reducing equivalents by diffusion to other dehydrogenases. A second difference is that the reactions of the flavin-linked dehydrogenases are not easily reversible, and hence the flavin coenzymes do not engage in the transfer of chemical reducing equivalents for reductive syntheses, as do the nicotinamide coenzymes.

Flavoproteins participate at several points where electrons are first funneled into the respiratory chain, in keeping with their relatively low reduction potenials (≤0 volts). Two electrons and two hydrogens are accepted by the FMN coenzyme from reduced NAD:

1. $NADH + H^+ + Enz\text{-}FMN \rightarrow NAD^+ + Enz\text{-}FMN\text{-}H_2$

Because there is a considerable difference in the reduction potentials between the NAD–NADH couple and the FMN–$FMNH_2$ couple ($\Delta E_0' \approx 0.3V$), there is more than enough energy released to provide for the generation of a mol of ATP ($\Delta F \approx -46 \times 0.3 \approx -14$ Cal). This reaction constitutes the *first* ***coupling site*** of oxidative phosphorylation, where the energy released from the oxidation is coupled to the formation of ATP from ADP and inorganic phosphate (P_i). Flavoproteins (containing FAD) also provide the entrée for electrons to the respiratory chain from succinate, fatty-acyl CoA, and α-glycerophosphate. In these instances, however, the E_0' values for the substrate couples are quite close to those for the corresponding Enz-FAD–Enz-$FADH_2$ couple, so that the difference, $\Delta E_0'$, is not great enough to promote the formation of ATP. With these substrates, as opposed to those that generate NADH, the first coupling site is bypassed,

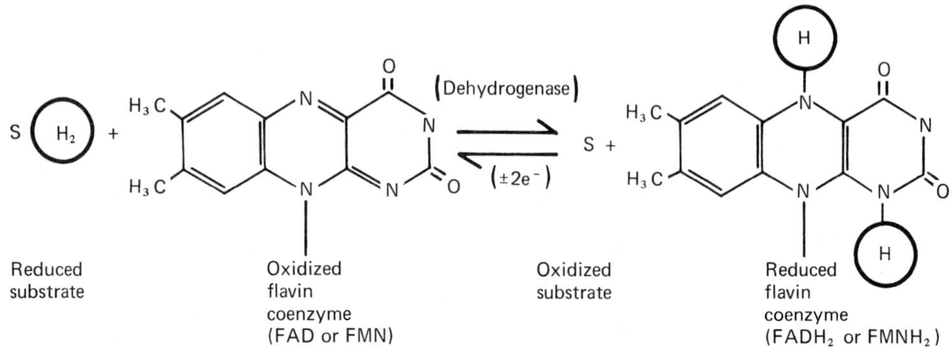

FIG. 3-8. Role of riboflavin coenzymes in dehydrogenation.

and hence their oxidation leads to one mole less of ATP per mole of substrate oxidized.

The second category of electron carriers comprises the benzoquinone compound, ***coenzyme Q*** (Fig. 3-9). Upon reduction, it accepts two electrons and two hydrogens, as do the flavoproteins, to become the hydroquinone compound, $CoQH_2$. The $CoQH_2$–CoQ couple has an E_0' value only slightly more positive than that of the flavoproteins, so the latter may transfer reducing equivalents to the quinone; the energy change is not large enough to generate a mol of ATP. It appears that the CoQ system acts as the focal point where all the various flavoproteins may discharge their electrons and hydrogens:

$$FP_1H_2 + CoQ \rightarrow FP_1 + CoQH_2$$

$$FP_2H_2 + CoQ \rightarrow FP_2 + CoQH_2, \text{ etc.}$$

In keeping with its multifarious role as a collecting point for the reducing equivalents from several different flavin enzymes, coenzyme Q is present in higher concentrations than most of the other electron carriers in the chain. Moreover, its relatively small molecular size and high solubility in the lipid phase of the mitochondrial membrane equip CoQ ideally as a mobile carrier between the fixed electron-transport proteins.

The third type of component of the respiratory chain is a diverse group of hemoproteins, the ***cytochromes.*** Varying chiefly with respect to the structure of their apoprotein portions, each of the cytochromes contains a conjugated tetrapyrrole ring coordinated to an iron atom as its prosthetic group (Fig. 3-10). Unlike the flavoproteins—which, as noted above, function *in parallel* to donate reducing equivalents to the common acceptor, CoQ—the cytochromes react in *series* to transfer those reducing equivalents from CoQ to the ultimate acceptor, oxygen. Furthermore, unlike all of the carriers discussed thus far, the cytochromes carry only electrons along the respiratory chain. This they accomplish by the reversible

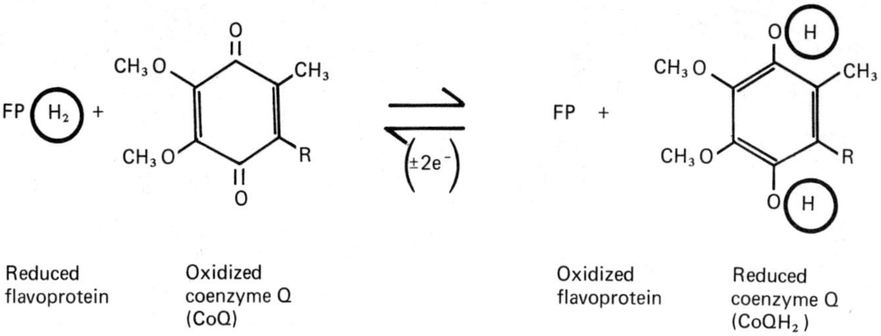

FIG. 3-9. Coenzyme Q.

oxidation and reduction of the iron atom in the prosthetic group between its ferrous and ferric states (Fe^{2+} or Fe^{3+}). Thus there is a sequence of reactions between cytochrome heme groups, with a single electron being transported at each step of the sequence.

The first step in this chain of reactions is the transfer of electrons from $CoQH_2$ to the cytochrome with the lowest E_0' value, *cytochrome b:*

$$CoQH_2 + 2Fe^{3+}\text{-cytochrome b} \rightarrow CoQ + 2Fe^{2+}\text{-cytochrome b} + 2H^+$$

It is noteworthy that the transfer of the two electrons from reduced CoQ requires 2 mols of the single-electron acceptor, cytochrome b; moreover, the hydrogen atoms, which cannot be transported by the reduction of the iron atom of the cytochromes, are released as protons. CoQ may be viewed as a mobile transport agent both for electrons within the membrane, and possibly for protons across the membrane. Both the CoQ–$CoQH_2$ and the cytochrome-b Fe^{2+}-Fe^{3+} couples have similar E_0' values, and little energy change is involved in this reaction. The reduced form of cytochrome b that is produced has a low enough E_0' value to allow it to act as the reductant for the heme groups of the next carriers in the chain, which are designated as the *c-type cytochromes.* In tracing the subsequent fate of the reducing equivalents, it is convenient to maintain the stoichiometry of electron pairs by representing the cytochrome reactions in pairs:

$$2Fe^{2+}\text{-cytochrome b} + 2Fe^{3+}\text{-cytochrome c} \rightarrow 2Fe^{3+}\text{-cytochrome b} + 2Fe^{2+}\text{-cytochrome c}$$

This particular reaction is important because it includes the *second coupling site* in oxidative phosphorylation; the E_0' values for cytochromes b and c are +0.04V and +0.25V; therefore $\Delta E_0' = 0.21$V and $\Delta F \approx -10$ Cal per two electrons, which is more than ample for the formation of ATP. An additional cytochrome, designated c_1, with an E_0' value very close to that of cytochrome c, is a bound intermediate in this reaction and carries out the actual transfer of electrons between cytochromes b and c.

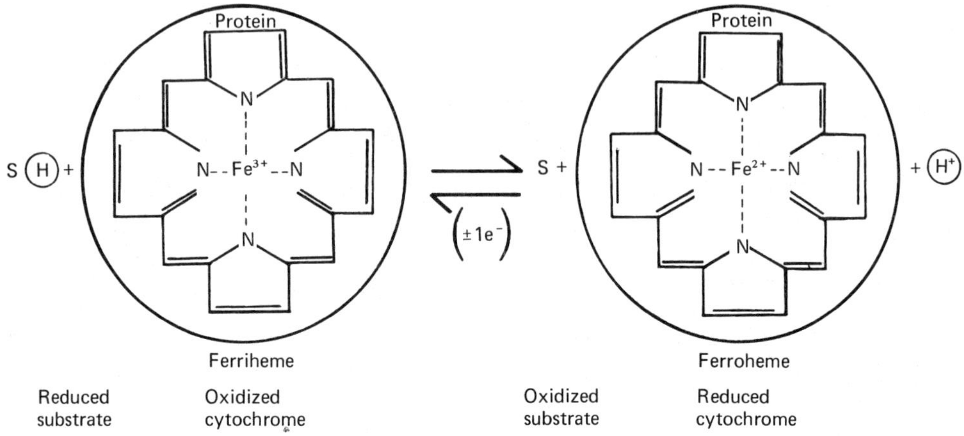

FIG. 3-10. Cytochromes.

The final stage in respiration is the reaction catalyzed by the enzyme ***cytochrome oxidase,*** a complex hemoprotein containing two chemically different heme groups that are designated *a* and a_3. The former reacts with cytochrome c to accept the electrons and transfer them to the a_3 component; the latter is unique among the cytochromes in its ability to react directly with oxygen, the final electron acceptor of the chain. The transfer of electrons sequentially from cytochrome c to a to a_3 occurs by reactions analogous to that detailed previously between cytochromes b and c. From an energy standpoint, the a and a_3 heme groups of cytochrome oxidase are at equilibirium since they have the same E_0' (+0.29V), and obviously there is little energy release during the transfer of electrons from cytochrome c ($\Delta E_0' \approx 0.04V$). It is during the subsequent reaction that sufficient energy may be generated to provide the *third and final coupling site:*

$$2Fe^{2+}\text{-cytochrome } a_3 + 2H^+ + \tfrac{1}{2}O_2 \rightarrow 2Fe^{3+}\text{-cytochrome } a_3 + H_2O$$

As noted previously, the H_2O–O_2 couple has a high E_0' (+0.82V), which is in keeping with our everyday experience of oxygen as a strong oxidizing agent. The avid capacity of oxygen to accept the electrons donated by the cytochrome-oxidase reaction provides a major driving force for the integrated sequence of electron transfers along the chain. This tendency of oxygen to act as an electron sink maintains each of the carriers in its oxidized state and thereby available to accept the reducing equivalents provided by the substrates of respiration (NADH, succinate, and so on). The energy produced at the third coupling site ($\Delta E_0' = 0.53V$; $\Delta F = -24$ Cal/2 e^-) is the largest of any component reaction of the chain. It would seem great enough to provide for the formation of several moles of ATP; nonetheless, the yield appears to be one mole of ATP at the third coupling

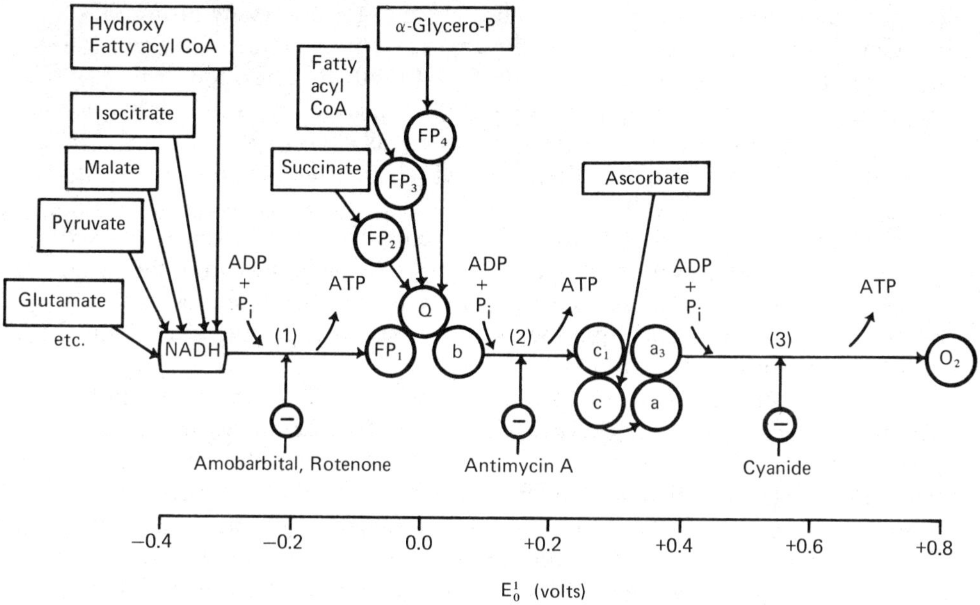

FIG. 3-11. The respiratory chain (*FP_1*, flavoprotein; *Q*, coenzyme Q; *b*, *c_1*, *c*, *a*, *a_3*, cytochromes). Reduction potentials (*E_0'*) are indicated by the positions of each of the carriers. The energy loss between carriers is great enough to support ATP formation at the three coupling sites, *1*, *2*, and *3*. The arrows indicate the direction of electron transfer from substrates to oxygen. Sites of action of respiratory inhibitors are represented by the circled minus signs.

site, as at the other two. It should also be noted that the cytochrome-oxidase reaction utilizes the two hydrogen ions from the substrates that were released to the medium during the oxidation of $CoQH_2$ by cytochrome b, and these hydrogen atoms ultimately become a source of *metabolic water* in the body.

Having considered the reactions of the respiratory chain in some detail, let's review the main features and principles of this most important energy-transducing system (Fig. 3-11). The first and most salient characteristic is the net yield of utilizable chemical energy transformed in the production of ATP. The yield and efficiency of the oxidative phosphorylation process is most commonly expressed as the ratio of phosphate esterified (P) to oxygen consumed (O), or the ***P:O ratio.*** A knowledge of the number of sites in the chain where phosphorylation is possible allows the calculation of this stoichiometry. Since the oxidation of NADH will involve the traverse of two electrons through each of the three coupling sites, we may assume that the oxidation of any of a number of substrates that pass their electrons to the respiratory chain by way of NADH will result in the production of 3 mols of ATP per electron pair transported to oxygen. Since two electrons will reduce ½ mol (½O_2) or one atom (O) of oxygen in the final cytochrome-oxidase step, we may express this relation by stating that three moles of phosphate (P) will be esterified with ADP to form ATP for every atom of oxygen (O) consumed by the respiratory chain; in shorthand notation, this is stated as P:O = 3. For the other common type of oxidizable substrate—which transfers its electrons into the chain at the level of the flavoproteins (*e.g.*, succinate, fatty-acyl CoA, or α-glycerophosphate)—only the last two coupling sites will be traversed and hence the oxidation of these substances will yield a ratio of P:O = 2. Finally, a substance such as ascorbic acid (vitamin C), which may directly reduce cytochrome c and thus bypass the first two coupling sites, would yield a value of P:O = 1 upon oxidation.

In terms of the efficiency of the process, the oxidation of one mol of NADH releases 52.4 Cal of energy, whereas the formation of three mols of ATP represents the capture of 3 × 7.5 = 22.5 Cal of useful chemical energy; therefore the efficiency will be (22.5/52.4) × 100, or approximately 40%. The remaining −30 Cal of energy released represents the driving force imparted to ensure that the respiratory chain operates essentially irreversibly in the direction: $SH_2 + O \rightarrow S + H_2O$. The energy loss is mainly in the form of heat energy, which cannot be used to perform useful chemical or mechanical work in our bodies, since we operate isothermally or at a fixed temperature. The heat generated by respiration, however, is a major factor in those *homeothermic mechanisms* that maintain our body temperature at a fixed value above that of our surroundings.

A number of poisons exert their toxic actions upon cells by selectively inhibiting some of the steps in the electron-transport chain. Best known among these ***respiratory inhibitors*** is hydrogen cyanide, the volatile gas liberated by the action of acid upon cyanide salts. The inhalation of HCN or the ingestion of KCN or NaCN salts in excess of a few milligrams leads to rapid death due to the avid binding of cyanide anion, CN^-, by the iron atom of cytochrome a_3. None of the other cytochromes in the chain is capable of forming such a complex with cyanide, but nonetheless the operation of the entire chain is brought to a standstill. By forming an inhibited compound with the heme a_3 component, cyanide blocks the ability of cytochrome oxidase to transfer electrons to oxygen from the other cytochromes below it in the chain. By damming up the outflow of electrons at the terminus, the cytochromes and other carriers of the chain accumulate in their reduced forms with no way of becoming reoxidized.

The antifungal antibiotic, antimycin A, produces toxic effects on mammalian cells by selectively inhibiting the transfer of electrons between cytochromes b and c_1. The consequence of this block in the respiratory chain is that all the carriers below cytochrome c_1 will remain reduced, while cytochromes c_1, c, a, and a_3 (*i.e.*, those on the oxygen side of the block) will remain oxidized.

A third class of inhibitors—which includes the barbiturate, amobarbital (Amytal), and the insecticide rotenone—includes those that block the chain lower down, at the level of the flavoprotein that transfers electrons from NADH to CoQ. Thus, amobarbital and rotenone selectively inhibit the oxidation of these substrates that are linked to the respiratory chain by NAD-coupled dehydrogenases, whereas the oxidation of substrates such as succinate, fatty-acyl CoA, and α-glycerophosphate, which transfer electrons directly to the flavoproteins, is not affected.

It is of interest that each of the above respiratory inhibitors acts at or near one of the coupling sites of oxidative phosphorylation. It should be recognized, however, that these inhibitors act primarily to prevent the oxidation–reduction reactions of the chain by blocking electron flow. Secondarily, they will also prevent ATP production by oxidative phosphorylation, and this explains their toxicity.

There are also inhibitory substances that act directly upon the formation of ATP by interfering with the coupling mechanisms of oxidative phosphorylation. An important group of such substances includes those known collectively as ***uncoupling agents,*** exemplified by the nitrophenols (*e.g.*, 2,4-dinitrophenol or DNP), the halophenols (*e.g.*, pentachlorophenol and thyroxine), as well as a number of seemingly unrelated organic compounds (salicylanilides and carbonylcyanide phenylhydrazones). Unlike the respiratory inhibitors, uncoupling agents such as DNP do not decrease the flow of electrons through the chain; they generally have the opposite effect of greatly accelerating the rate of cellular respiration. This accelerated respiration, however, does not lead to the production of ATP. In short, the processes of phosphorylation and oxidation are no longer coupled together. Usually the three coupling sites along the chain are equally affected. This uncoupling of oxidative phosphorylation results in uncontrolled, nonproductive respiration, where the energy released is dissipated entirely as heat. When ingested accidentally, toxic nitrophenols like DNP accelerate metabolism and heat production, causing an elevated body temperature. The increased metabolic rate seen in humans with hyperactive thyroid glands is also related to such a toxic uncoupling action, which, in this case, is due to greatly elevated concentrations of the thyroid hormones, thyroxine and triiodothyronine.

The second type of agent that acts directly upon the phosphorylation reactions associated with respiration is the ***coupling inhibitor***—exemplified by the antibiotic, oligomycin—which acts directly upon the enzyme system that converts ADP to ATP. Thus, both phosphorylation and the electron flow that is coupled to the phosphorylation process are prevented by such agents. When the electron flow is not coupled to ATP formation, oligomycin has no inhibitory effect on respiration; the presence of an uncoupling agent like DNP will therefore completely overcome the respiratory inhibition by oligomycin.

RESPIRATORY CONTROL

The understanding and significance of these seemingly esoteric observations may best be appreciated in the context of the phenomenon termed ***respiratory control.*** This is an important example of the mechanisms whereby the rates of metabolic processes may be regulated in the body and is illustrated by common experience. A person at rest utilizes only a small amount of oxygen, but when galvanized into physical action, his respiratory system must heave and pant to provide the additional oxygen needed. In quantitative terms, we can assert that the oxygen utilized by actively contracting skeletal muscle may be as much as 100 times greater than at rest. How does this come about? Probably the best way to approach this problem is not so much to try to understand what causes the maximal activation of the respiratory system, but rather to understand what constrains it from running at greater rates when the body is at rest. Maximal respiratory activity simply represents the top capacity for the passage of electrons through the chain when it is operating in a runaway, uncontrolled manner. To appreciate the way in which respiration becomes switched on requires a knowledge of the control system that turns it off when not required.

The primary factor involved in maintaining low rates of oxidation in the resting state is the low requirement for energy. As a consequence, most of the cellular adenine-nucleotide content is in the form of ATP, with very low concentrations of AMP and ADP. It is ADP that acts as the principal regulatory substance as a result of the obligatory coupling between electron flow and the phosphorylation of ADP in the respiratory chain. Between the energy loss accompanying electron transport and the energy conservation of ATP formation, there exists an energy-transducing system that links the two processes as rigidly as would a system of meshing gears at each of the coupling sites (Fig. 3-12). Thus, despite the provision of excess substrate (NADH, succinate, and so forth) as well as excess oxygen for the respiratory reactions, because of the tight coupling to phosphorylation there will be no electron flow through the coupling sites if there is no ADP and inorganic phosphate present to remove the energy that builds up in the transducing linkage. Normally, inorganic phosphate is present in excess, so the limiting factor becomes the ADP concentration (Fig. 3-12A). It is noteworthy that although the electrons may flow through parts of the chain other than the coupling sites, in the absence of ADP a blockage of electron flow at coupling sites 1, 2, and 3 will block the entire respiratory process just as inexorably as would the presence of amobarbital, antimycin A, or cyanide.

When a resting cell is stimulated into activity—for example, to cause mechanical contraction in the muscle, to produce an electrical impulse in the nerve, to carry out active transport in the kidney, or to promote other ATP-requiring processes such as biosyntheses—it will break down its re-

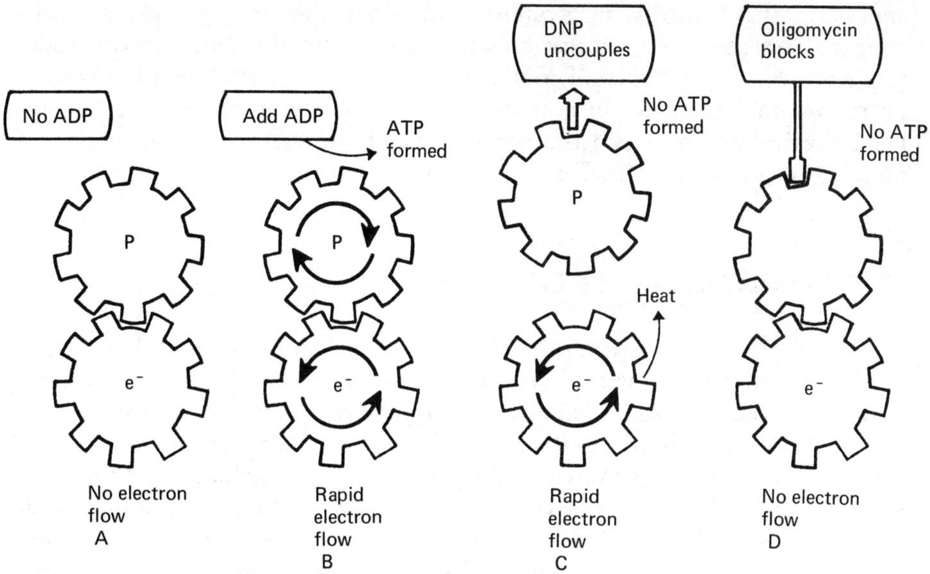

FIG. 3-12. Respiratory control. The energy transduction of oxidative phosphorylation is represented by the coupling of the electron-transport system *(e^-)* with the phosphorylation system *(P)* as intermeshing gears. **A.** In the absence of ADP; **B.** in the presence of ADP; **C.** in the presence of 2, 4-dinitrophenol (DNP), with or without ADP; **D.** in the presence of oligomycin, with or without ADP.

serves of creatine phosphate and ATP, depleting the former and converting the latter into inorganic phosphate and ADP. Now the electrons will be able to pass through the coupling sites at maximal rates because the availability of ADP for the phosphorylation system will dissipate the backpressure of energy in the transducing system and allow the "gears" to turn freely (Fig. 3-12B). This rapid respiration will continue until the demand for energy from activity ceases and the ADP becomes entirely reconverted to ATP, once again instituting the respiratory control of the resting state. Thus, the electron-transport chain is finely attuned to the cell's energy demands by the balance between ATP and ADP.

The mode of action of uncoupling agents such as DNP is explained by their ability to disengage the linkage of the intermediate transduction process. Thus, in the presence of DNP, no ATP is formed and no energy backpressure results on the electron carriers, since the energy that is built up is dissipated as heat. In short, the uncouplers act *between* the electron transfer and the phosphorylation process so that the "gears" no longer mesh, allowing respiration to proceed at its maximum, uncontrolled rate (Fig. 3-12C). The coupling inhibitors like oligomycin, on the other hand, prevent the interaction between the adenine nucleotides and the phosphorylation system so that ADP is no longer able to dissipate the pent-up energy in the transducing system, and, since the two systems are still coupled together, the energy backpressure halts the electron-transport process, even in presence of ADP (Fig. 3-12D). When DNP is added, however, maximal respiration is resumed, because the uncoupler acts to release the linkage between the respiratory chain and the site of action of oligomycin at the terminal process of ATP formation.

THE CITRIC-ACID CYCLE

The sources of the electrons that feed into the oxidative phosphorylation system are many and varied. We have seen, for example, that the carbohydrate intermediates, succinate and α-glycerophosphate, as well as the fatty acids, through the oxidation of their CoA derivatives, can donate electrons directly to flavoproteins of the respiratory chain, yielding P:O ratios of two. Other carbohydrate derivatives (pyruvate, malate, isocitrate, or α-ketoglutarate), lipid derivatives (hydroxy forms of fatty-acyl CoA or β-hydroxybutyrate), and amino acids (glutamate or proline) transfer their electrons in the form of NADH, yielding a P:O ratio of three. This list is by no means complete, and many other dietary substances as well as tissue components may serve to provide fuel for the respiratory chain, depending upon the particular conditions of surplus and demand. One particular source, however, performs such a key role in stoking the matabolic furnace of oxidative phosphorylation that it requires separate and special consideration. This is the process whereby the seemingly simple conversion of two-carbon compounds (*e.g.*, acetate) to 2 mols of CO_2 is consummated. As in the case of the seemingly simple cleavage of hexose to two three-carbon fragments, nature has endowed this process with considerable complexity. In this case, we have a multiple-enzyme system that operates in a wheel, or cycle, of reactions. This process is commonly named according to one of its principal reactants as the ***citric-acid cycle*** or, more generally, the *tricarboxylic-acid cycle,* or, after its discoverer, as the *Krebs cycle.*

Acetate enters this cycle in its activated form as the thio ester of coenzyme A, or *acetyl CoA*. The importance of the citric-acid cycle in metabolism stems from the fact that acetyl CoA plays such a major role as the end product of catabolism, deriving from carbohydrates by the decarboxylation of pyruvate, from lipids by the stepwise removal of two-carbon units from the long alkyl chains during fatty-acid oxidation, as well as from the catabolism of certain amino acids such as lysine, phenylalanine, tyrosine, leucine, and isoleucine (Fig. 3-13). Moreover, many other amino acids contribute to intermediates of the cycle upon undergoing deamination: glutamate, proline, histidine, arginine (to α-ketoglutarate), aspartate (to oxaloacetate), and alanine, cysteine, serine, and glycine (to pyruvate). It is no exaggeration to refer to the citric-acid cycle as the

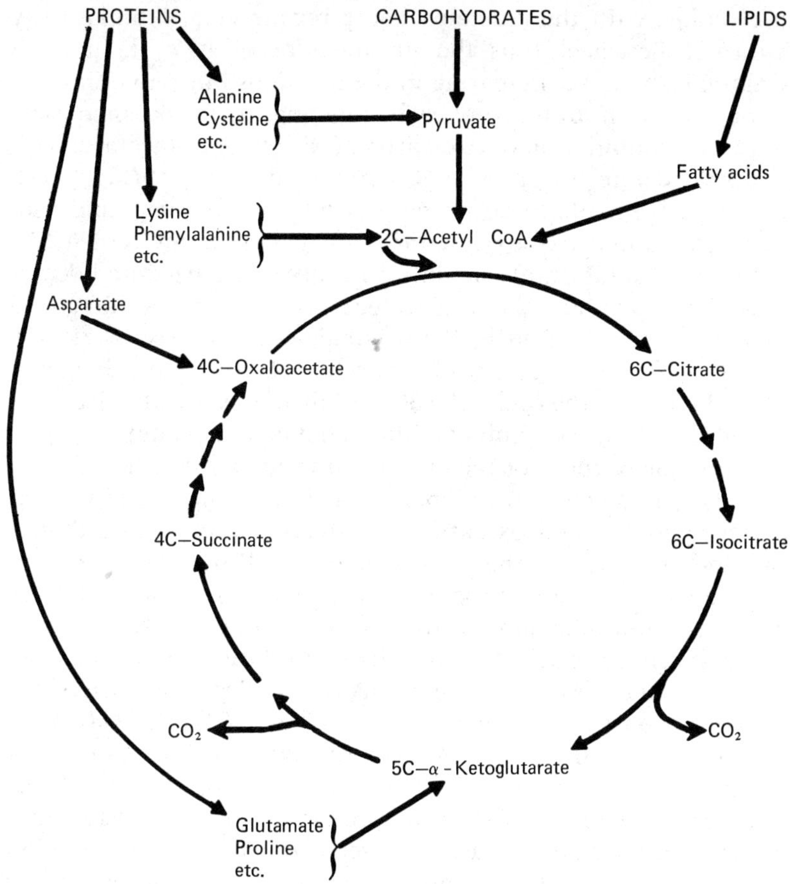

FIG. 3-13. Outline of the citric-acid cycle, showing its role as the final common pathway for catabolism of proteins, carbohydrates, and lipids.

final common pathway for the catabolism of carbon compounds in the body. Another feature that renders the cycle of great importance in the production of energy is the fact that all its component enzymes, coenzymes, and intermediates are located *inside* the mitochondria of the cell. Thus, NADH and the other electron donors of the citric-acid cycle may pass their electrons freely and directly to the respiratory chain in the mitochondrial membrane system.

The rationale of the citric-acid cycle is to provide a stepwise series of reactions, as in glycolysis, whereby the energy and electrons liberated during the oxidation of the acetate molecule are released in manageable packets. Unlike glycolysis, however, the process is unidirectional; no capability exists in the human to reverse the cycle for the regeneration of acetate from carbon dioxide. Also unlike glycolysis, the sequence is not linear but circular, with the reactants being regenerated automatically with each turn of the wheel; thus, the intermediates of the cycle function in true catalytic fashion, participating in the reactions but remaining unaltered in concentration. In terms of carbon balance, the process consists essentially of the combination of two-carbon (2C) moieties (acetate) with 4C groups (oxaloacetate) to produce 6C compounds (citrate) (Fig. 3-13). After some rearrangements to an isomeric form, the 6C compound (isocitrate) is oxidized and decarboxylated to a single-carbon molecule (CO_2) and a 5C unit (α-ketoglutarate). The latter in turn is oxidized and decarboxylated to 1C (CO_2) and a 4C unit (succinate) that, after oxidations and rearrangements, is reconverted to the original 4C group (oxaloacetate), thus regenerating the starting point of the cycle. Acetate is the only monocarboxylic acid found in the cycle; the 6C intermediates are all tricarboxylic acids, and the 5C and 4C units are dicarboxylic compounds.

Since the major purpose of the citric-acid cycle is the generation of CO_2 and energy, let us concentrate upon those important steps involved in the decarboxylations and oxidations. It should be emphasized that there is only one reaction in the cycle whereby ATP may be generated directly; the bulk of the energy production is dependent upon the extraction of the hydrogen atoms and electrons from acetate and transferring them to the respiratory chain. The first decarboxylation is catalyzed by *isocitrate dehydrogenase* (a), which, as its name implies, also carries out an oxidative function by transferring a pair of electrons to NAD, thus producing CO_2 and NADH (Fig. 3-14). The second decarboxylation acts upon the product of this reaction; in this step, *α-ketoglutarate dehydrogenase* (b) also forms CO_2 and NADH as end products, but by a considerably more complex process than that used for isocitrate. The decarboxylation of α-ketoglutarate requires the B vitamin, thiamine, as its pyrophosphate coenzyme, the oxidation step requires bound FAD to transfer electrons ultimately to free NAD, and the carbon chain remaining after the abstraction of CO_2 and electrons requires the disulfhydryl coenzyme, lipoic acid, to transfer the succinate to CoA as its thio ester. As a result, part of the energy from the oxidation of α-ketoglutarate is retained in the product, succinyl CoA; the latter transforms this high-energy thio ester to high-energy phosphate as guanosine triphosphate (GTP), which may be interconverted to ATP inside the mitochondrion. This is the only step where ATP may be directly generated by the cycle. The other oxidative steps of the cycle involve the FAD-linked *succinate dehydrogenase* (c), which, as noted previously, feeds electrons directly to CoQ of the respiratory chain, and finally *malate dehydrogenase* (d), which produces NADH and regenerates the oxaloacetate.

To determine the energy yield from the cycle, we may assume that each of the NADH and $FADH_2$ molecules generated from the oxidative reactions could be oxidized directly by electron transport through the coupling sites of oxidative phosphorylation. Thus, for each mole of acetyl

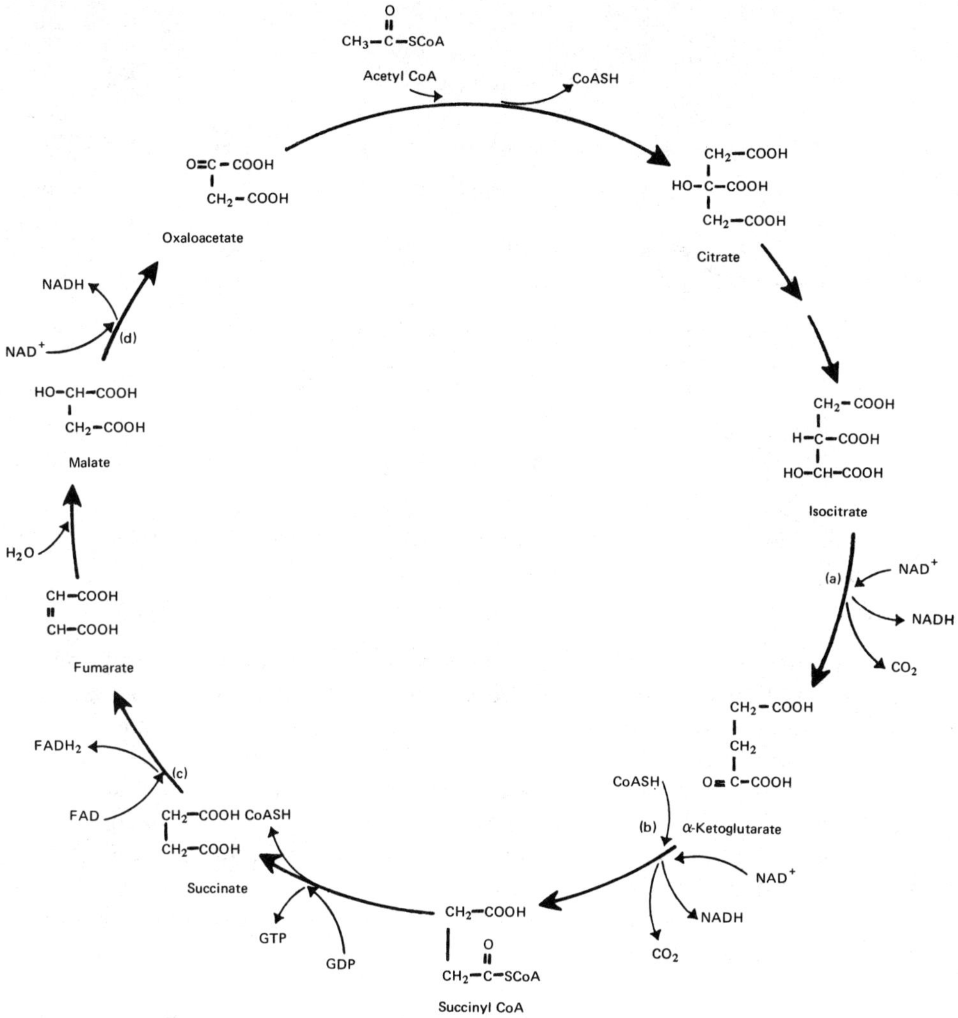

FIG. 3-14. Details of reactions in the citric-acid cycle.

CoA converted to citrate and then to isocitrate, the following yields result in each revolution of the cycle:

a.	Isocitrate →	α-Ketoglutarate + CO_2 + NADH	3 ATP
b.	α-Ketoglutarate →	Succinyl CoA + CO_2 + NADH	3 ATP
c.	Succinyl CoA →	Succinate + GTP	1 ATP
d.	Succinate →	Fumarate + $FADH_2$	2 ATP
e.	Malate →	Oxaloacetate + NADH	3 ATP
			12 ATP

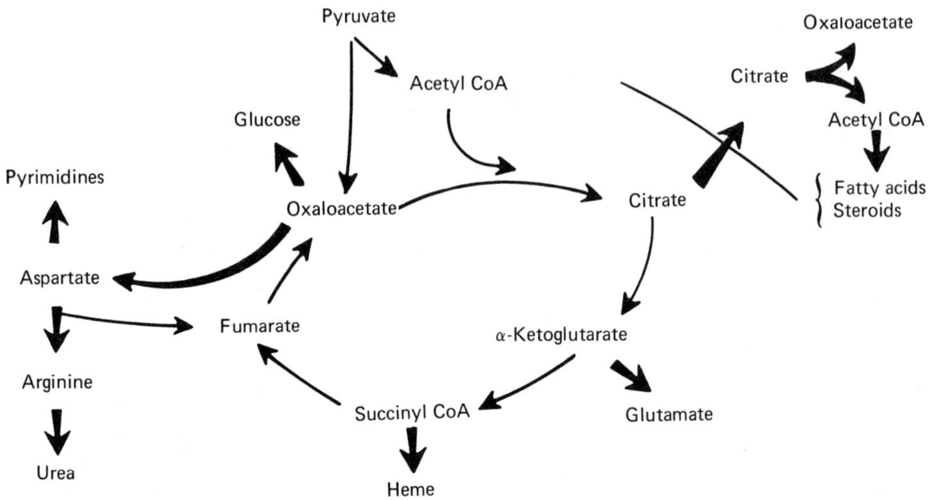

FIG. 3-15. Biosynthetic functions of the citric-acid-cycle intermediates.

In addition to providing a mechanism for the oxidation of acetyl CoA, the cycle is the path for the catabolism of a number of amino acids by way of α-ketoglutarate and oxaloacetate. Moreover, since deaminations of glutamate and aspartate are reversible in the human body, the two keto-acid intermediates of the cycle can contribute to amino-acid synthesis as well as breakdown (Fig. 3-15). This is but one example of the biosynthetic functions of the Krebs cycle that link it to carbohydrate, lipid, and amino-acid metabolism. Provision of one of the nitrogen atoms of arginine, and hence of urea, is dependent upon operation of that portion of the Krebs cycle between fumarate and oxaloacetate. Citrate, which is formed inside the mitochondrion, can permeate the membrane and be broken down to acetyl CoA in the cytoplasm; this will provide the substrate for the synthesis of fatty acids and steroids. The biosynthesis of the heme prosthetic groups of the cytochromes and hemoglobin requires, as its first step, a mitochondrial enzyme that utilizes succinyl CoA from the citric-acid cycle as one of its substrates.

Oxaloacetate obviously occupies a primary position in these biosynthetic functions of the cycle. It contributes to the formation of aspartate, which in turn may act as a precursor for the formation of the purines and pyrimidines as well as of urea and arginine. It plays a key role in providing carbon atoms from a variety of amino acids and other noncarbohydrate sources for gluconeogenesis. Moreover, oxaloacetate provides the chief means of replenishing the reactants of the cycle for these various biosynthetic processes, as will be subsequently described.

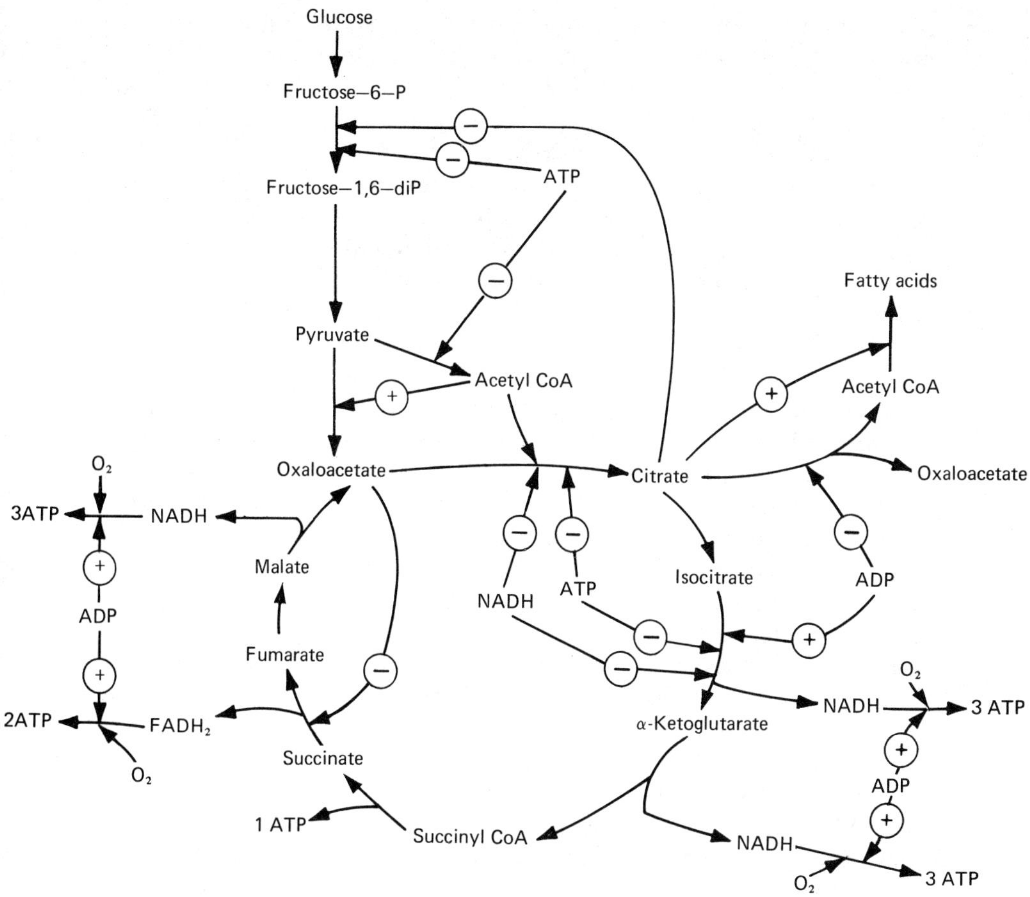

FIG. 3-16. Regulation of the citric-acid cycle. Compounds exerting effects as negative modifiers are shown by arrows with circled minus signs indicating the reactions that are inhibited; positive modifiers are represented by circled plus signs indicating the reactions that are stimulated.

REGULATION OF THE CITRIC-ACID CYCLE

The control of the activity of the citric-acid cycle is of cardinal significance in the adaptation of the individual to changing circumstances in the environment and to varying physiologic conditions. There are several regulatory factors and they may be divided into two general categories: regulation by the carbon substrates and intermediates in the cycle itself and regulation by the adenine nucleotides that serve as coenzymes and end products (Fig. 3-16).

First, let us consider the function of the cycle intermediates and, prin-

coupled event is the expulsion of H^+ ions, whereas the Cl^- ions probably follow passively to maintain electroneutrality. Farther down in the gastrointestinal tract, the absorption of sugars and amino acids occurs against concentration gradients across the microvilli of the jejunum at the expense of respiratory energy. The uptake of Na^+ invariably accompanies the uptake of the organic molecules, which suggests a link between the energy transduction systems for ion pumps and those for the migration of charged and neutral metabolites. The absorption of Ca^{2+}, mainly as calcium phosphate, is another energy-linked transport system of the intestines. In the epithelial cells of the ileum, a specific protein carrier, which requires vitamin D for its synthesis, functions in transluminal uptake. Respiratory energy is essential to enable the Ca^{2+} uptake to occur against a concentration gradient: first, at the cell membrane to provide ATP for a Ca^{2+},Mg^{2+}-ATPase transport system analogous to that of Na^+, K^+-ATPase; second, to promote uptake into the "sink" of the mitochondrial matrix.

In the kidney, the nature and direction of solute movement are governed *inter alia* by architectural considerations, by the differential permeability of different regions, and by superimposed hormonal controls. Despite the complexity of these processes, there is a fundamental unity at the level of the energetics of the active transport events. The retention of essential components by reabsorption through the epithelial cells lining the renal proximal tubules is intimately associated with a sodium pumping mechanism that operates by means of the outward-oriented, ouabain-sensitive, Na^+,K^+-ATPase system found in other cells. The primary motive force is the active transport of Na^+ ions, which results in their retention, together with the uptake of anions (such as Cl^-) and neutral molecules (such as glucose), which are passive fellow travelers that utilize the energy derived from Na^+ movement. As in other cells, K^+ ions move in the opposite direction to Na^+ and thus are actively secreted into the urine. Foreign substances that are actively secreted (such as *p*-aminohippuric acid) probably utilize the same energy system as do those that are actively absorbed, since they show competitive effects with one another when present simultaneously.

ENERGY METABOLISM IN NEURAL TISSUE

As any student will readily attest, the operations of the brain are demanding in terms of energy expenditure. In fact, almost one fourth of the oxygen consumption of the body is due to respiration in the brain. *Anoxia,* the deprivation of O_2, lasting for a few seconds will result in a loss of consciousness, and extended anoxia produces irreversible damage to brain function. The energy requirements of the tissue are high and continuous,

cipally, the regulatory role of oxaloacetate. It is well established that a depletion of the tissue levels of oxaloacetate can shut down the operation of the cycle. The consequences for the cell are dire, not only as a result of the elimination of one of the cell's main energy sources, but also because in the absence of adequate levels of oxaloacetate to accept the acetyl CoA generated by fat and carbohydrate catabolism, the excess two-carbon fragments may condense to produce an excessive accumulation of acetoacetate and other "ketone bodies," a pathologic condition known as *ketosis*. Ketosis will occur whenever the production of acetyl CoA greatly exceeds the capacity of oxaloacetate to condense with it and remove it by oxidation through the cycle. In short, ketosis results from a *relative deficiency* of oxaloacetate—relative, that is, to the levels of acetyl CoA—and it generally accompanies an imbalance of fat oxidation relative to carbohydrate catabolism. The reason for this is the ability of carbohydrates to regenerate oxaloacetate by the carboxylation of pyruvate. This ATP-requiring reaction is catalyzed in the mitochondria by the biotin enzyme, pyruvate carboxylase, and it is the main mechanism in the body for the fixation of carbon dioxide into carbohydrates. The oxaloacetate formed by this fixation of CO_2 may be used to replenish the catalytic capabilities of the citric-acid cycle, as well as to furnish the carbons for gluconeogenesis, for the multifarious metabolic functions of aspartate, and for the biosynthetic functions of the cycle that require a net production of citrate, α-ketoglutarate, and succinyl CoA. The reactions within the cycle leading to the formation of oxaloacetate are subject to negative feedback control by virtue of the fact that oxaloacetate, a structural analog of succinate, acts as a competitive inhibitor of succinate dehydrogenase. The latter thus becomes a regulatory site within the cycle, since an accumulation of oxaloacetate will block the cycle at the stage of succinate oxidation until an influx of additional acetyl CoA removes the excess oxaloacetate.

Acetyl CoA plays another major role in the autoregulation of the cycle because it is a powerful activator of pyruvate carboxylase; in fact, acetyl CoA is an obligatory allosteric modifier of the enzyme, so that negligible production of oxaloacetate occurs in the tissues in the absence of acetyl CoA. This system of checks and balances ensures, within the normal limits of metabolic supply and demand, that the cycle's operations will be controlled so neither deficiencies nor excesses of the intermediates will occur. Imbalances generally occur when fat oxidation swamps the tissues with acetyl CoA (as in starvation) and when carbohydrate catabolism is inadequate to provide the necessary pyruvate for oxaloacetate formation (as in diabetes mellitus).

Accumulations of citrate are prevented by several feedback operations and diversionary tactics. In the latter category is the process mentioned previously in which the excess citrate in the mitochondrion will diffuse to the cytoplasm, where its cleavage to acetyl CoA will result in the diversion of the two-carbon fragments to lipid synthesis. Moreover, the first step in fatty-acid biosynthesis—the carboxylation of acetyl CoA—is catalyzed by the allosteric biotin enzyme, acetyl-CoA carboxylase, which requires either citrate or isocitrate as positive activators. Hence, a buildup of either of these tricarboxylic acids will further accelerate the diversion of acetyl CoA from oxidation in the citric-acid cycle to fat synthesis. As an additional feedback mechanism to curtail its own accumulation, citrate functions as a negative modifier of phosphofructokinase, thereby blocking carbohydrate catabolism at an early stage in glycolysis.

In the second category of regulators of the activity of the citric-acid cycle are the nucleotide systems of ATP–ADP–AMP and NADH–NAD^+. The first site of control is at the point of entry of acetyl CoA, where the enzyme citrate synthase is subject to negative feedback inhibition by the products of the cycle, NADH and ATP. Thus, when the further production of either reducing or phosphorylating energy is no longer required, the operation of the citric-acid cycle will stop due to the allosteric inhibition of the first committed enzyme reaction. The reaction of the enzyme isocitrate dehydrogenase provides the other major regulatory step within the cycle. This enzyme is converted to its catalytically active form only in the presence of ADP, whereas the binding of either NADH or ATP will inactivate it. Thus, this step will also control the cycle's activity on the basis of cellular energy demands.

As noted previously, the energy state of the cell will control the overall rate of respiration requiring ADP for oxidative phosphorylation. This respiratory control produced by low ADP concentrations will also slow down the citric-acid cycle, since a depressed capacity of the respiratory chain for the reoxidation of the NADH and $FADH_2$ generated by the cycle will depress the activity of the dehydrogenases because of a deficiency of oxidized carriers, NAD^+ and FAD.

Other indirect actions of the nucleotides include the inhibition of the extramitochondrial citrate-cleavage enzyme by ADP; this ensures that when cellular energy levels are low, the citrate inside the mitochondria will be used for the production of ATP by the cycle, rather than for diversion to acetyl-CoA formation in the cytoplasm. The synthesis of fats from carbohydrate will thus occur only when the levels of ATP are high. High levels of ATP in the cytoplasm will also act to enhance glucose synthesis from noncarbohydrates and to inhibit glycolysis. On one hand, ATP inhibits phosphofructokinase and pyruvate dehydrogenase, thereby preventing glucose utilization, and, on the other, it serves as an allosteric activator of fructose diphosphatase in the liver, thus promoting glucose formation. Both ADP and AMP antagonize the effects of ATP so that when energy levels are low (*i.e.*, high ADP and AMP but low ATP levels), more glucose will be broken down to pyruvate to provide increased amounts of acetyl CoA for the cycle, whereas an increase in cellular energy (low ADP and AMP but high ATP levels) will inactivate glycolysis and accelerate its reversal in gluconeogenic tissues such as the liver.

ROLE OF MITOCHONDRIA IN ENERGY METABOLISM

The mitochondria are responsible for the generation of the bulk of metabolic energy in all body cells except the erythrocytes. Erythrocytes lack the organelles found in other human cells and hence are totally dependent upon anaerobic mechanisms, such as glycolysis. The aerobic mechanisms associated with respiration—a much more potent source of ATP—are confined to the mitochondria, which contain, in addition to the electron carriers and associated energy transducers of oxidative phosphorylation, the enzymes for the citric-acid cycle, for pyruvate oxidation and carboxylation, for fatty-acid oxidation, and for key reactions in amino-acid metabolism. Because all these processes in the mitochondria are separated from the rest of the cytoplasm by a membrane boundary, they constitute a distinct metabolic domain, almost as a cell within a cell. Moreover, the membranes of the mitochondria perform a unique function inasmuch as they contain many of the enzyme systems of oxidative metabolism as bound, integral components.

As noted in Chapter 1, the mitochondrion has two different membrane systems enclosing an internal space termed the *matrix;* this intramitochondrial fluid compartment contains the enzymes of the citric-acid cycle and those for fatty-acid oxidation as well as their necessary coenzymes and intermediate substrates. Surrounding the matrix is the mitochondrial *inner membrane,* an extensively invaginated membrane with regular arrays of spherical particles that project from the inward surface facing the matrix (Fig. 3-17). The invaginations are termed the *cristae,* and they greatly increase the surface area of the inner membrane. The particles are known as the *inner membrane spheres.* The inner membrane contains, embedded within its framework, all the carriers in the respiratory chain; it also acts as a permeability barrier to the substances passing between the cytoplasm and the matrix. The coupling of the electron flow within the membrane to oxidative phosphorylation is provided by the inner membrane spheres, which catalyze the synthesis of ATP from ADP and inorganic phosphate. This functional apparatus of the mitochondrion—the inner membrane plus the matrix—is encompassed by a smooth *outer membrane.* The latter has but few enzymatic activities, is freely permeable to all except large macromolecules, and acts mainly as a protective envelope, or corset, to confine the involuted inner membranes within a limited volume.

The passage of substrates between the mitochondrial matrix and the cytoplasm is of vital significance for cellular energy metabolism. The inner membrane, for example, is not freely permeable to nucleotides, yet ADP must penetrate from the cytoplasm to the inner membrane spheres for the formation of ATP, and the latter must pass back out to meet the cell's energy demands. To maintain the balance, the inner membrane contains a specific *translocase system,* which possesses receptors with a high affinity

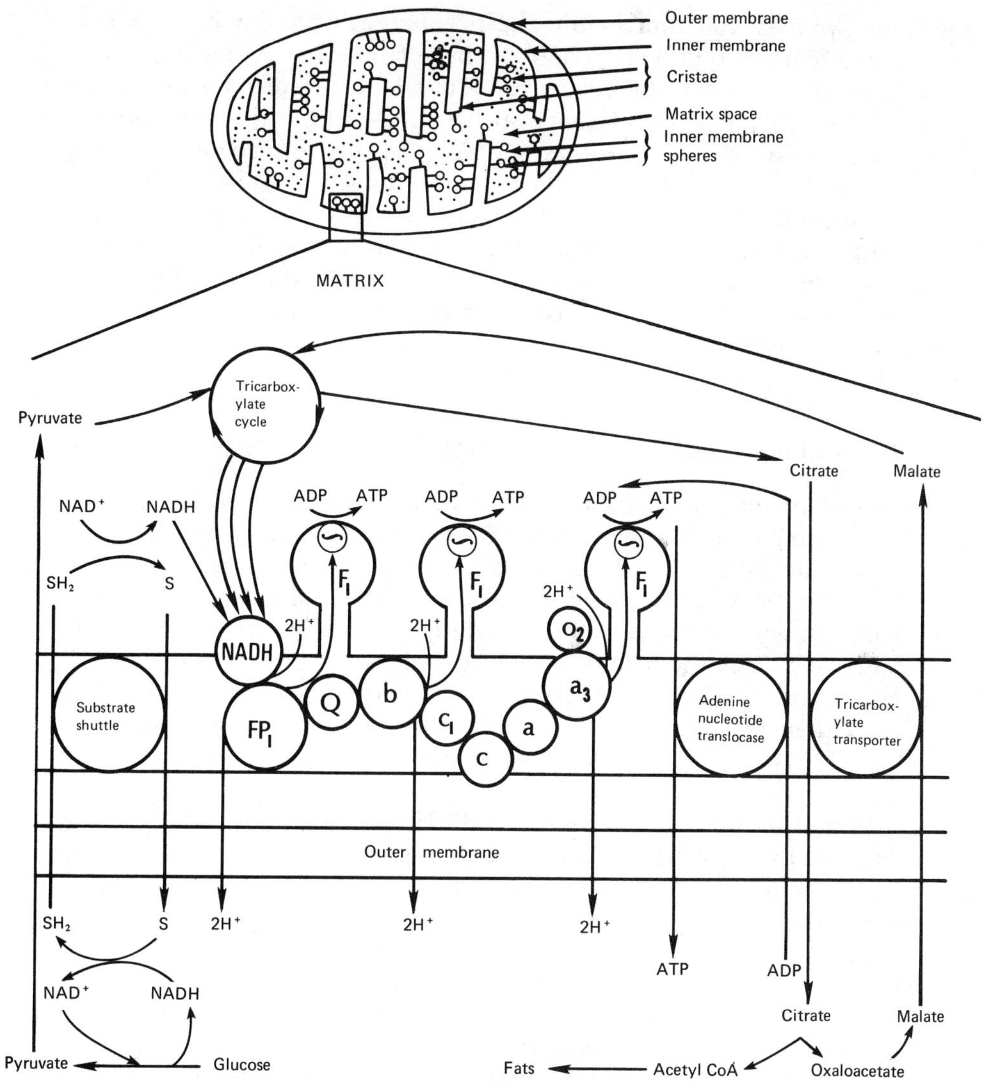

FIG. 3-17. The mitochondrion. The section below is an enlarged representation of the inner membrane, with the electron carriers depicted as in Figure 3-11, to show the movements of metabolites between the external cytoplasm and the intramitochondrial space of the matrix.

for ADP on the external surface and for ATP on the internal surface. By employing a "flip-flop" of the two receptors, the translocase will promote the exchange of newly synthesized ATP molecules moving out to the cytoplasm for an equivalent number of ADP molecules moving inward. In this way, the total adenine-nucleotide content inside the mitochondrion will always stay constant, while the production of ATP will keep pace with the breakdown of ATP to ADP in the surrounding cytoplasm.

The movement of reducing energy equivalents between the cytoplasm and the intramitochondrial space utilizes an entirely different type of translocating system, the *substrate shuttle.* Since the nicotinamide nucleotides, NAD^+ and NADH—like the adenine nucleotides—do not diffuse readily through the inner mitochondrial membrane, they transfer their reducing equivalents to substrates that will pass through the membrane and will hence act as the shuttles. All that is required is the presence of the appropriate NAD-linked dehydrogenases, E_1 and E_2, on *both* sides of the inner mitochondrial membrane:

1. *Cytoplasm:* $NADH + H^+ + S \overset{E_1}{\rightleftarrows} NAD^+ + SH_2$
2. *Intramitochondrial:* $SH_2 + NAD^+ \overset{E_2}{\rightleftarrows} S + NADH + H^+$

Proceeding in the forward sense of the reactions as written from left to right, reducing equivalents from the NADH that is produced in the cytoplasm by such processes as glycolysis may enter the mitochondria for oxidation by the respiratory chain in the form of SH_2. In the reverse sense, reducing energy generated within the mitochondria by activity of the citric-acid cycle or by the oxidation of fatty acids may be transferred out to the cytoplasm in the form of SH_2; the latter, by the reversal of reaction (1), may then form the NADH needed for reductive syntheses, such as reversal of glycolysis for glucose synthesis in the cytoplasmic compartment.

The nature of the substrates S and SH_2 of the shuttle are still to be established. In the liver, SH_2 is probably malate, a metabolite that crosses the inner membrane by a carrier system. Malate dehydrogenase satisfies the criterion of existing both inside and outside the mitochondrion and thus can catalyze the functions exemplified by reactions (1) and (2). Oxaloacetate—S in the malate shuttle—may not diffuse rapidly enough to complete the shuttle system *per se*, but it may require intermediate transformation, for example, to aspartate. Other S-SH_2 pairs that have been proposed as the shuttles include dihydroxyacetone-phosphate–α-glycerophosphate and acetoacetate–β-hydroxybutyrate, but their functions have not been proved to be of general significance for all tissues.

The movement of substrates such as pyruvate or malate between the mitochondria and the cytoplasm are facilitated by important carrier systems or *transporters*. These carriers of organic acids act—as does the adenine-nucleotide transport system—by exchange with other anions, on a mol-for-mol basis, that move in the opposite direction. Thus, malate, succinate, and other dicarboxylates interact with a specific transporter that moves phosphate as the counter ion. Citrate, isocitrate, and other tricarboxylates diffuse by exchange with malate moving in the opposite direction and use another specific carrier system. The importance of the dicarboxylate movement has been alluded to in connection with the transfer of reducing equivalents by means of the malate shuttle. The movements of citrate and isocitrate from the mitochondrial to the cytoplasmic compartment are of key significance in fatty-acid biosynthesis, with citrate furnishing acetyl CoA in the cytoplasm and isocitrate serving as a source of NADPH through the cytoplasmic, NADP-specific, isocitrate dehydrogenase. Other transporters exist for glutamate, aspartate, and ornithine to provide the essential linkage between mitochondrial and cytoplasmic amino-acid metabolism as well.

A special carrier mechanism exists to facilitate the movement of long-chain fatty acids, which must cross from the cytoplasmic compartment, where they originate from lipolytic reactions, into the matrix space that houses the enzymes of fatty-acid oxidation. The transfer of the fatty-acyl group across the mitochondrial inner membrane involves the formation of a diffusible ester of the compound carnitine, a reaction that is catalyzed by an enzyme—fatty-acyl carnitine transferase—embedded in the membrane and having outward and inward catalytic functions. The outward function of the enzyme is to transfer the fatty-acyl group from its activated CoA ester to carnitine on the cytoplasmic side:

$$\text{Fatty-acyl CoA} + \text{Carnitine} \rightleftharpoons \text{Fatty-acyl carnitine} + \text{CoA}$$

Its inward function is to catalyze the reverse of this process, reconverting the fatty-acyl carnitine—which passes freely across the membrane—into fatty-acyl CoA on the matrix side. It is important to note that fatty-acyl CoA—which does not penetrate the inner mitochondrial membrane—has been transferred effectively by this carrier mechanism from the cytoplasmic to the intramitochondrial compartment without any change in the coenzyme-A levels of the two compartments.

Uptake of cations into the mitochondrial matrix is intimately associated with the energy-transducing system of the inner membrane. In fact, it appears likely that the mechanism of energy conservation in oxidative phosphorylation may be understood eventually in terms of ion movements and the concomitant electromotive forces across the membrane. In the first place, it should be noted that the mitochondrial inner membrane is quite impermeable to most cations in general and to hydrogen ions in particular. Therefore, it is possible to establish an ionic gradient between the interior of the mitochondrion and the cytoplasm. If this were effected by the movement of H^+ in a certain direction, it would produce a *p*H gradient across the membrane or a ***protonmotive force.*** It is quite clear, moreover, that the electron carriers of the respiratory chain are organized within the inner mitochondrial membrane in an asymmetric manner such that they are exposed to their substrates on one side of the membrane or the other, but not both. NADH, for example, may interact with the respiratory chain only from the matrix or *M side,* whereas cytochrome c functions on the outer cytoplasmic surface or *C side.* Such an arrangement of carriers across the inner membrane results in a vectorial movement of electrons during the oxidation–reduction steps of the chain.

Consequently, the process of electron transport will result in charge separations and ionic gradients between the mitochondrion and the cytoplasm in a way analogous to that in which the controlled combustion of hydrogen in a fuel cell produces electrical energy by separating the charged H^+ and OH^- ions generated in the oxidation reaction:

$$2H_2 + O_2 \rightarrow 2H^+ + 2OH^-$$

During respiration, protons are ejected from the mitochondrion, leaving the inside of the inner membrane negatively charged and the matrix space alkaline in relation to the outside or cytoplasmic compartment. This H^+ ejection implies a vectorial arrangement of carriers such that the release of H^+ is outward, while H^+ consumption occurs on the matrix side during the cytochrome-oxidase reaction:

$$2Fe^{2+} - \text{cytochrome } a_3 + 2H^+ + \tfrac{1}{2}O_2 \rightarrow 2Fe^{3+} - \text{cytochrome } a_3 + H_2O$$

Thus, the energized state of the mitochondria that is produced by respiratory-chain oxidations may be viewed in terms of an electrochemical gradient of H^+, and, in effect, the energy inherent in the substrate's *reduction potential* is converted into an *electrochemical potential.*

The consequence of the electronegativity of the mitochondrial interior in the energized state is that cations other than protons—K^+ and Ca^{2+}, for example—may move down the electrical potential gradient to enter the matrix space, even against a chemical concentration gradient. In so doing, the movement of cations will discharge the energized state that was obtained by proton extrusion coupled to respiration; in other words, the energy normally to be conserved by oxidative phosphorylation is dissipated in the active transport of ions, so that ATP formation and cation uptake become competing alternative demands on the respiratory energy pool.

Calcium ions provide an exception to the general rule that the inner membrane is impermeable to cations. In this case, there are specific receptors in the membrane that facilitate the uptake of Ca^{2+} across the membrane barrier. The significance of this system for the cell is not clear except in the cells of the intestinal mucosa, where Ca^{2+} absorption from the gut may be accelerated due to the rapid entry of Ca^{2+} into mitochondria, thereby providing a sink for the transfer of these ions into the cell from the intestinal lumen. Because of the key regulatory actions of Ca^{2+} in the cytoplasm, the capability of mitochondria to take up and store Ca^{2+} may play an important part in controlling metabolic activity.

In the case of other ions, such as K^+, their accumulation by mitochondria is very slow unless the cation is caged by a complexing agent, such as valinomycin, which encases the K^+ selectively within a large, nonpolar, ring structure that may permeate the hydrophobic interior of the membrane. Once a mechanism such as that noted for the lipid-soluble, ***"ionophore"*** agents like valinomycin is introduced to permit passage across the membrane barrier, however, the energetics of all cation uptake processes (Ca^{2+}, K^+, or whatever) are strictly parallel and utilize an identical energy-transduction system.

The participants in the coupling mechanism that utilizes the energized state for ATP production may be described but not readily explained. The physical basis of the reaction ADP + $P_i \rightarrow$ ATP + H_2O resides within the inner membrane spheres. When separated from the inner membrane, a protein factor that is essential for oxidative phosphorylation is obtained from the spheres. This protein catalyzes the reverse of ATP synthesis—that is, it is an ATPase—and, since this was the first coupling factor to be identified, it is termed *F_1-ATPase*. This is the step in oxidative phosphorylation that is sensitive to inhibition by the antibiotic, oligomycin. Why does isolated F_1 catalyze ATP hydrolysis, whereas the native particle promotes ATP synthesis?

In terms of the linkage of energy transduction, the answer probably relates to asymmetry of the spheres in their orientation on the inner membrane, and the importance of vectorial reactions relative to the membrane is again emphasized. In the isolated state, F_1 has free access to H_2O, an essential reactant for the hydrolytic function of ATPase; accordingly, in an aqueous solution, F_1 acts in the thermodynamically favored direction, namely, toward ATP breakdown. When present in the functional state, however, F_1 is bound to the hydrophobic inner membrane and may be specifically oriented so that it is accessible to ADP and ATP only from the intramitchondrial or M side. It is probable that the enzyme is quite limited in its access to H_2O due to the hydrophobic environment, and it has been suggested that the ionization products of H_2O—that is, H^+ and OH^-—may be the actual reactants at the active site of F_1. If the accessibility of H^+ and OH^- to the enzyme were also asymmetric with respect to the two sides of the inner membrane, then the true activity of H_2O at the catalytic site would be proportional to the ion product $[H^+]_M \cdot [OH^-]_C$, where the subscripts designate the matrix (M) and cytoplasmic (C) sides. In this formulation, a movement of protons from the M to the C side—as is observed in respiration—will effectively decrease $[H^+]_M$ while simultaneously lowering $[OH^-]_C$, and this movement will thereby bring about a reduction in the ion product and the activity of H_2O at the F_1 catalytic site. If the effective H_2O concentration is lowered in the reaction:

$$\text{ADP} + P_i \rightarrow \text{ATP} + H_2O$$

then the reaction will proceed toward the right. In essence, ATP synthesis as depicted by this mechanism is a dehydration process, with the electrochemical energy required for water removal being supplied by the pumping of protons unidirectionally across the membrane during coupled respiration.

The phenomenon of uncoupling is explained simply by such a mechanism. Lipophilic uncoupling agents such as 2,4-dinitrophenol can permeate the mitochondrial inner membrane, either as the phenolate ion (ΦO^-) or the associated phenol (ΦOH). In the latter form, they may carry a proton into the mitochondrial matrix and dissociate to release the H^+ and phenolate ions; the phenolate can diffuse outward again to repeat the proton transfer. In this sense, uncouplers are ionophores rather like valinomycin, and they act to break down the proton gradient produced by vectorial electron transfer that is essential for ATP formation. The result is that energy conservation is disrupted, and respiration proceeds at an uncontrolled, maximal rate that is not dependent upon ADP, as discussed previously.

Although many of these notions about the mechanism of ATP production are incompletely delineated, it is quite clear that the processes involved in high-energy compound formation and utilization, respiratory energy-trapping systems, the pumping of ionic molecules across mitochondrial membranes, and many associated systems involved in the chemical, electrical, and osmotic work in the cell are interconnected by a common linkage through vectorial metabolism and charge separation.

ENERGY-REQUIRING TRANSPORT SYSTEMS

The cell membrane serves as a protective barrier to prevent the entry of noxious or unnecessary substances from the environment and to maintain the cytoplasmic contents by impeding the leakage of essential constituents. In addition to these rather passive restrictions, which are based upon the impermeability properties of the membrane, a number of active processes are involved in the facilitated uptake of essential nutrients, the output of waste products, and so on, that require energy expenditure. Since the external membrane of the cell—unlike that of the mitochondria—lacks its own energy-generating system, it must rely upon the ATP produced by catabolic reactions elsewhere in the cytoplasm. The ***active transport*** systems in the cell membrane couple the energy-requiring movements of ions and metabolites to the hydrolysis of ATP, and thus, these systems—like the F_1 factor of mitochondria—are commonly characterized and designated on the basis of their ATPase activity.

The most widely distributed active transport system of this type is the ***Na^+,K^+–ATPase*** system, so-named because it requires both Na^+ and K^+ for maximal ATP hydrolysis. The enzymatic activity is intimately associated with the process whereby cells maintain their internal ionic composition. Whereas blood plasma and other extracellular fluids have high

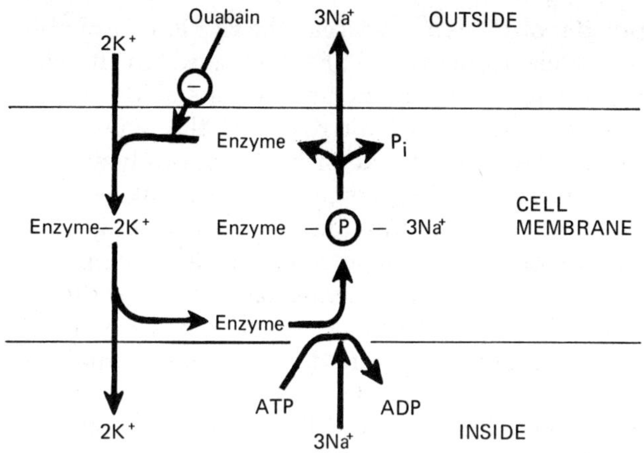

FIG. 3-18. The Na^+, K^+-ATPase system for the active transport of cations across cell membranes.

concentrations of Na^+ and low K^+, the intracellular spaces have the opposite, namely, high K^+ and low Na^+ contents. The cells must continuously expend energy to preserve these ionic conditions against the concentration gradients. This is accomplished by the Na^+,K^+–ATPase complex, which consists of an enzymatic protein situated asymmetrically in the membrane in association with phospholipids.

The ATP-hydrolyzing capacity of the complex, as well as the ability of the cell to transport Na^+ and K^+, are both selectively inhibited by the cardiac glycoside, *ouabain*. The ATPase enzyme acts vectorially, combining with ATP in the presence of Na^+ only on the inner surface of the membrane. In the process, the enzyme is phosphorylated and moves three Na^+ ions outward across the membrane for each molecule of ATP utilized (Fig. 3-18). The phosphorylated enzyme is capable of undergoing hydrolysis to yield inorganic phosphate only when exposed to K^+ from the outer surface of the membrane. During this dephosphorylation step, two K^+ ions migrate inward across the cell membrane for each mole of P_i liberated. It is this latter K^+-associated reaction that is blocked by ouabain. The net effect of this "ion pump" is that for each ATP hydrolyzed to ADP and P_i, three Na^+ are expelled from the cell and two K^+ are taken in from the extracellular fluid. This ouabain-sensitive, Na^+,K^+-ATPase activity is present in the outer membranes of all cell types—muscles, brain cells, erythrocytes, and so on—where the active transport of Na^+ and K^+ occurs.

Many different classes of cells and tissues may be specially differentiated with respect to their absorptive and secretory functions. In the gastric mucosa, for example, the parietal cells carry on the production of HCl for acidification of the stomach contents. The process is respiration-dependent and possesses obvious parallels to the acid release from respiring mitochondria; in this case, too, it appears that the primary energy-

and they may be met effectively only by the aerobic processes of oxidative phosphorylation.

As in many body tissues, much of the brain's energy consumption is related to the active transport of substances across the cell membranes. In the case of neural tissue, the processes are similar to those found to operate in other cell types, but the demands are much greater. The erythrocyte, for example, has a low capacity for ATP generation and uses about 30% of the total produced to maintain its internal ionic composition. The brain, on the other hand, with its tremendous oxidative activity in relation to its size, requires almost 70% of its large ATP production in order to maintain the ionic gradients between the inside of neural cells and the extracellular environment. The reason for this large energy requirement in neural tissue is that there is an extensive flux of ions across the neural membranes during the transmission of nerve impulses, and a continuous necessity exists for metabolism to restore the electrical potential of the membranes after the nerves have discharged. It is in the latter recovery stage, that is, after the ionic movements of nerve activity, that the uniquely heavy demands are placed upon the energy metabolism of the neural tissues.

Neurons, like other cells, maintain high intracellular levels of K^+ but low concentrations of Na^+ and Cl^- relative to the external space. In the resting state, the cell membrane is many times more permeable to K^+ and Cl^- than it is to Na^+. Hence, there is a tendency for K^+ to escape from the interior and for Cl^- to penetrate from the exterior, respectively, by diffusion down their concentration gradients. The Na^+, however, moves inward much more slowly. The result is a greater abundance of positive charges on the outside of the membrane, which is manifested in an electrical potential difference of about +70 millivolts on the outside relative to the inside of the cell (Fig. 3-19A). This is termed the *resting potential* of the neuron, and it is maintained in the resting nerve by a small, continuous energy utilization so that the sodium pump may reverse the slow trickle of Na^+ that enters the cell. In this sense, resting nerve cells are not unlike other cell types. The characteristic that distinguishes nerve and muscle cells is their ***excitability,*** a response to stimuli that changes the distribution of ions and hence induces an electrical change, or *action potential,* that may be transmitted along the length of the axonal membrane of the cell.

Details of the stimuli and the factors involved in the propagation of nerve action potentials are beyond the scope of this discussion. From the standpoint of the energetics of nerve conduction, it suffices to consider the following molecular events that accompany the electrical changes at the membrane surface. First, the primary response to stimulation of the nerve, regardless of the stimulus or receptor, is a dramatic local alteration in the neuronal membrane so that its permeability to Na^+ is enhanced several

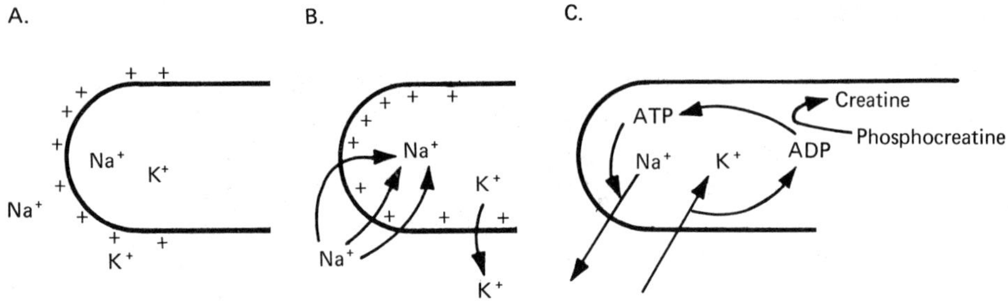

FIG. 3-19. Energy requirements of nerves in relation to ion movements.

hundredfold (Fig. 3-19B). Because external Na^+ concentrations are much higher than those within the nerve and because the outside space bears a net positive charge, Na^+ is rapidly drawn into the neuron by both the concentration gradient and the electrochemical gradient. As a result, the resting potential is abolished—that is, *depolarization* occurs—and the large influx of Na^+ reverses the potential difference so that it becomes positive on the inside of the cell, with a value of about +50 mV. After passage of the stimulus, the cell membrane becomes once more impermeable to Na^+, and K^+ diffuses outward down its concentration and electrochemical gradients to restore the original potential difference.

It is noteworthy that these ionic movements occur by passive movement *with* the existing gradients, and hence the actual electrical disturbance characterizing the transmitted action potential along the nerve is not itself an energy-requiring process. Having inverted the normal K^+ and Na^+ gradients, however, the neuron is refractory to subsequent stimuli; it is the restoration of excitability to the nerve by the transport of Na^+ back out and of K^+ back into the cell that is dependent upon ATP.

The movements of Na^+ and K^+ during recovery from excitation occur by mechanisms analogous to those found in other cells. Neural tissue contains a highly active, ouabain-sensitive Na^+,K^+-ATPase that has the properties required for an outward-directed Na^+ pump with concomitant inward-directed K^+ movement (Fig. 3-19C). The energy demands for the reestablishment of resting ionic levels are met by a chain of metabolic events, the first of which is the direct utilization of ATP. Upon stimulation of the nerve, there is a transitory drop in ATP, reflecting this role in the ionic pump, but the ATP level is rapidly restored to normal. This is due to the reservoir of energy storage in the form of *phosphocreatine,* which maintains the ATP level by the action of creatine phosphokinase:

$$\text{Phosphocreatine} + \text{ADP} \rightleftarrows \text{Creatine} + \text{ATP}$$

During sustained stimulation, the ATP level in the nerve does not fall, but it remains quite steady at the expense of phosphocreatine, which declines subsequently if the energy demand outstrips the supply.

At the time when the phosphocreatine reserve is bolstering the ATP levels, respiration in the mitochondria is stimulated by the momentary rise in ADP concentration. This is observable in isolated brain segments, whose rate of respiration is quite low in the resting state but may be stimulated manyfold by the application of electrical impulses. The oxygen consumption by human brain tissue *in vivo* continues at near maximal rates even when "at rest," because the living brain cells continue to discharge and regenerate their ionic gradients at all times. Enhanced energy metabolism and hence increased oxygen consumption by the brain are detectable *in situ* only under conditions of massive external stimulation, as, for example, during electroshock therapy when the respiratory rates may rise twofold. Under normal circumstances, energy consumption under conditions of rest versus that under intense, conscious mental activity differs surprisingly little, despite the fatigue engendered frequently by the latter state. In terms of energy metabolism, one may draw a parallel between the resting brain and the random noise produced by the uncoordinated performances of a roomful of individual musicians; it requires no greater energy expenditure, relative to the basal activity, for the concerted and orchestrated actions of the players under the baton than it does during their period of tuning up. In both cases, there is an absolute requirement for energy, but the organization of that energy expenditure to a defined end is subject to a higher level of control.

ENERGY METABOLISM IN MUSCLE

The utilization of metabolic energy by muscle tissues exhibits several important characteristics. First is an obvious quantitative consideration: the contractions of skeletal muscle, of smooth muscle, and of cardiac muscle proceed to varying degrees dependent upon the individual activity levels, but they proceed nonetheless continuously and, like the nervous system, their metabolic machinery requires a continuous source of fuel. The type of fuel may also vary (as will be discussed in later sections) but whether it is obtained by oxidation of ketone bodies in heart muscle, from fatty acids in resting skeletal muscle, or by anaerobic catabolism of glycogen stores in vigorously acting skeletal muscle, the ultimate driving force for the mechanical work in all cases comes from the breakdown of ATP.

A second significant feature is the existence of *myoglobin* in many muscles. This hemoprotein, like the hemoglobin of blood, combines reversibly with O_2 and thereby serves as an intracellular oxygen reservoir to sustain aerobic energy production when the tissue O_2 becomes depleted. These *red muscles* characteristically contain abundant mitochondria, which are packed tightly with highly infolded membranes (indicating their high respiratory capacity) and which are oriented in close proximity to the contractile myofibrils that utilize the ATP from mitochondrial oxidative phosphorylation. This class of skeletal muscle typically is slow to contract but is capable of maintaining contracture for long periods. On the other hand, muscles that are required to function in short, rapid bursts of activity are deficient in myoglobin and mitochondria, and hence they are sometimes designated as *white muscles*. The latter contain large deposits of glycogen in their cytoplasm and depend predominantly upon anaerobic glycolysis for ATP generation.

Finally, the energy demands of muscles are quite complex by virtue of the fact that a muscle cell is excitable as well as contractile; accordingly, there are three, separate ATP-requiring processes that operate during a cycle of contraction and relaxation in a muscle: (1) a sodium-ion pump at the cell membrane, (2) a calcium-ion pump within the cell itself, and (3) the process that converts the chemical energy from ATP hydrolysis into the mechanical energy of contraction (Fig. 3-20).

The muscle sodium pump, like that of neural tissues, is mediated by the Na^+,K^+-ATPase system at the cell membrane, or *sarcolemma*, of the muscle. The inward migration of K^+ and the outward transport of Na^+ against their concentration gradients utilize a significant portion of the energy output of resting muscles (Fig. 3-20A). As in the case of the nerve-cell membrane, the sarcolemma exhibits a membrane potential, approximately +50 mV on the exterior, due to these ionic gradients. Stimulation of the muscle occurs by means of a wave of depolarization that moves along the muscle membrane, like the propagation of a nerve impulse along the axonal membrane but at a slower rate. The result is that the whole muscle will respond as a unit to the contractile stimulus. The restoration of the ionic gradients and the repolarization of the sarcolemma are thus significant, ATP-requiring processes in the recovery period after muscle activity, in addition to the energy demands involved in the mechanical aspects of contraction (Fig. 3-20C).

The calcium-ion pump of the muscle resides in the intracellular membrane system known as the *sarcoplasmic reticulum*. This interconnected series of tubules and cisternae encloses a fluid space that is isolated from the cytoplasm which surrounds the contractile elements of the muscle. At rest, the bulk of muscle Ca^{2+} is sequestered within this space, with only very low concentrations ($<10^{-7}$ *M*) in the cytoplasm. A stimulation of the muscle leading to the depolarization of the outer cytoplasmic membrane of the sarcolemma causes a secondary release of Ca^{2+} from the intracellular membranes that enclose the sarcoplasmic reticulum. The elevation in Ca^{2+} concentrations (to $\sim 10^{-6}$ *M*) in the cytoplasmic fluid space that bathes the contractile fibers is fundamental to the initiation of contraction, and it also activates the breakdown of glycogen (Fig. 3-20B). When the muscle is no longer under stimulation, the Ca^{2+} rapidly disappears from the cytoplasm. Some is probably taken up by the mitochondria at the expense of the proton gradient generated by respiration. The bulk of

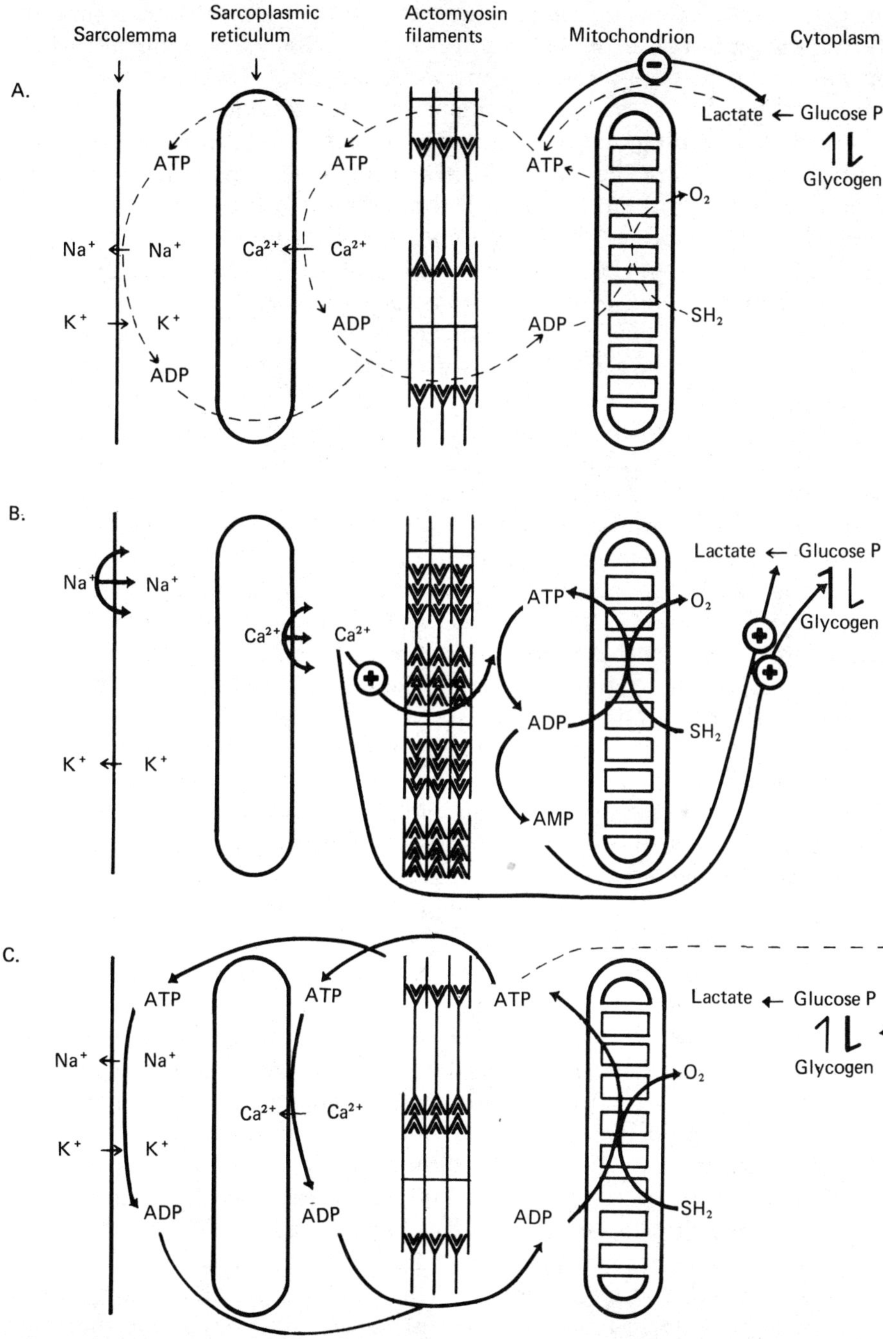

FIG. 3-20. Energy metabolism of muscle in response to stimulation of the contractile process.

the Ca^{2+}, however, is returned to the spaces of the sarcoplasmic reticulum during the period of recovery from contraction. Since Ca^{2+} concentrations may become extremely low in the cytoplasm in a very short time, it is apparent that uptake by the sarcoplasmic reticulum must be activated by a metabolic pumping mechanism to counteract the unfavorable Ca^{2+} concentration gradient. The energy for this process is provided by ATP, and it is mediated by an enzyme system bound within the membrane of the sarcoplasmic reticulum that will hydrolyze ATP only in the presence of Ca^{2+}, that is, Ca^{2+}-ATPase. For each mole of ATP hydrolyzed, two Ca^{2+} ions are transported into the sarcoplasmic reticulum.

The molecular basis of the contractile process depends upon the interaction between two filamentous proteins, *actin* and *myosin,* that are characteristic of muscle tissues. In the tissue, they form a complex structure, *actomyosin,* which has collective properties that differ from those of either of the individual proteins considered alone. Both molecules react with adenine nucleotides: actin in its filamentous form with ADP; myosin with ATP. When combined as actomyosin, however, the complex dissociates to its component proteins in the presence of ATP. Only the subsequent addition of Ca^{2+} allows the reassociation of actin with myosin, and, in the process, the ATP is hydrolyzed to ADP and inorganic phosphate. It seems probable that these molecular events, together with a detailed understanding of the architecture of the molecules, will provide an explanation for the mechanism of the contractile process. The molecular model that, at present, best satisfies these criteria involves the rupture and reformation of the linkages between the actin and the myosin filaments. These bonds are continually broken and reestablished in such a way that the two molecules slide past each other and hence shorten the array of longitudinally oriented filaments that is present within a muscle fiber during contraction (Fig. 3-20B). From the standpoint of energetics, the principal feature is the fact that ATP must be hydrolyzed each time a connection is made and broken between the two proteins. The presence of Ca^{2+} is essential for the splitting of ATP and hence for the initiation of the contractile process. Because of the large number of connections or bridges that are made and broken between the two functional proteins, the amount of ATP used to provide this mechanical energy requirement for contraction greatly outstrips the amounts required for the restoration of ionic gradients in actively functioning muscles.

It may be calculated that at maximal rates of work, the total ATP content of muscle will be depleted in a second or less. Sustained activity at maximal rates may still proceed, however, by virtue of two reserve systems for high-energy phosphate. The first involves the creatine-phosphokinase enzyme system. As in neural tissue, this system can maintain ATP levels fairly constant despite the continuous, rapid hydrolysis of the nucleotide. Phosphocreatine levels are generally five times the ATP concentration in resting muscle, and hence maximal activity may be sustained for several more seconds on the basis of these sources. As the phosphocreatine becomes depleted, a second reaction, which is catalyzed by the *adenylate-kinase* enzyme system, comes into play:

$$2\ ADP \rightleftarrows ATP + AMP$$

This reaction utilizes the second reactive phosphate group to generate additional ATP from ADP; the reaction is important in that it serves to raise the concentration of AMP in the muscle tissue. As noted previously, AMP serves as a key signal to warn of declining cellular energy reserves, and it acts as a powerful positive stimulus to several crucial enzymes involved in the catabolism of carbohydrates to produce more ATP.

The elevation in ADP concentrations from continuous ATP hydrolysis activates the oxidative phosphorylation mechanism, as well as the regulatory enzymes of the citric-acid cycle and glycolysis (Fig. 3-20B). Thus, during the recovery phase from contraction, an accelerated oxidative and glycolytic metabolism will ensue so long as the tissue levels of AMP and ADP are elevated. Even though the muscle has ceased working, catabolic processes must catch up with the deficit in high-energy compounds that results from the previous period of activity. This is frequently referred to as the ***metabolic debt*** incurred by the muscle, and it is characterized by accelerated oxygen uptake or increased lactate production, or both, above the resting levels. The time taken to repay the debt may be several times the duration of the actual activity. As in the case of many biologic systems, the muscular contractile apparatus is poised and ready to function at top capacity for periods in excess of its reserves, but ultimately the debt must be repaid.

SUGGESTED READINGS

Adelstein RS, Eisenberg E: Regulation and kinetics of the actin–myosin–ATP interaction. Annu Rev Biochem 49:921–956, 1980

Burg MB: The nephron in transport of sodium, amino acids, and glucose. Hosp Pract 13, No. 10:99–109, 1978

Bygrave FL: Mitochondria and the control of intracellular calcium. Biol Rev 53:43–79, 1978

Christensen HN: Biological Transport, 2nd ed. Reading, Mass, WA Benjamin, 1975

Cohen C: The protein switch of muscle contraction. Sci Am 233, No. 5:36–45, 1975

Cross RL: The mechanism and regulation of ATP synthesis by F_1-ATPases. Annu Rev Biochem 50:681–714, 1981

Fillingame RH: The proton-translocating pumps of oxidative phosphorylation. Annu Rev Biochem 49:1079–1113, 1980

Hays RM: Principles of ion and water transport in the kidney. Hosp Pract 13, No. 9:79–88, 1978

Ikemoto N: Structure and function of the calcium pump protein of sarcoplasmic reticulum. Annu Rev Physiol 44:297–317, 1982

Knowles JR: Enzyme-catalysed phosphoryl transfer reactions. Annu Rev Biochem 49:877–919, 1980

LaNoue KF, Schoolwerth AC: Metabolite transport in mitochondria. Annu Rev Biochem 48:871–922, 1979

Lemberg R, Barrett J: Cytochromes. New York, Academic Press, 1973

Margaria R: The sources of muscular energy. Sci Am 226, No. 3:84–91, 1972

Mitchell P: Biochemical mechanism of proton motivated phosphorylation in FoF_1 adenosine triphosphatase molecules. In Lee CP, Schatz G, Dallner G (eds): Mitochondria and Microsomes, pp 427–457. Reading, Mass, Addison–Wesley, 1981

Munn EA: The structure of mitochondria. London, Academic Press, 1974

Murray JM, Weber A: The cooperative action of muscle proteins. Sci Am 230, No. 2:59–69, 1974

Racker E: A New Look at Mechanisms in Bioenergetics. New York, Academic Press, 1976

Sokoloff L: Circulation and energy metabolism of the brain. In Siegel GJ, Albers RW, Katzman R, Agranoff BW (eds): Basic Neurochemistry, 2nd ed, pp 388–413. Boston, Little, Brown & Co, 1976

Trachtenberg MC, Packey DJ, Sweeney T: *In vivo* functioning of the Na^+,K^+-activated ATPase. Curr Top Cell Regul 19:159–217, 1981

Wallick ET, Lane LK, Schwartz A: Biochemical mechanism of the sodium pump. Annu Rev Physiol 41:397–411, 1979

Wikström M, Krab K, Saraste M: Proton-translocating cytochrome complexes. Annu Rev Biochem 50:623–655, 1981

Wilson DB: Cellular transport mechanisms. Annu Rev Biochem 47:933–965, 1978

Chapter Four Carbohydrates

The ***carbohydrates*** include the sugars and their derivatives; their general chemical formulation is that of the combination of carbon with water $[C(H_2O)]_n$, or hydrated carbon, as the name implies. The most abundant sugar in the human body is ***glucose,*** a hexose ($n = 6$), but other hexoses such as galactose and fructose are also important, as are certain pentoses ($n = 5$) such as ribose and deoxyribose (Fig. 4-1). The trioses ($n = 3$), tetroses ($n = 4$), and heptoses ($n = 7$) are minor components quantitatively, but they play key roles as metabolic intermediates.

When the sugars are not combined with other sugars, they are termed *monosaccharides* (Fig. 4-1B). When two sugars are in combination—as in the case of the milk sugar lactose, which contains glucose combined with galactose—the compound is termed a *disaccharide* (Fig. 4-1C). A large number of monosaccharides joined head-to-tail in a chain constitute a *polysaccharide* (Fig. 4-1D). Carbohydrates are essential metabolites for a wide range of processes in the body. Free sugars, particularly glucose, are important energy fuels in the nervous system, muscles, and many other tissues. In combined form the pentose sugars contribute to structures of nucleotide coenzymes and polynucleotides of DNA and RNA. Sugars may also be combined with lipids forming ***glycolipids,*** which are of particular significance in myelin and other membranes of nerves, and in the erythrocytes where they determine the blood-group reactions. Combined with protein as ***glycoproteins,*** the sugars play a role in secretion and the external recognition properties of cell membranes. Finally, the sugars may polymerize among themselves to produce the storage energy substance, ***glycogen,*** or they may form heteropolymers, such as the glycosaminoglycans of connective-tissue ground substance. Anomalies in carbohydrate metabolism include the common medical problem of diabetes mellitus, as well as several less frequent genetic disorders such as the glycogen-storage diseases.

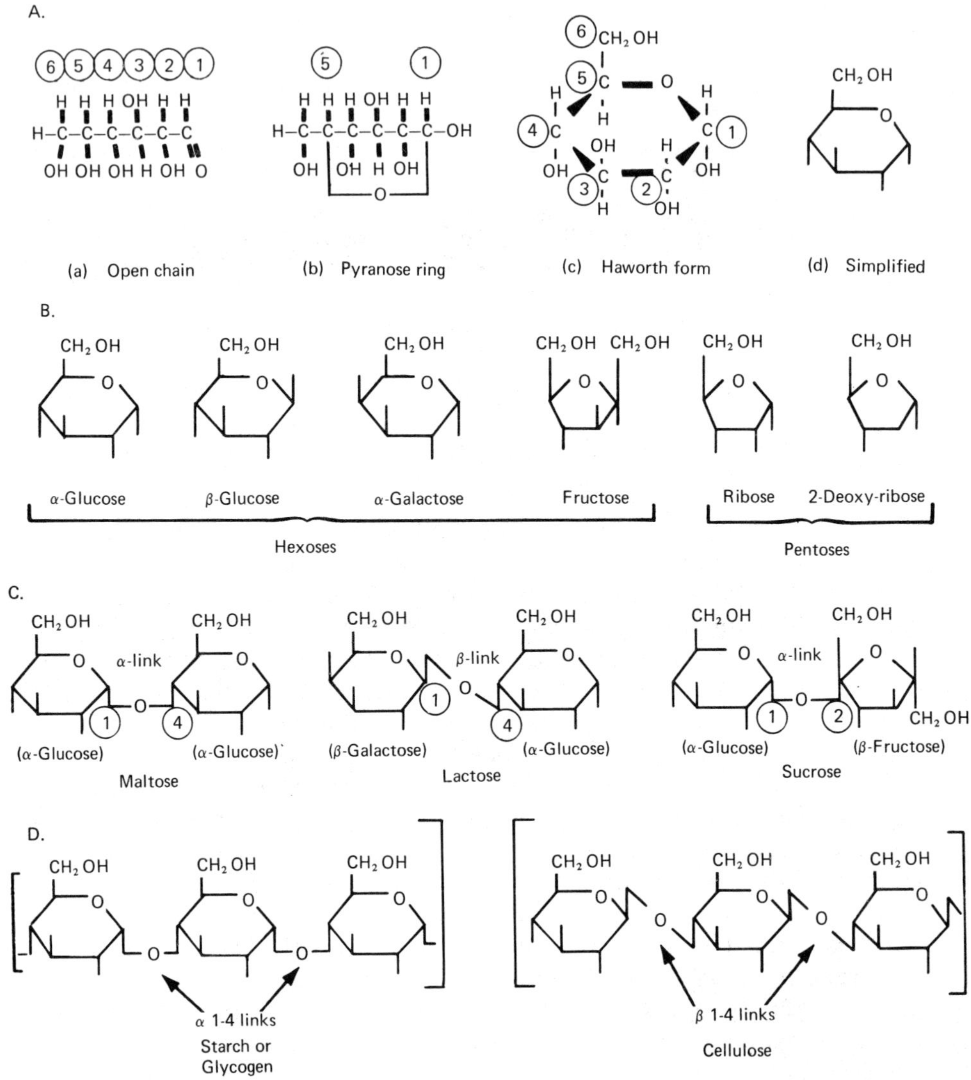

FIG. 4-1. Some important carbohydrates. **A.** Glucose is represented in four ways: **(a)** As the open-chain compound, with a free aldehyde in position 1; **(b)** as the pyranose ring, formed by a hemiacetal link between the aldehyde in position 1 with the hydroxyl in position 5; **(c)** as the Haworth form, which shows the pyranose ring as a flat hexagon, with substituents projecting above or below the plane of the ring; **(d)** in the simplified form used throughout this book, which depicts the ring with only the –OH projections shown (H atoms are not shown). **B.** Structures of major monosaccharides are depicted in the simplified form. The hydroxyl group at position 1 of the ring may project down (α) or up (β). **C.** Disaccharides are designated as α or β linked, depending upon the configuration at position 1 of the linkage. **D.** Polysaccharides may be joined by α or β linkages.

DIETARY CARBOHYDRATES

Most of the carbohydrate ingested in a natural diet consists of *starch,* the plant storage polysaccharide that is similar in structure to glycogen. Cane sugar or *sucrose*—a disaccharide of glucose with another hexose, fructose (Fig. 4-1C)—is abundant in "civilized" diets, and lactose is a significant component in a high milk diet, such as that of infants. Small amounts of free hexoses (glucose and fructose) are found in fruits and honey.

There is no specific carbohydrate requirement for the human, but in a balanced diet, about 50% of the energy requirement of the body should be provided from this source. The only carbohydrate derivative required in the diet is ascorbic acid, or *vitamin C.* Virtually all members of the animal kingdom are capable of synthesizing this six-carbon compound by three enzymatic conversions of glucose. Man, however, in common with the other primates and the guinea pig, lacks one of the enzymes required for this metabolic sequence. This represents a type of inborn error of metabolism that arose by a mutation during primate evolution. Other important carbohydrates, such as galactose and the pentoses, may be synthesized in the body and are not required for human nutrition.

A few persons in the population possess an inherited inability to utilize fructose from the diet. Apart from an abnormal excretion of fructose in the urine (which could be misinterpreted in the diagnosis of diabetes by undifferentiated sugar tests), such fructosuria is benign and unimportant clinically. More serious is the intolerance to galactose in the diet, experienced by infants with the inborn error known as galactosemia. The sequelae are mental retardation and cataracts, which may be prevented by the avoidance of milk and other sources of galactose in the diet. In terms of oral health, large amounts of free sugars in the diet should be avoided because they are excellent substrates for bacteria in the mouth and hence contribute to dental caries. Other than these exceptions, there are few strictures on the types of carbohydrates ingested by humans.

DIGESTION AND ABSORPTION

Only the monosaccharides are absorbed to any significant extent from the gastrointestinal (GI) tract. Disaccharides, such as sucrose, are broken down in the small intestine by specific *hydrolases,* enzymes that add the elements of water to split the disaccharide linkage and yield the constituent monosaccharides. The polysaccharides, such as starch from plant sources and glycogen from animal foods, are broken down in digestion by another group of hydrolases known as the *amylases,* which specifically cleave the linkages of these long-chain polymers (Fig. 4-2). The amylases are produced by both the salivary glands and the pancreas, thus ensuring continual and complete hydrolysis of polysaccharides during the passage of food down the gut. The plant structural polysaccharide, *cellulose* (Fig.

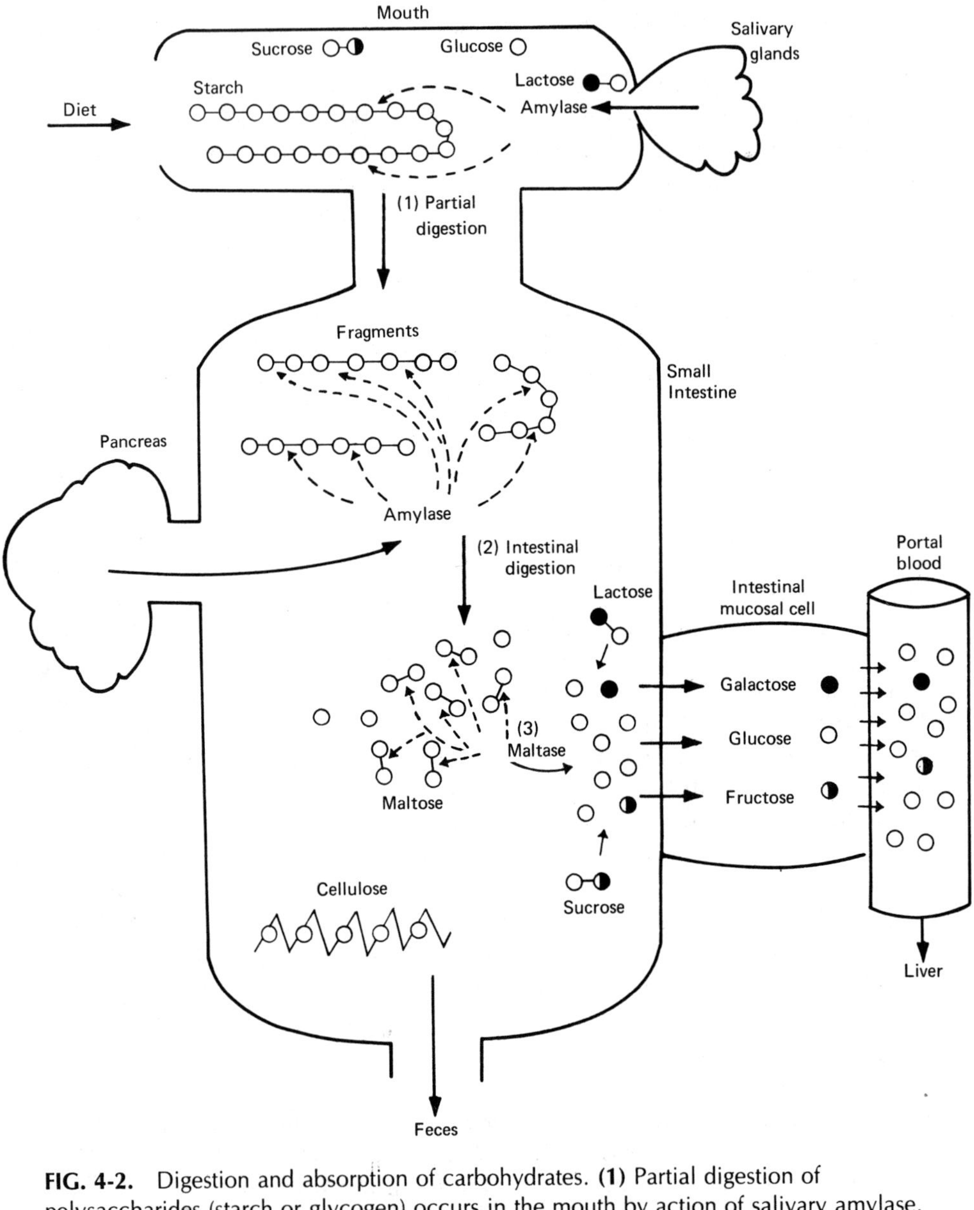

FIG. 4-2. Digestion and absorption of carbohydrates. **(1)** Partial digestion of polysaccharides (starch or glycogen) occurs in the mouth by action of salivary amylase. **(2)** Intestinal digestion completes the process under the action of pancreatic amylase. **(3)** Disaccharide hydrolases of the intestinal secretion break down maltose, lactose, and sucrose. Resulting monosaccharides are absorbed via mucosal cells into the portal bloodstream.

4-1D), has a different configuration of the linkages between the glucose monomers and is not cleaved by the amylases. Thus, cellulose cannot constitute a food source of carbohydrates for humans, but it may serve as the semisolid support, or bulk, of the intestinal contents. Amylases will attack starch and glycogen randomly along the length of the chain and will break down the shorter fragments until only disaccharides of two glucose units, *maltose,* remain (Fig. 4-1C). The maltose molecules are broken down to free glucose by a specific hydrolase known as *maltase,* which is found in the brush-border surface of the mucosal cells that form the villi of the small intestine.

Galactose, glucose, fructose, and the pentose monosaccharides ingested or released from other saccharides in digestion are absorbed by the mucosal cells lining the small intestine in the order given. The differences in the rates of absorption reflect different mechanisms of transport of the sugars across the membranes of the cells of the intestinal tract. Thus, glucose and galactose may be absorbed even when their concentrations in the cells are higher than in the intestinal lumen (*i.e.,* against a concentration gradient). This is a characteristic property of living cells known as *active transport,* and it requires an expenditure of energy to accomplish the osmotic work involved. Small amounts of the disaccharides sucrose and lactose may be absorbed if a superabundance of these is provided in the diet. They cannot, however, be utilized as such in the body and will be excreted unchanged in the urine.

THE LIVER

The hepatic cells of the liver serve as a major retailoring site and exert a regulatory influence upon the dietary carbohydrates (Fig. 4-3). Once absorbed, the sugars are taken from the intestinal-lining cells into the blood capillaries that drain into the portal vein. The latter vessel then conveys the monosaccharides derived from the diet directly to the liver. Here the sugars are rapidly taken up by the hepatic cells, where glucose may be formed from the fructose and galactose. Thus, the major carbohydrate that emerges in the general circulation after passage of the dietary sugars through the liver is glucose. One of the principal metabolic functions of the liver is to act as a blood *glucostat,* converting excess incoming glucose into storage reservoirs in times of plenty and reconverting the stores to glucose during food restriction, in order thereby to maintain an adequate level of glucose in the bloodstream.

As in other tissues of the body, the energy needs of the liver itself may be met by the intracellular catabolism of incoming glucose. Two distinct processes are involved in glucose catabolism: (1) the ***glycolytic pathway,*** which converts one mole of glucose to two of lactate and also yields two

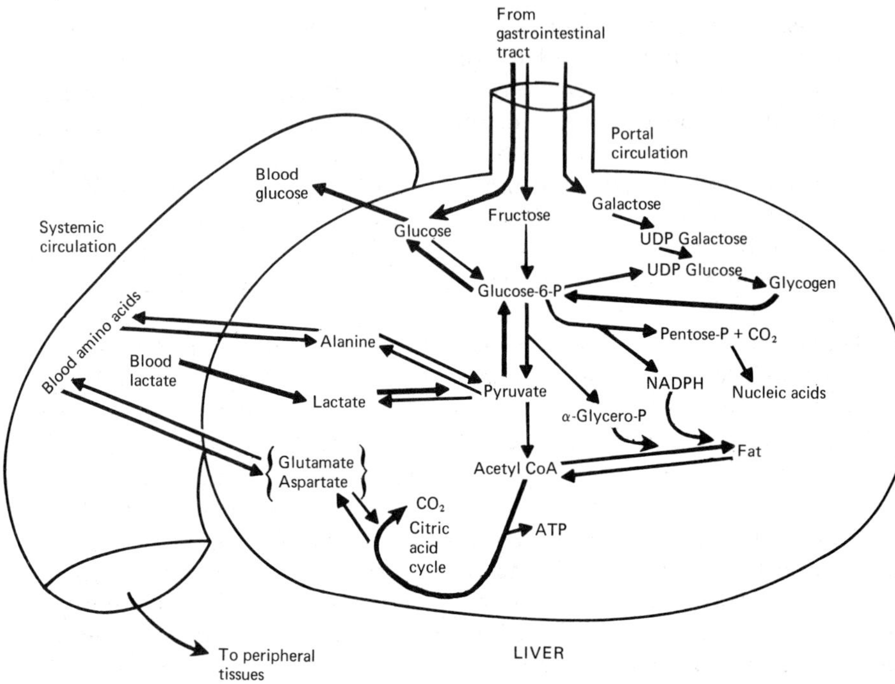

FIG. 4-3. Role of the liver in carbohydrate metabolism.

moles of ATP; and (2) the ***pentose–phosphate pathway,*** which converts one mole of glucose to six of CO_2 and also yields 12 mols of NADPH. Both processes are anaerobic, both enzyme systems are contained in the soluble cytoplasm, and both pathways require the prior phosphorylation of glucose to glucose-6-phosphate by the ATP-requiring enzyme, glucokinase. Here, however, the similarities end. Whereas glycolysis provides the energy for cellular phosphorylations, protein synthesis, and other processes that require ATP, the pentose–phosphate pathway is a primary source of reducing energy for biosynthetic reactions such as the steps in fatty-acid and steroid syntheses that require NADPH. The pentose–phosphate pathway breaks down the hexose molecule completely to CO_2 under anaerobic conditions, but glycolysis effects only the partial catabolism of glucose to three-carbon fragments; for complete oxidation of the latter, the presence of oxygen and mitochondria is required. Under aerobic conditions, the combined actions of glycolysis in the cytoplasm and of the citric-acid cycle and oxidative phosphorylation in the mitochondria will result in a maximum yield of 38 mols of ATP per mol of glucose.

The two pathways differ also in the intermediates formed: the pentose–phosphate pathway involves the intermediate production of pentose sugars, which may be tapped off to provide the cell's requirements for ribose in the formation of nucleotides and nucleic acids. The glycolytic pathway can be diverted to produce α-glycerophosphate for fat synthesis by the reduction of triose intermediates. It also produces pyruvate, which may be used in the net syntheses of alanine, aspartate, and various other substances through oxaloacetate formation, as well as of the myriad cellular constituents formed from acetyl CoA. Although reactions within the pentose–phosphate pathway may be used to form glucose in photosynthetic organisms, in mammalian tissues the process is essentially irreversible in the direction of glucose breakdown. In contrast, the reactions of glycolysis may be reversed to allow the synthesis of glucose by gluconeogenesis. In the liver, approximately one third of the glucose oxidized follows the pentose–phosphate pathway, the remaining two thirds being metabolized by the glycolytic reactions.

GLYCOLYSIS

In outline form (Fig. 4-4), the process of glycolysis is seen to involve the isomerization of glucose-6-phosphate to the ketohexose, fructose-6-phosphate, and the phosphorylation of the latter by a specific kinase, phosphofructokinase. The resulting six-carbon sugar with a phosphate esterified to each end, fructose-1,6-diphosphate, is now cleaved in the middle to produce two mols of triose phosphate; by a reversible isomerization, these three-carbon sugars become metabolically equivalent. The oxidative reaction of glycolysis catalyzed by triose-phosphate dehydrogenase transfers reducing equivalents to NAD^+ from the triose molecule, forming NADH. The energy liberated from the oxidation step is conserved by the uptake of inorganic phosphate in the highly reactive acyl phosphate of the product, 1,3-diphosphoglycerate. The latter transfers this phosphate grouping to ADP, forming ATP and 3-phosphoglycerate. After mutation of the phosphate to position 2, the product is dehydrated by the enzyme enolase to the second high-energy intermediate of glycolysis, phosphoenolpyruvate. The latter donates its phosphate to ADP, forming ATP and pyruvate. Thus, each of the two triose molecules formed by the cleavage of one molecule of hexose will generate two ATP molecules, and the glycolysis of 1 mol glucose will generate 4 mol ATP. Since 2 mol ATP are necessary to convert glucose to fructose-1,6-diphosphate, the net energy yield of glycolysis is 2 mol ATP/mol glucose utilized.

The oxidative function in glycolysis is dependent upon the conversion of NAD^+ to NADH and therefore requires a means of regenerating NAD^+ from the NADH thus formed. In the absence of oxygen, this is accomplished by lactate dehydrogenase, which accepts hydrogens from NADH and transfers them to the pyruvate formed as the end product of the sequence of reactions below triose phosphate. The products thus formed are NAD^+, which can then be recycled for the oxidative reaction of triose-phosphate dehydrogenase, and lactate, which is the ultimate endpoint of

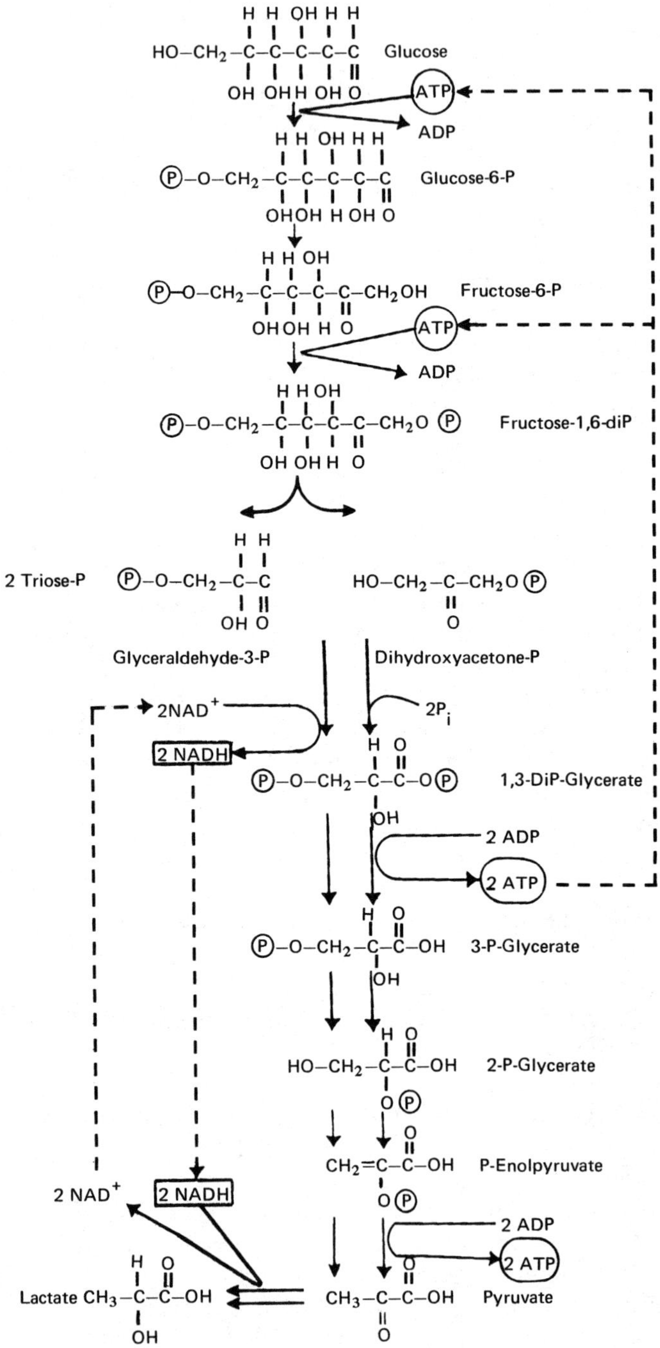

FIG. 4-4. Glycolysis. Splitting of hexose yields 2 mols of triose; therefore all steps below are doubled (twin arrows). NADH is recycled anaerobically by reduction of pyruvate. The net reaction therefore is 1 Glucose + 2 ADP + 2 $P_i \rightarrow$ 2 Lactate + 2 ATP.

anaerobic glycolysis. Thus, there is no net change in the NAD^+-NADH couple, since the oxidation–reduction reactions of triose-phosphate dehydrogenase and lactate dehydrogenase are linked together to form an internally compensated system.

In a sense, the process of glycolysis is basically an intermolecular oxidation of the polyhydroxylic hexose molecule, with one end of the product, lactate, becoming more highly oxidized, (—COOH) and the other end becoming more highly reduced (—CH_3). Aerobic glycolysis produces much less lactate; in the presence of O_2, most of the pyruvate is converted instead to acetyl CoA, and most of the NADH is oxidized by the electron-transport chain.

GLUCONEOGENESIS

The reverse of glycolysis, ***gluconeogenesis,*** is a particularly important process in the liver, which is the major organ involved in glucose formation from carbohydrate and noncarbohydrate sources. Although the energy changes in most of the glycolytic reactions are small enough that they may be reversed, there are three enzymatic steps that proceed to such a small extent in the direction of glucose synthesis that they may be considered as essentially irreversible (Fig. 4-5). These are *(a)* glucokinase (glucose → glucose-6-phosphate); *(b) phosphofructokinase* (fructose-6-phosphate → fructose-1,6-diphosphate); and *(c) pyruvate kinase* (phosphoenolpyruvate → pyruvate).

Special gluconeogenic enzymes are used to bypass these barriers to the reversibility of the glycolytic pathway. The first two kinases are simply circumvented by the provision of two specific phosphohydrolases: *(a′) glucose-6-phosphatase* (glucose-6-phosphate → glucose) and *(b′) fructose-1, 6-diphosphatase* (fructose-1,6-diphosphate → fructose-6-phosphate). It is important to note that these phosphatases do not promote the reverse reactions of the corresponding kinases, which would involve the thermodynamically unfeasible resynthesis of ATP from ADP. These are hydrolytic reactions in which H_2O reacts to release inorganic phosphate, and the energy of the phosphate ester is released as heat.

Surmounting the barrier at the pyruvate-kinase level requires more strenuous metabolic efforts and must be accomplished in stages involving two enzymes: *(c′) pyruvate carboxylase* (pyruvate + CO_2 → oxaloacetate) and *(c″) phosphoenolpyruvate carboxykinase* (oxaloacetate → phosphoenolpyruvate + CO_2). Both reactions require energy; the first, as noted previously, is accompanied by the hydrolysis of ATP to ADP, while the second is driven by the dephosphorylation of GTP to GDP (the latter may be reconverted to GTP at the expense of ATP). Since a third mol of ATP is needed to reverse the 3-phosphoglycerate to 1,3-diphosphoglycerate step, the total energy expenditure for the synthesis of glucose from 2 mols of pyruvate will be six equivalents of high-energy phosphate. In addition, reducing energy in the form of NADH is required to reverse the triose-phosphate dehydrogenase step.

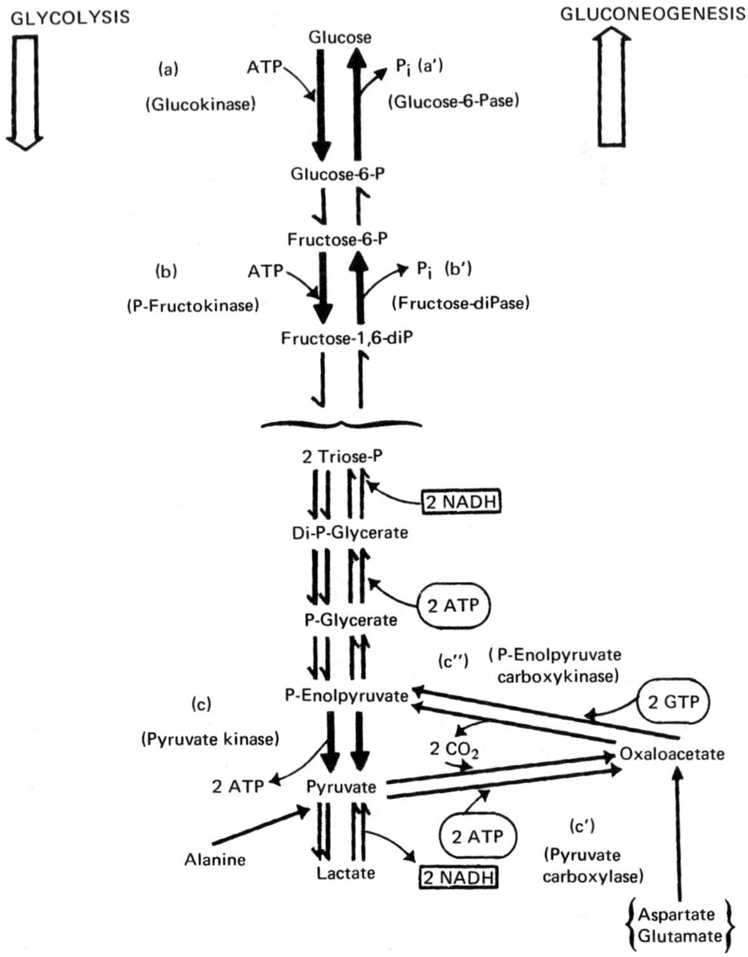

FIG. 4-5. Gluconeogenesis, showing the three irreversible steps of glycolysis at **(a)**, **(b)**, and **(c)** as well as the reactions **(a′)**, **(b′)**, **(c′)**, and **(c″)** of gluconeogenesis that circumvent these steps.

Thus, the stoichiometry for gluconeogenesis from pyruvate (or its precursors such as alanine) will be

$$2\ \text{Pyruvate} + 2\ \text{NADH} + 6\ \text{ATP} \rightarrow 1\ \text{Glucose} + 2\ \text{NAD}^+ + 6\ \text{ADP} + 6\ \text{P}_i$$

Starting with oxaloacetate (which may originate, for example, from the catabolism of amino acids such as aspartate, glutamate, and so on), the overall reaction is:

$$2\ \text{Oxaloacetate} + 2\ \text{NADH} + 4\ \text{ATP} \rightarrow 1\ \text{Glucose} + 2\ \text{CO}_2 + 2\ \text{NAD}^+ + 4\ \text{ADP} + 4\ \text{P}_i$$

Note that only 4 mols of ATP are needed here, because one of the energy-requiring steps is not used.

Another extremely important starting point for liver gluconeogenesis is the lactate that originates in other tissues, such as the muscles, and arrives at the liver in the bloodstream. In this case, the reducing requirements for the reversal of the triose-phosphate dehydrogenase reaction will be provided by coupling it to the reversal of the lactate-dehydrogenase reaction, and the two enzymes will provide an internally compensated oxidation–reduction system as in anaerobic glycolysis:

$$2 \text{ Lactate} + 6 \text{ ATP} \rightarrow 1 \text{ Glucose} + 6 \text{ ADP} + 6 \text{ P}_i$$

REGULATION OF GLYCOLYSIS AND GLUCONEOGENESIS

The balance between the catabolism and the anabolism of glucose in liver cells is subject to an elaborate series of regulatory influences, including the concentrations of metabolites and nucleotides as well as the effects of hormones (Fig. 4-6). It should be noted that the liver cells and, to a lesser extent, those of kidney and intestinal mucosa are unique in that they contain the full complement of enzymes to allow the reversal of glycolysis. Other tissues have control factors that are solely involved with turning the glycolytic process on or off in parallel with their own cellular energy demands. Hepatic cells, on the other hand, have larger responsibilities to the rest of the body; they must sense the requirements for glucose elsewhere and switch on the gluconeogenic reactions accordingly. Moreover, the regulatory system must also ensure that the catabolic reactions of glycolysis are "off" while gluconeogenesis is taking place; otherwise, the cell might engage in a futile, energy-dissipating cycle:

$$\text{Glucose} \rightarrow 2 \text{ Lactate} + 2 \text{ ATP}$$

$$2 \text{ Lactate} + 6 \text{ ATP} \rightarrow \text{Glucose}$$

Intracellular Regulators

Substrates and intermediates play an important role in determining whether glycolysis or gluconeogenesis will predominate. As in the case of the regulation of the citric-acid cycle, the oxaloacetate concentration will be rate-limiting for gluconeogenesis. To provide an effective source of substrate for phosphoenolpyruvate formation, the oxaloacetate concentration must exceed that required for the cycle. The sources of oxaloacetate production, apart from the carboxylation of pyruvate, derive mainly from the catabolism of proteins and the deamination of the free amino acids to form such keto acids as oxaloacetate and α-ketoglutarate. Consequently, any physiologic factors that stimulate the breakdown of tissue proteins, that prevent their biosynthesis from tissue amino acids, or that enhance the activity of aminotransferases and other enzymes which promote the

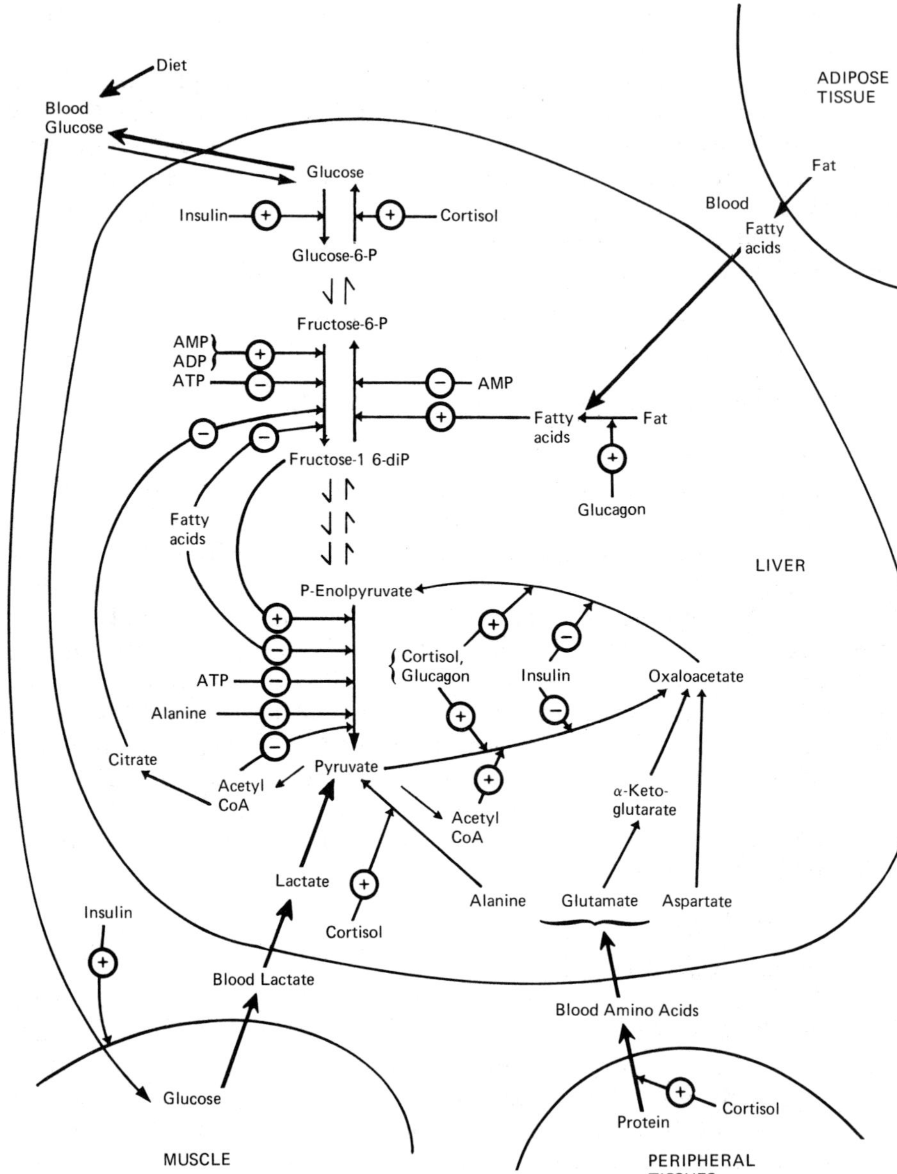

FIG. 4-6. Regulation of glycolysis and gluconeogenesis in liver. Negative modifiers are represented by the circled minus signs, with the arrow indicating the reaction that is inhibited or repressed; positive modifiers are shown by circled plus signs, with the arrow indicating the reaction stimulated or induced.

conversion of aspartate and glutamate to their keto-acids, will enhance gluconeogenesis indirectly by providing more substrate.

The amino acid, alanine, plays an important role, too, since it furnishes pyruvate upon deamination and acts as a negative allosteric modifier of pyruvate kinase, thus providing a source of carbon for gluconeogenesis and preventing the operation of a futile cycle:

Pyruvate + ATP + GTP → Phosphoenolpyruvate

Phosphoenolpyruvate → Pyruvate + ATP

Citrate, as noted previously, acts as a negative feedback agent by inhibiting phosphofructokinase to block glycolysis at an early stage. Thus, when excessive amounts of the citric-acid-cycle intermediates are produced in the liver from amino-acid deamination, the brakes will be applied to the glycolytic pathway, and glucose synthesis will be favored.

Excess acetyl-CoA production will also favor gluconeogenesis, because pyruvate carboxylase requires acetyl CoA as a positive modifier. Like alanine, acetyl CoA will also inhibit pyruvate-kinase activity.

It is of interest that although the alkyl side-chains of fats are not able to provide carbon for the synthesis of carbohydrates, the long-chain fatty acids will promote gluconeogenesis by acting as allosteric modifiers. Thus, physiologic conditions and hormonal agents that stimulate the breakdown of fats to free fatty acids or that inhibit tissue fatty-acid utilization will stimulate glucose synthesis in the liver. The major site of this stimulation is the gluconeogenic enzyme, fructose-1,6-diphosphatase, which is increased in activity manyfold by the binding of free fatty acids. At the same time, pyruvate kinase and phosphofructokinase are both inhibited by fatty acids, which may thus prevent the operation of glycolysis while gluconeogenesis proceeds.

Fructose-1,6-diphosphate exhibits a unique type of controlling influence upon the enzyme, pyruvate kinase. The latter exists under normal conditions in an inhibited state in the liver, and it requires fructose diphosphate as a positive allosteric modifier. This is an example of an early intermediary in the metabolic sequence of glycolysis acting as a positive feedforward agent to ensure not only that it will not accumulate excessively, but also that its own further catabolism will be promoted.

Another example of the precise regulation through interaction between countervailing influences is the control exerted through cellular energy changes as reflected in the adenine-nucleotide balance. When the energy levels in the liver are low—that is, when low ATP, high AMP, and high ADP concentrations are present—glycolysis will be turned on, and gluconeogenesis, off. The opposite situation also applies, ensuring that glucose synthesis can occur only when the cell has a surfeit of energy, as indicated by high ATP, low AMP, and low ADP concentrations. Apart from the obvious fact that the gluconeogenic process demands energy input in the form of ATP as a reactant, there are other, more subtle controlling influences exerted by the adenine nucleotides acting in concert as allosteric modifiers. The major shutdown site for gluconeogenesis is the fructose diphosphatase reaction, which is turned off by AMP binding and acting as a negative allosteric modifier. Thus, when cellular energy reserves are low, all the fructose diphosphate produced by phosphofructokinase may be channeled into ATP production by glycolysis, and the presence of AMP will prevent the nonproductive, energy-dissipating cycle:

Fructose-6-phosphate + ATP → Fructose-1,6-diphosphate

Fructose-1,6-diphosphate → Fructose-6-phosphate + P_i

Moreover, elevated concentrations of AMP or ADP will activate phosphofructokinase, whereas elevations of the ATP concentration will inhibit both phosphofructokinase and pyruvate kinase. Thus, a decreased energy level switches gluconeogenesis off and glycolysis on by the effects of AMP and ADP at the level of the fructose-6-phosphate–fructose-diphosphate interconversion. On the other hand, an increased energy charge provides the driving force for gluconeogenesis to occur while simultaneously switching off the kinase reactions unique to glycolysis.

Extracellular Regulators

Superimposed upon these intracellular control factors, which relate primarily to the capabilities and requirements of the liver itself, are the extracellular factors that signal the needs of the body as a whole. *Substrate availability* from the bloodstream is one such factor. During strenuous muscular activity, for example, the production of lactate will exceed the capacity of the muscle mitochondria for pyruvate and NADH oxidation, and anaerobic glycolysis will predominate over aerobic catabolism by means of the citric-acid cycle and the electron-transport chain. The resulting excess of lactate will diffuse into the bloodstream and, upon reaching the liver, will provide the carbon source for gluconeogenesis. The glucose produced will then pass back through the circulation to the exercising muscle, thereby constituting the so-called ***Cori cycle:***

Muscle: Glucose → 2 Lactate (+2 ATP)

Liver: 2 Lactate → Glucose (−6 ATP)

In this case, the blood acts as intermediate carrier of the metabolites. Stated another way, the muscle can derive two ATPs for the mechanical work of contraction at the expense of six ATPs in the liver. Other but similar means of providing substrates for gluconeogenesis include the extrahepatic sources of free amino acids, which are produced by the catabolism of muscle and other tissue proteins, and the mobilization of fat stores from depot tissues, which may provide elevated concentrations of free fatty acids in the liver to act as energy sources for gluconeogenesis and as modulators of the enzymes involved.

Another factor from outside the liver that will regulate the balance between hepatic glucose production and utilization is the *blood glucose concentration* itself. First, glucose equilibrates readily and rapidly across liver cell membranes by a process of facilitated diffusion. Thus there is no detectable concentration gradient between liver cells and the surrounding medium, and the intracellular concentrations will reflect those in the blood. Obviously, then, in times of plenty, such as following a high-carbohydrate meal, the increased blood glucose concentration will promote glycolysis in the liver by providing increased levels of intracellular substrate. Conversely, during a fast, the low extracellular glucose concentration will enhance the ability of the intrahepatic glucose produced from gluconeogenesis to diffuse into the circulation. A second characteristic that renders the liver especially sensitive to changes in blood glucose concentrations is the presence of the unique enzyme, glucokinase. Although other tissues possess nonspecific hexokinases with very low K_m values for glucose (10^{-6} to 10^{-5} M), the liver enzyme is highly specific and has a relatively high K_m for glucose ($K_m = 2 \times 10^{-2}$ M). Since the latter value corresponds to a concentration of 360 mg per 100 ml (whereas the normal fasting levels of glucose in blood are about 90 mg per 100 ml or 5×10^{-3} M), we can readily calculate, using the Michaelis–Menten equation, that liver glucokinase in normal fasting persons is operating at only a fraction of its capacity:

$$v = \frac{V_m[\mathrm{S}]}{K_m + [\mathrm{S}]}$$

Therefore,

$$\frac{v}{V_m} = \frac{[\mathrm{S}]}{K_m + [\mathrm{S}]} = \frac{90}{360 + 90} = \frac{90}{450} = 20\%$$

When the blood glucose concentration doubles, say, following a high-carbohydrate meal, v/V_m will rise markedly to 33% of capacity, whereas a fall to half the normal level, as in hypoglycemia, will reduce glucose phosphorylation rates correspondingly to about 10% of capacity. The combination of free glucose diffusion through the cell membrane and the high K_m of glucokinase makes the extent of utilization of glucose by the liver highly sensitive to the glucose concentration in the blood.

The other group of extracellular regulators of gluconeogenesis and glycolysis in the liver consists of the *endocrine agents.* ***Insulin,*** for example, indirectly enhances the utilization of glucose by causing an increased synthesis of the critical enzyme, glucokinase; increased amounts of the latter will remove glucose from the circulation by converting it to nondiffusible glucose-6-phosphate within the liver cells. Hormones from the adrenal cortex with an oxygen function at position 11 of the steroid ring have profound effects upon carbohydrate metabolism, and hence they are termed the ***glucocorticoids.*** Glucocorticoids, such as cortisol, act upon the liver in the opposite direction to insulin, promoting gluconeogenesis. The mechanism of this effect is due to the induction of synthesis of several enzymes required for the gluconeogenic process, including certain transaminases, pyruvate carboxylase, phosphoenolpyruvate carboxykinase, fructose-1,6-diphosphatase and glucose-6-phosphatase. The effects of the glucocorticoids are counteracted by insulin, and thus the balance between these antagonistic hormones will determine the extent of these effects. Although they promote the biosynthesis of the enzymatic proteins in the liver, the glucocorticoids generally enhance protein catabolism in other tissues as well, thus providing increased amounts of free amino-acid substrates for glucose synthesis by liver cells. ***Glucagon,*** which, like insulin, is a pancreatic hormone, stimulates the enzyme reactions that convert pyruvate to phosphoenolpyruvate. It also indirectly increases gluconeogenesis by activating hepatic lipase, which in turn causes elevated fatty-acid levels in the cells. As in the case of the effects produced by the glucocorticoids, the actions of glucagon in promoting gluconeogenesis are also antagonized by insulin.

Apparently, it is difficult to isolate the functions of individual endocrine agents upon these processes; rather, it would seem that the critical determinant will be the ratios of the hormones to one another, with the effects of excesses or deficiencies of one agent being considered in relation to the concentrations of its antagonists.

PYRUVATE METABOLISM

Complete oxidation of the pyruvate produced in glycolysis is by the final common pathway shared by carbohydrates, fats, and proteins—namely, the citric-acid cycle (Chap. 3). Before entering the cycle, pyruvate first must be carried from the cytoplasm into the mitochondria; here the pyruvate-dehydrogenase enzyme system acts by exactly analogous mechanisms to those described earlier for the oxidative decarboxylation of α-ketoglutarate within the cycle (Fig. 4-7). Thus, the complex multienzyme system first uses thiamine pyrophosphate to decarboxylate pyruvate to acetaldehyde; the latter, while still bound to the enzyme, is oxidized to an acetyl group; the acetyl group is transferred to lipoic acid and the hydrogens from the oxidation are transferred to NAD^+ through FAD; and finally, this activated acetyl group, with the energy of the thio ester retained, is transferred to a free molecule of coenzyme A. The overall reaction is

$$\text{Pyruvate} + \text{CoASH} + \text{NAD}^+ \rightarrow \text{Acetyl}{\sim}\text{SCoA} + \text{NADH} + \text{CO}_2$$

The liver, as all tissues in the human body, has no mechanism to convert the energy of the thio ester of acetyl CoA to ATP that is parallel to the mechanism for succinyl CoA; rather, the energy is used for condensation

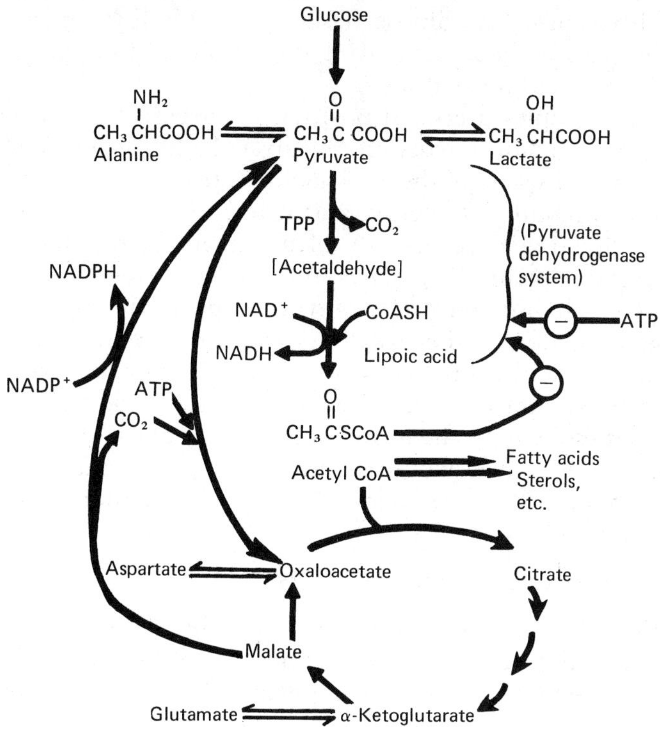

FIG. 4-7. Metabolism of pyruvate. (*TPP,* thiamine pyrophosphate.)

reactions, such as the citrate synthetase reaction, or for converting the acetyl groups to fatty acids, sterols, and other important constituents of the liver.

The oxidative decarboxylation of pyruvate is essentially non-reversible, and the pyruvate dehydrogenase reaction will lead to depletion of the body pool of glucose because there is no process in humans that can convert acetyl CoA into glucose. Thus the control of pyruvate oxidation is important for conservation of glucose in liver and other tissues. Pyruvate dehydrogenase is subject to feedback regulation at two levels by the end-products of its reaction: (1) acetyl CoA and NADH will inhibit the reaction by competing with CoASH and NAD^+ respectively; (2) ATP will inactivate the enzyme by phosphorylating it, and acetyl CoA and NADH will favor the phosphorylation. When there is abundant acetyl CoA for citric-acid-cycle oxidations (*e.g.*, from fat catabolism), and when there is ample cellular energy as indicated by high NADH and ATP levels, pyruvate dehydrogenase will be turned off, and glucose will be conserved for other requirements. As noted in Chapter 5, pyruvate dehydrogenase is also a key control point of lipogenesis in adipose tissue. Insulin will favor dephosphorylation of the enzyme, hence activating it to produce more acetyl CoA from glucose for fatty-acid biosynthesis.

Pyruvate obviously occupies a crucial position in carbohydrate metabolism: it may generate lactate to sustain glycolysis in the absence of O_2; it may be fully oxidized by way of the citric-acid cycle in the presence of O_2 to yield enormous amounts of energy; and it may be carboxylated to keep the cycle operational or to reverse glycolysis. In addition to these important functions relating primarily to carbohydrate metabolism, there are numerous ways in which interconversions involving pyruvate impinge upon other metabolic pathways. Either by direct transamination to alanine or through other keto acids (through which it is precursor to aspartate, glutamate, and others), pyruvate contributes and is itself regenerated through amino-acid metabolism. In the form of acetyl CoA, pyruvate contributes to the anabolism of a variety of lipids and other important compounds such as acetylcholine. Moreover, as we shall note in Chapter 5, the action of malic enzyme involves pyruvate:

$$\text{Malate} + \text{NADP}^+ \rightleftarrows \text{Pyruvate} + CO_2 + \text{NADPH} + H^+$$

This reaction may constitute an important source of reducing power for those cellular anabolic paths, such as fatty-acid and sterol biosynthesis, that require a source of reduced NADP. Thus, pyruvate truly represents a crossroads where the manifold processes of carbohydrate, fat, and protein metabolism converge and interlock with one another.

PENTOSE–PHOSPHATE PATHWAY

The pentose–phosphate pathway involves a complex series of transformations that will not be dealt with in detail here. The main oxidative reactions are catalyzed by two different NADP-linked dehydrogenases: (1) glucose-6-phosphate dehydrogenase oxidizes the aldehyde function of glucose to a carboxylic-acid group, thus forming 6-phosphogluconic acid and NADPH, and (2) 6-phosphogluconate dehydrogenase acts on the product of the preceding enzyme by an oxidative decarboxylation mechanism similar to that catalyzed by isocitrate dehydrogenase, liberating the carboxyl group as CO_2 and leaving behind a five-carbon sugar, ribulose-5-phosphate, and NADPH (Fig. 4-8). The chief importance of this pathway for the cell, in addition to the provision of pentose sugars, is the reductive synthetic energy generated as NADPH. This coenzyme also serves as the main regulatory factor in the pathway, since both of the dehydrogenases are strongly inhibited by the reduced form of NADP. If allowed to accumulate in the cell, the NADPH will cause nearly total inhibition of the dehydrogenases, which will be relieved only when some reductive process utilizes the NADPH, reconverting it to $NADP^+$. Thus, the pentose-phosphate pathway will respond to the cellular NADPH requirements that

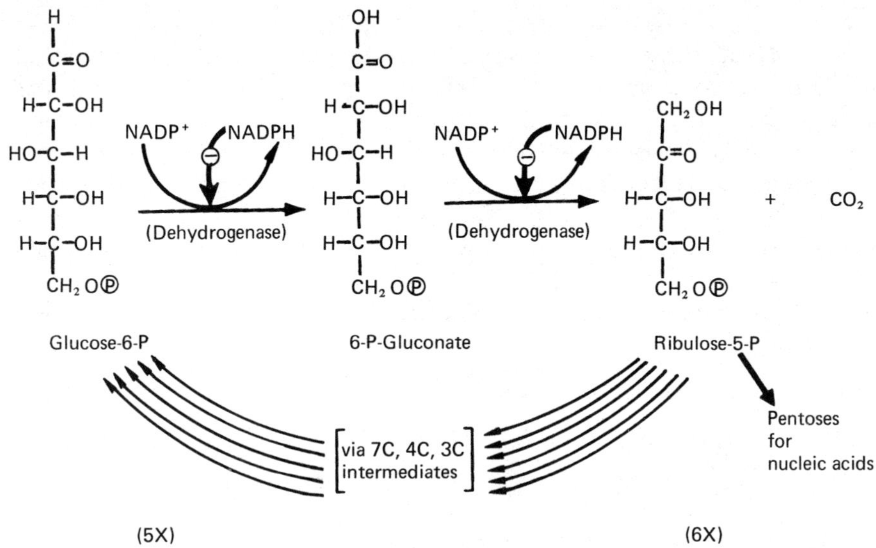

FIG. 4-8. The pentose–phosphate pathway.

arise, for example, in the synthesis of fatty acids. The metabolic economy of the cell will be preserved, however, since a decreased requirement for NADPH will result in the shutdown of the pentose–phosphate pathway in its initial reactions, thereby conserving glucose-6-phosphate for other purposes.

The complete oxidation of glucose-6-phosphate to CO_2 in the pentose–phosphate pathway involves a series of interconversions of the pentoses (that are produced by the dehydrogenations) to re-form hexoses. At first sight, this would seem impossible without the refixation of the CO_2 that is initially lost, but in reality it occurs by a piece of metabolic legerdemain whereby bits of one pentose molecule are juggled about with additions and subtractions from other pentoses. The overall stoichiometry may be seen if we start with 6 mols of glucose-6-phosphate. In the first two dehydrogenase steps

1. 6 Glucose-6-phosphate + 12 $NADP^+$
$\rightarrow$ 6 Ribulose-5-phosphate + 6CO_2 + 12 NADPH

The remaining steps of the process are concerned with the net reaction:

2. 6 Ribulose-5-phosphate $\rightarrow$ 5 Glucose-6-phosphate + P_i

Summing (1) and (2),

Glucose-6-phosphate + 12 $NADP^+$ $\rightarrow$ 6CO_2 + 12 NADPH + P_i

The individual reactions of the six pentose sugars in reaction (2) may be indicated by considering their arithmetic: in the first reaction, a two-carbon portion of *one pentose* is transferred to a *second pentose* (5C + 5C → 7C + 3C). Next, the seven-carbon sugar transfers a three-carbon portion to the three-carbon sugar (7C + 3C → 4C + 6C), forming *one hexose molecule.* The four-carbon sugar remaining receives a two-carbon portion donated by a *third pentose* (4C + 5C → 6C + 3C), forming a *second hexose molecule* plus a triose. It may be seen that doubling each of the above processes would therefore utilize *six pentose molecules* to produce *four hexose molecules* and two trioses; the latter could condense by a reversal of glycolysis to produce the *fifth hexose molecule.* Thus, only the substances in italics are net reactants in the pathway, and these account for the overall balance:

$$6 \times 5\text{C sugars} = 5 \times 6\text{C sugars}$$

The other reactants—that is, the 7C, 4C, and 3C sugars—are only transient intermediates of the pathway (Fig. 4-8).

GLYCOGEN METABOLISM IN THE LIVER

Since glucose in its monosaccharide form is freely diffusible through the hepatic cell membrane, it cannot be stored as such within this tissue in order to provide a reservoir. The glucose phosphate and fructose phosphate esters, however, do not pass through the membrane readily, and hence the phosphorylation of these sugars serves to trap them, effectively sequestering the reactions of glycolysis and the pentose–phosphate pathway within the intracellular spaces. These sugar phosphates, though, could not accumulate extensively within the liver, since, due to osmosis, a large accumulation of water would accompany the buildup of such solutes. To provide a nondiffusible storage form of glucose without the attendant osmotic swelling of the cells, the liver has the capacity to convert its excess carbohydrate into the large, insoluble polymer *glycogen.*

After a heavy intake of carbohydrates, as much as one tenth of the liver mass may consist of this storage form of glucose, whereas in starvation, the glycogen reserve may be almost depleted. Under customary dietary fluctuations, the content of glycogen in liver will vary between 2% and 8% of the weight of the organ, with an average or normal value of about 5%. The importance of this rather small energy store, which is equivalent to 70 g to 90 g glucose in a normal, well-fed adult, is that it constitutes a rapidly mobilized source of sugar, both for the blood that nourishes other tissues and for the needs of the liver itself. In the first 12 to 24 hr after withdrawal of food, for example, most of the body's carbohydrate needs will be met by the breakdown of liver glycogen. After more prolonged fasting, the slower response of the tissue to endocrine factors will lead to an increased synthesis of the enzymes required for amino-acid catabolism, phosphoenolpyruvate formation, and so forth, so that gluconeogenesis will then operate maximally after the glycogen has become depleted.

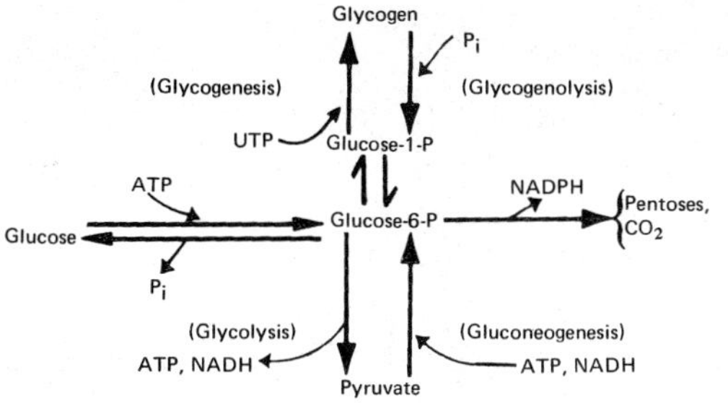

FIG. 4-9. Metabolism of glycogen and glucose-6-phosphate in the liver.

Glycogenesis

The process of glycogen synthesis or ***glycogenesis*** is an anabolic sequence that requires energy expenditure both from ATP and from uridine triphosphate (UTP). The latter is a common reactant in the activation of glucose, galactose, and so on for other enzyme reactions, and the uridine nucleotides may be considered the universal coenzymes of sugar metabolism. Glycogenesis starts with glucose, which must first be converted to glucose-6-phosphate by glucokinase at the expense of ATP. Thus, it may be seen that glucose-6-phosphate occupies a key position as a common intermediate in carbohydrate metabolic pathways: it may be converted reversibly to pyruvate by glycolysis or gluconeogenesis; it may be converted to pentoses or to CO_2 irreversibly by the pentose–phosphate pathway; and it may be converted reversibly to glycogen (Fig. 4-9).

The first unique step in the synthesis of glycogen is a simple enzymatic mutation of the phosphate group of glucose-6-phosphate from position 6 to the position 1. The next step, the activation of glucose-1-phosphate, utilizes the uridine nucleotide. In the process, the two terminal phosphates are removed from UTP as inorganic pyrophosphate, while the remaining UMP portion is joined by a pyrophosphate bridge to glucose-1-phosphate to form *UDP-glucose* (Fig. 4-10). This highly reactive, coenzyme form of glucose is capable of participating in numerous enzyme reactions, either in joining sugar molecules to other substances (*e.g.*, to proteins or lipids to form glycoproteins or glycolipids) or in interconverting sugars (*e.g.*, glucose to galactose).

In the reaction catalyzed by ***glycogen synthetase,*** a glucose monomer from UDP-glucose is joined to the free 4-hydroxy group of the terminal glucose unit of an existing glycogen chain. This reaction is repeated with more incoming UDP-glucose donors, each of which attaches one more glucose residue through its one-position and releases UDP. This UDP may re-form UTP at the expense of ATP. The resulting glycogen polymer contains varying numbers—ranging from several thousand to hundreds of thousands—of glucose monomers in 1–4 linkage.

The molecule does not extend indefinitely in a linear chain, because there is a second enzyme that transfers part of the growing chain back to link up through a free hydroxyl in position 6 of a glucose monomer within

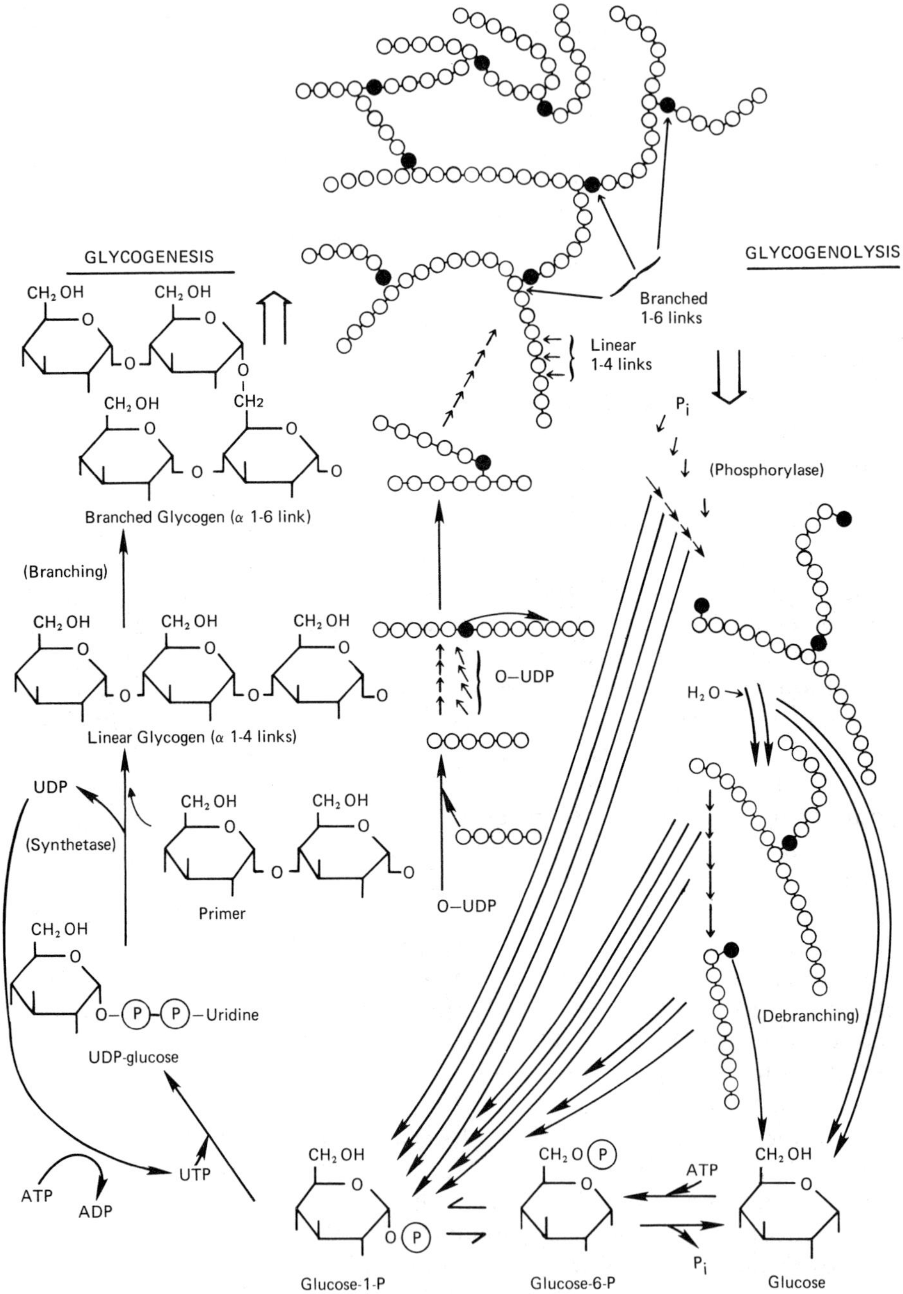

FIG. 4-10. Glycogenesis and glycogenolysis. Actions of the synthetase and branching enzymes are shown diagrammatically in the central portion of the figure, beside the structural formulas on the left. Actions of phosphorylase to release glucose-1-phosphate from terminal 1–4 linkages and of the debranching enzyme to release glucose from branched 1–6 linkages are shown on the right.

an existing chain. The concerted action of this *branching enzyme* with glycogen synthetase produces linear chains of glucose monomers with 1–4 linkages and side branches with 1–6 linkages at the branch points. As a result, glycogen has a structure rather like a tree, with a branch occurring at approximately every tenth glucose residue. This highly branched structure serves two major purposes in the cell: first, it provides numerous ends for glucose molecules to be attached or removed rapidly; second, it forms a dense, compact storage particle in the cell.

Electron microscopy reveals that these particles in the liver cytoplasm are joined together into clusters or rosettes about 20 nanometers in diameter. The particles contain, in addition to the stored glycogen, the enzymes both for glycogen synthesis (the synthetase and branching enzymes) and for glycogen breakdown in a bound, multienzyme complex.

Glycogenolysis

The breakdown of liver glycogen occurs in response to the accelerated requirements for glucose by the hepatic tissue itself or by other parts of the body. The enzyme ***phosphorylase,*** as its name implies, is responsible for breaking the linkages between glucose monomers of the glycogen chains by adding inorganic phosphate, instead of water as in the hydrolytic action of amylase upon the polysaccharides. Phosphorylase also differs from amylase in its actions by cleaving off the glucose units one at a time from the tips of the glycogen tree, rather than by random attack. Phosphorylase can act only upon the linear portions of the polymer down to the branch points; a separate *debranching enzyme* hydrolyzes the 1–6 linkage, yielding free glucose, and renders the rest of the "tree" susceptible to further pruning by phosphorylase. Because the latter adds phosphate at the linear 1–4 linkages, the products of phosphorylase action will be glucose-1-phosphate molecules, which will outnumber the free glucose molecules from the debranching process by roughly 10:1.

The glucose-1-phosphate molecules may be reconverted by phosphoglucomutase to glucose-6-phosphate, and the latter may be utilized directly for energy within the liver through either the glycolytic or pentose–phosphate pathways. When used in this fashion for intracellular energy, the convenience of glycogen storage costs the cell one high-energy equivalent per molecule of glucose-6-phosphate passing through the cycle:

Glucose-6-phosphate → Glucose-1-phosphate

Glucose-1-phosphate + UTP → UDP-glucose + Pyrophosphate

UDP-glucose → Glycogen + UDP

UDP + ATP → UTP + ADP

Glycogen + P_i → Glucose-1-phosphate

Glucose-1-phosphate → Glucose-6-phosphate

In order to provide energy for other tissues, the glucose-6-phosphate must be hydrolyzed to free glucose to permit diffusion through the bloodstream, thereby dissipating an additional high-energy equivalent:

Glucose + ATP → Glucose-6-phosphate + ADP

Glucose-6-phosphate → Glucose + P_i

Thus, the cycle, blood glucose → liver glycogen → blood glucose, that provides an energy storage for extrahepatic tissues will require two high-energy equivalents per mol of glucose.

Regulatory Factors in Glycogenesis and Glycogenolysis

As in the case of the glycolysis–gluconeogenesis system, the synthesis and degradation of glycogen do not proceed as independent processes in the liver. Glycogenesis and glycogenolysis also operate together as a closely integrated system that is subject to strict regulation, based upon both intracellular and extracellular glucose supply and demand (Fig. 4-11). The controls, however, are not exerted in this instance by the slow induction of enzyme synthesis, and the allosteric effects of metabolites are superimposed upon the primary conformational changes produced by ***cyclic AMP.*** The effects of cyclic AMP are such as to ensure that glycogen metabolism in the liver will respond *rapidly* to extracellular requirements and that glycogen formation and degradation may not occur simultaneously by the futile, energy-dissipating cycle:

Glucose + 2 ATP → Glycogen

Glycogen → Glucose

In the "resting" state, a normal liver cell will contain infinitesimal quantities of cyclic AMP, and the enzymes of glycogen metabolism will exist in their unmodified forms, termed *phosphorylase b* and *glycogen synthetase I* (Fig. 4-11A). The I, or *insensitive,* form of the synthetase is so designated because its function is insensitive to allosteric control by metabolites. In the resting state, the synthetase will be maximally active, and, provided that there is enough UDP-glucose substrate, glycogen synthesis will be turned "on." On the other hand, the b form of phosphorylase is inactive as such; the conformation of phosphorylase b will be converted to a catalytically active state only by allosteric interaction with AMP. Since the binding of AMP is inhibited by ATP, the phosphorylase b of the unstimulated liver cell will be active only when energy reserves within the cell are low enough to require mobilization of glycogen. When ensuing glucose catabolism restores adequate levels of ATP, the AMP concentration will fall and glycogenolysis will be turned "off." Thus, in the absence of external stimulation, the rate of breakdown of liver glycogen will respond solely to the intracellular energy state.

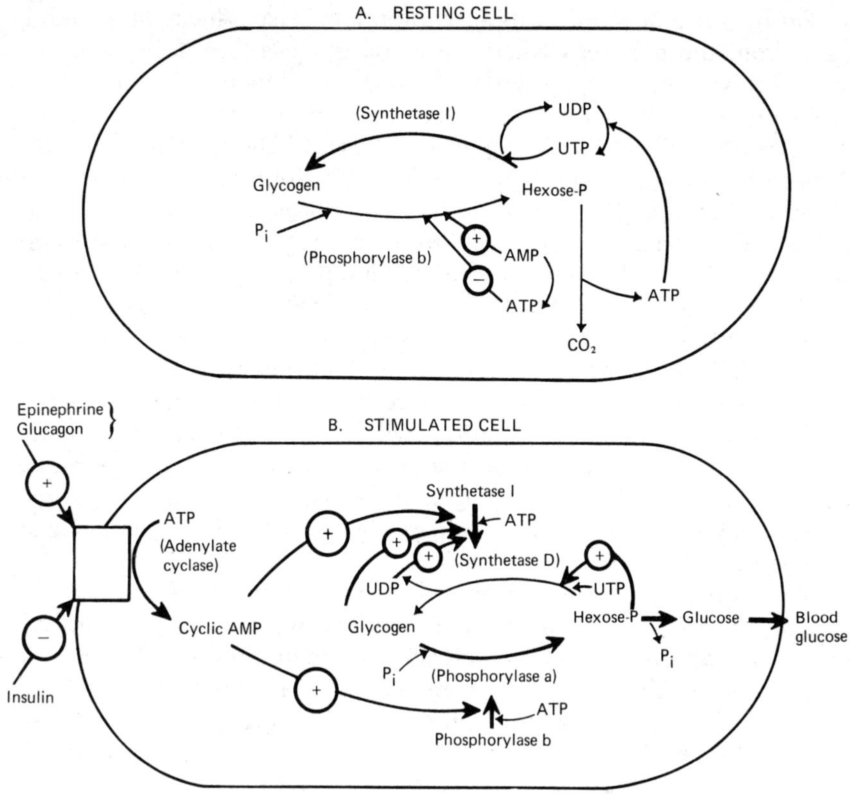

FIG. 4-11. Regulation of glycogen metabolism in the liver. **A.** Resting state. Glycogen synthesis predominates because the I form of the synthetase is active while the b form of phosphorylase has low activity. **B.** Stimulated state. Glycogen breakdown predominates because the D form of the synthetase has low activity while phosphorylase a has high activity.

The response of glycogen metabolism in the liver to extracellular needs is mediated by *adenylate cyclase,* the enzyme that sits like Janus with two faces at the portal of the cell membrane (Fig. 4-11B). The outward-looking face of the cyclase possesses specific receptors for certain endocrine hormones; the latter are unable to penetrate the hepatic cell membrane, but they alter the cell's metabolism by either positive or negative modification of the cyclase through binding to the *external receptor sites.* The introspective face of adenylate cyclase is the *internal catalytic site,* which will generate cyclic AMP from ATP intracellularly only when a positive modifier reacts with the extracellular receptors.

Two different classes of endocrine agents may interact with the external receptors of adenylate cyclase: epinephrine, which is released in the blood in times of stress, will activate the cyclase; glucagon, which is released from the pancreas when the blood glucose level falls, will also activate the cyclase, a response that is antagonized by insulin. Thus, both epinephrine and glucagon will raise the hepatic cyclic-AMP levels, whereas insulin will cause the opposite effect. In the presence of cyclic AMP, both the glycogen synthetase and the phosphorylase enzymes will be phosphorylated at the expense of ATP by the *protein kinases* that are normally present in an inactive state in the resting cell. These kinases are maximally activated by the very small amounts of cyclic AMP, and it should be recognized, too, that the number of molecules of enzyme phosphorylated will be miniscule. The consequences are twofold: first, the amount of ATP consumed in cyclic-AMP production and in protein-kinase action will be negligible in terms of the cellular energy economy; second, the effect of such a cascade of reactions will be to amplify enormously the effect of the attachment of a molecule of hormone at the cell membrane. In a sense, the metabolism of glycogen at rest is poised upon the verge of an explosion, with hormone effectors serving solely to trip the hair-trigger mechanism of adenylate cyclase.

The phosphorylation of phosphorylase b converts it to the fully active form, *phosphorylase a*. The latter will demonstrate its maximal catalytic activity in the absence of AMP, and it will not be inhibited by ATP. In this way, it resembles the I form of synthetase, since it is insensitive to allosteric modifiers and will rapidly catabolize liver glycogen regardless of the energy state of the tissue itself. Under hormonal influence, the metabolism of glycogen in the liver subserves the needs of the body as a whole, for example, to raise the glucose concentrations in the blood. Upon phosphorylation of glycogen synthetase, the enzyme will no longer function catalytically unless it is in the presence of high concentrations of glucose-6-phosphate. Because of its dependence upon the latter allosteric modifier, the phosphorylated synthetase is referred to as the *D form* or dependent form.

The implication of this type of control is that a hormonal signal to break down glycogen indicates that there is a short supply of glucose and its phosphate esters. Low glucose-6-phosphate concentrations in conjunction with high cyclic-AMP levels will thus ensure that glycogen synthesis will be turned off while its degradation is maximally activated. In reinforcement of this effect, both of the products of glycogen synthetase—UDP and glycogen—will act as negative modifiers to keep the synthetase in its inhibited D form. Otherwise, the response to hormonal control will normally be discrete and delimited in time; upon removal of the stimulus, cellular hydrolytic enzymes will act rapidly to degrade the cyclic AMP to ordinary AMP and to dephosphorylate the enzymes of glycogen metabolism, thereby reconverting phosphorylase a to inactive b and glycogen synthetase D to active I, the forms found in the resting state.

The phosphatase enzyme that is responsible for the conversion of glycogen synthetase into its active state is inhibited by the active a form of phosphorylase. The synthesis of glycogen, therefore, may not proceed so long as it is subject to breakdown. A separate phosphatase that converts phosphorylase into its inactive b state is in turn stimulated by the presence of high concentrations of glucose. Thus in the fed state, when glucose levels in the liver cell become elevated, a coordinated sequence of reactions follows: (1) stimulation of phosphorylase a phosphatase converts it quantitatively to the b form; (2) the consequent lowering of phosphorylase a results in relief of the inhibition imposed on glycogen–synthetase-D-phosphatase; and (3) the latter therefore converts the synthetase enzyme to the fully active I form. While the breakdown of glycogen is turned off its synthesis is turned on simultaneously. In effect the elevation of glucose returns the liver from the stimulated (B) condition to the resting (A) condition depicted in Fig 4-11.

Hepatic Glycogenoses and Galactosemia

A number of genetically transmitted diseases are known in which a failure to carry out normal enzymatic reactions in the liver causes an abnormal deposition of glycogen; these are referred to collectively as the hepatic ***glycogenoses.*** Excessive hepatic glycogen storage in some patients will result in such massive enlargement of the liver tissue that it is readily palpable and may grossly distend the abdomen. This *hepatomegaly* is a characteristic feature of such disorders.

The first described and most common of the hepatic glycogen-storage diseases is referred to as *type I* or, after its discoverer, *von Gierke's disease.* In actuality, the disorder affects glycogen metabolism secondarily, for the primary lesion is a recessively inherited deficiency of the enzyme glucose-6-phosphatase. Accordingly, the patients have low blood glucose concentrations in the fasting state. This hypoglycemia results in convulsions in acute conditions and in growth retardation due to chronic suppression of insulin production. The glycogen accumulates in the liver and kidney as a result of the decreased removal of glucose-6-phosphate, which is not only a precursor of the substrate for glycogen synthetase, but also an activator of the D form of the enzyme.

Type III or *Cori's disease* is characterized by increased deposits of glycogen with very short outer branches in the liver and other tissues. The cause is an absence of the debranching enzyme. In contrast, *type IV* or *Andersen's disease* is due to a deficiency of the branching enzyme, resulting in the accumulation of glycogen with abnormally long chains. Although the amounts of glycogen deposited in type III disease are much greater than those in type IV, the former is a fairly benign condition that is compatible with survival to adult life. On the other hand, possibly because of a reaction of the body to the abnormal unbranched glycogen, type IV disease invariably leads to liver cirrhosis and death in infancy.

Another glycogenosis that affects the liver is *type VI* or *Hers' disease,*

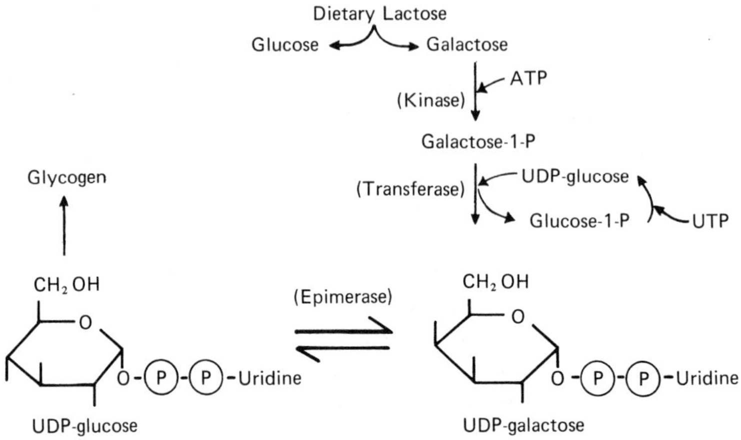

FIG. 4-12. Galactose metabolism in the liver. The lesion in galactosemia is a deficiency of the tranferase enzyme that forms UDP-galactose.

which is characterized by the absence of phosphorylase and hence by the excessive deposition of glycogen (with normal structure) in the liver. The symptoms are similar to but milder than those of type I disease. Apart from the type I form, these glycogen-storage diseases appear to be quite rare.

Another metabolic disorder affecting the carbohydrate metabolism of the liver is *galactosemia,* the result of a recessive trait in which a failure to metabolize dietary galactose may lead to cataract formation and mental retardation. Normally, galactose is converted in the liver to glucose by the uridine intermediate, UDP-galactose (Fig. 4-12). A deficiency of the transferase enzyme that produces the sugar nucleotide results in the accumulation of galactose-1-phosphate, which is toxic to the brain, and in an overflow of galactose in urine. The galactosuria, together with liver enlargement and milk intolerance, may lead to a ready diagnosis in an infant who fails to thrive. Early detection is essential, because the simple avoidance of milk, the major dietary source of galactose, will allow normal development and an excellent prognosis for survival. In adults, accessory pathways for galactose metabolism develop so that some milk may then be tolerated in the diet.

THE MUSCLES

In contrast to the liver, which must integrate the needs of other organs and tissues, the muscles are strictly self-serving from the standpoint of carbohydrate metabolism. Muscle cells have a rather single-minded purpose in this context, namely, to use the incoming glucose most effectively for production of the ATP necessary to perform the mechanical work of contraction. At rest, considerable amounts of glucose may be stored as glycogen, which will be induced to break down upon stimulation of the muscle to contract. The muscle cytoplasm contains high concentrations of glycolytic enzymes, and an abundance of mitochondria ensures effective catabolism of the products of glycolysis by the citric-acid cycle and the electron-transport chain. Only under extremes of exertion will the latter aerobic processes be unable to prevent the accumulation of lactate.

The muscle performs few synthetic functions apart from glycogenesis; the key enzymes for gluconeogenesis are absent, and glycolysis is virtually irreversible. There is no requirement for NADPH for reductive syntheses, and the pentose–phosphate pathway is almost nonexistent. There is no significant utilization of carbohydrate end products for biosynthetic reactions, except of pyruvate for alanine formation. For example, there is little conversion of acetyl CoA → fatty acids.

In summary, carbohydrate metabolism in the muscles is geared to provide storage reserves of glycogen when at rest and to utilize these stores and incoming glucose when working strenuously; the basal energy requirements of all muscle types are met chiefly by the oxidation of products derived from the metabolism of fats. In fact, neither slowly contracting smooth muscle tissue nor cardiac muscle utilize glucose extensively. During heavy exertion, the heart, for example, will derive considerable substrate for its numerous mitochondria by oxidation of the lactate provided by way of the circulation from exercising skeletal muscles. The points discussed below will pertain to the latter category of striated muscles, the main muscle mass of the peripheral limbs and skeleton.

GLUCOSE UPTAKE AND PHOSPHORYLATION

Glucose entry into skeletal muscle cells is facilitated at the membrane surface by a transport system that is enhanced in activity by either the presence of insulin or increased muscular work (Fig. 4-13). Once it has entered the cell, glucose is effectively trapped by phosphorylation to glucose-6-phosphate. The hexokinase catalyzing this reaction has a much lower K_m value for glucose than does the glucokinase of liver (10^{-5} *M* versus 10^{-2} *M*). Other differences between the two enzymes relate to their physiologic regulation: the muscle enzyme is strongly inhibited by negative feedback from its product glucose-6-phosphate whereas the liver enzyme is unaffected; liver glucokinase synthesis is enhanced by insulin but muscle hexokinase is not. Thus, the muscle enzyme is responsive solely to intracellular control, and it will phosphorylate glucose only so long as the glucose-6-phosphate can be used in the tissue for glycolysis or glycogen formation.

Another major distinction between liver and muscle tissue is the com-

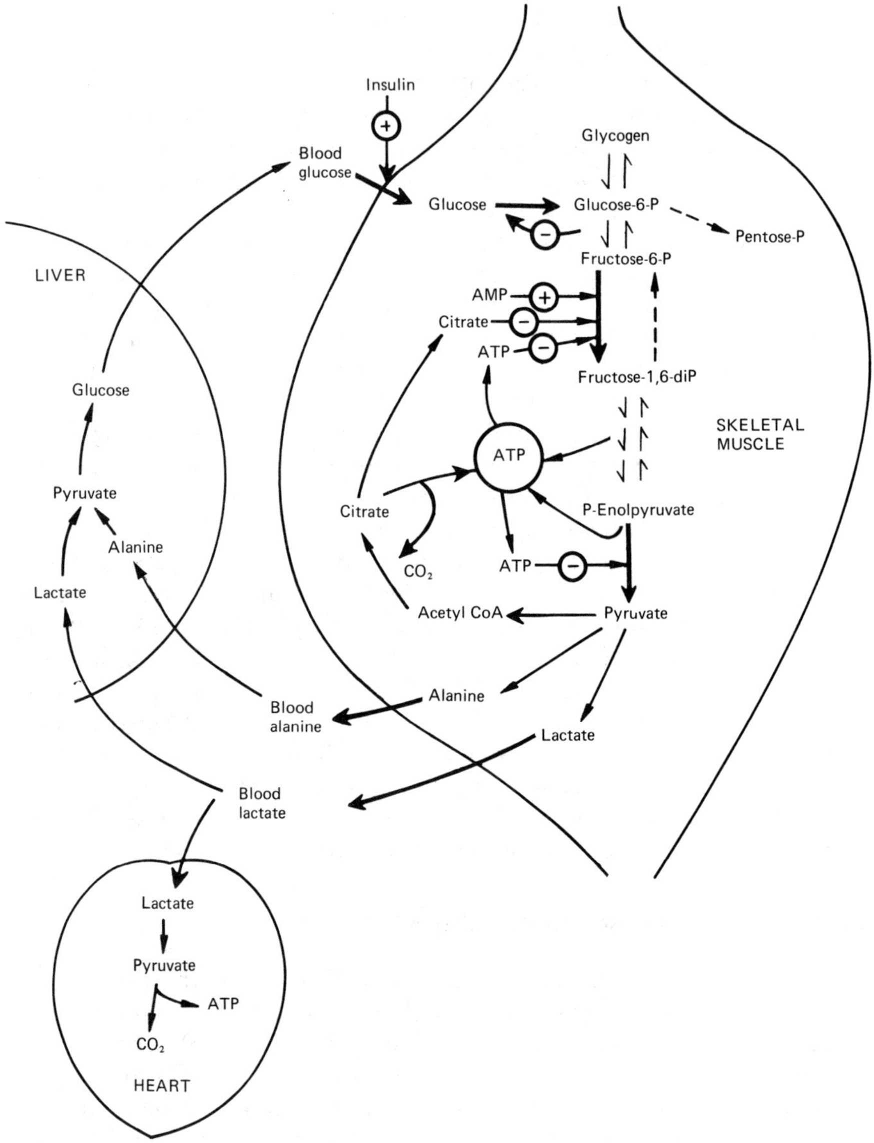

FIG. 4-13. Carbohydrate metabolism in muscle.

plete absence in the latter of the enzyme glucose-6-phosphatase. This is also true of most extrahepatic tissues, except those of the kidney and small intestine, which have minor roles in glucose production. For the muscles and other tissues, the conversion of glucose to glucose-6-phosphate is a one-way process that sequesters carbohydrate solely for intracellular metabolism.

GLYCOLYSIS AND ITS REGULATION

Glycolytic reactions in the muscles are similar to those of the liver, but there are differences in the quantities and regulatory properties of the individual enzymes involved. Muscle phosphofructokinase activity is more than 20 times greater than that of the liver, whereas the fructose diphophatase of liver is 20 times more active than that of muscle. These ratios reflect the situation under normal physiologic conditions where glycolysis predominates in muscle in the catabolic direction, while in the liver these reactions are poised to favor glucose synthesis. Phosphofructokinase is of major importance in the regulation of muscle glycolysis, just as in the liver. The muscle enzyme exhibits the same type of allosteric negative feedback by ATP and citrate, with AMP reversing the inhibition; however, it is not affected by long-chain fatty acids. The effect of citrate is particularly important in fasting states. The shift from carbohydrate to fat catabolism in starvation will increase the circulating acetoacetate levels (see Chap. 5). The resulting metabolism of acetoacetate to acetyl CoA in muscle will produce excess citrate, which inhibits phosphofructokinase and shuts down glucose utilization, favoring instead the oxidation of ketone bodies as an energy source. The control of the very low fructosediphosphatase activity is probably of marginal significance in muscle.

The pyruvate kinase activity in muscle is also 20 times greater than the corresponding activity in liver, whereas pyruvate carboxylase and phosphoenolpyruvate carboxykinase, which together effectively produce the shunt to reverse this reaction, are virtually undetectable in muscle. The muscle pyruvate-kinase system shows lesser negative allosteric effects with ATP than does the liver enzyme and the effects of fructose-1,6-diphosphate and other metabolites are not pronounced. Thus, the major factors controlling muscle glycolysis relate to the ATP concentrations, a situation that is compatible with the primary function of glucose metabolism in maintaining the energy needs for intracellular work.

Lactate Dehydrogenases

A basic difference exists between skeletal and cardiac muscle with respect to their utilization of pyruvate and the production of lactate. The heart is essentially an aerobic organ, and its mitochondria are so abundant that cardiac muscle not only oxidizes the pyruvate produced by its own glycolysis but also the lactate generated by other tissues. In skeletal muscle by contrast, the potential maximum activity of the glycolytic enzymes that is evoked by rapid exertion will occasionally outstrip the oxygen delivery system and the mitochondrial oxidative capacity to reoxidize NADH; in such periods of pronounced activity, the skeletal muscles will function anaerobically, converting pyruvate quantitatively to lactate. The lactate thus produced will diffuse to the liver for glucose resynthesis or to the heart and other tissues for oxidation.

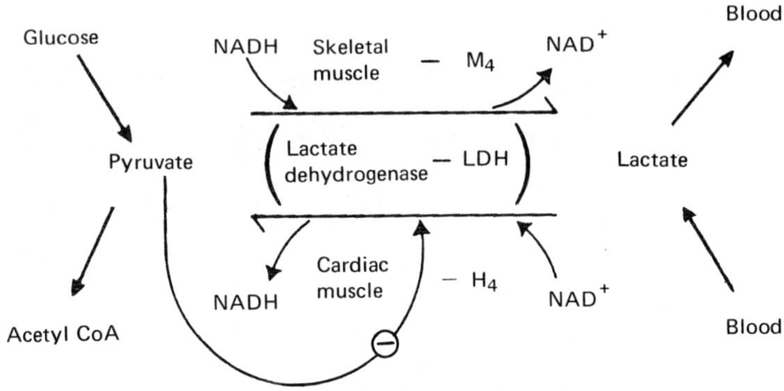

FIG. 4-14. Lactate dehydrogenase isoenzymes. In skeletal muscle, the M_4 form favors conversion of pyruvate to lactate, whereas the H_4 form of heart muscle favors lactate oxidation to pyruvate.

These differences in metabolism relate to the different *lactate-dehydrogenase* ***isoenzyme*** units, designated either *H* for heart or *M* for skeletal muscle. The active enzyme contains any combination of four of these subunits (H_4, M_4, H_3M_1, H_2M_2, or H_1M_3). Heart and other aerobic tissues contain a preponderance of H forms (H_4, H_3M_1, and so on), whereas skeletal muscle possesses only M_4 (Fig. 4-14). The H_4 isoenzyme and those combinations with H subunits predominating operate much less effectively with pyruvate than they do with lactate as substrate; in addition, the H_4 isoenzyme is inhibited by high concentrations of pyruvate. Thus, the cardiac muscle and other tissues with H_4 will preferentially utilize pyruvate by reactions other than the lactate dehydrogenase reaction, which will in fact serve to oxidize lactate to pyruvate in these aerobic tissues. The M_4 isoenzyme shows no inhibition by pyruvate, and it is poised toward the conversion of pyruvate to lactate when other means of utilizing NADH and pyruvate are inadequate, as will arise during the rapid contraction of skeletal muscle. The H_4 form is found in elevated amounts in blood plasma after damage to heart tissue, and in conjunction with the assay for creatine-phosphokinase is used as a diagnostic indicator in patients with suspected myocardial infarction.

GLYCOGEN METABOLISM IN MUSCLE

Muscle glycogen metabolism involves enzyme systems and control mechanisms analogous to those found in the liver. Since the muscle glycogen store serves for its own requirements only, the concentration of the polysaccharide is less responsive to fasting than it is in the liver, and it does not reach such high levels upon resumption of feeding. The glycogen content in well-nourished, rested muscle rarely rises above 1% of the muscle weight, resulting from inhibition of glycogen synthetase by glycogen concentrations in excess of this amount. Because of the large total mass of muscle in the body, the total muscle stores of glycogen will exceed those of the liver roughly fivefold. Despite this large potential glucose store, muscle glycogen cannot be utilized to replace depleted blood sugar, since the tissue lacks glucose-6-phosphatase. Accordingly, muscle glycogen metabolism is strictly an intracellular affair, sensitive to the cell's own internal energy requirements. The glycogenesis-glycogenolysis cycle in muscle comprises these steps:

Glucose-6-phosphate → Glycogen

Glycogen → Glucose-6-phosphate

The energy cost to the muscle cell is therefore only one high-energy equivalent per mol of glucose for the convenience of storage, as calculated previously for intracellular energy utilization in the liver.

Regulation of Glycogenolysis and Glycogenesis

The phosphorylase b of resting muscle, like that of liver, is inactive, and under physiologic conditions, it is quite unresponsive to cellular concentrations of ATP and AMP. Phosphorylation to the phosphorylase a form by kinase is the primary activation response, and it relates to early events in the contractile process itself (Fig. 4-15). The phosphorylase kinase of muscle is strongly activated by Ca^{2+}, which is in low concentration in the resting tissue; stimulation of the nerves leading to the muscle induces the release of Ca^{2+} from its bound form into the cytoplasm (Chap. 3, Fig. 3-20). This Ca^{2+} release initiates contraction and simultaneously promotes the conversion of phosphorylase b to a and of glycogen synthetase from the I to D form. Thus, the same signal that instructs the muscle to work also provides the impetus for glycogen breakdown to ensure an adequate energy supply. Upon cessation of the stimulus, Ca^{2+} is removed from the cytoplasm and the phosphatases cause glycogenolysis to be switched off, while converting the synthetase enzyme back to its more active I form.

In addition to this intracellular regulation by Ca^{2+} concentrations, the muscle glycogen system responds through its adenylate cyclase system to extracellular epinephrine; unlike the liver, the muscle cell membrane has no analogous receptors for glucagon and hence is insensitive to this hormone. The muscle cyclase is more sensitive than the liver enzyme to epinephrine, which serves as the general alarm signal to induce maximal performance from the muscles in times of supreme effort. The elevation of intracellular cyclic-AMP concentrations by activation of adenylate cyclase converts the phosphorylase kinase into a form that is active even in the presence of very low Ca^{2+} concentrations. Thus, phosphorylase b may be activated to the a form independently of the events surrounding contraction, thereby producing a steady, increased output of glucose-6-phosphate for glycolytic energy production.

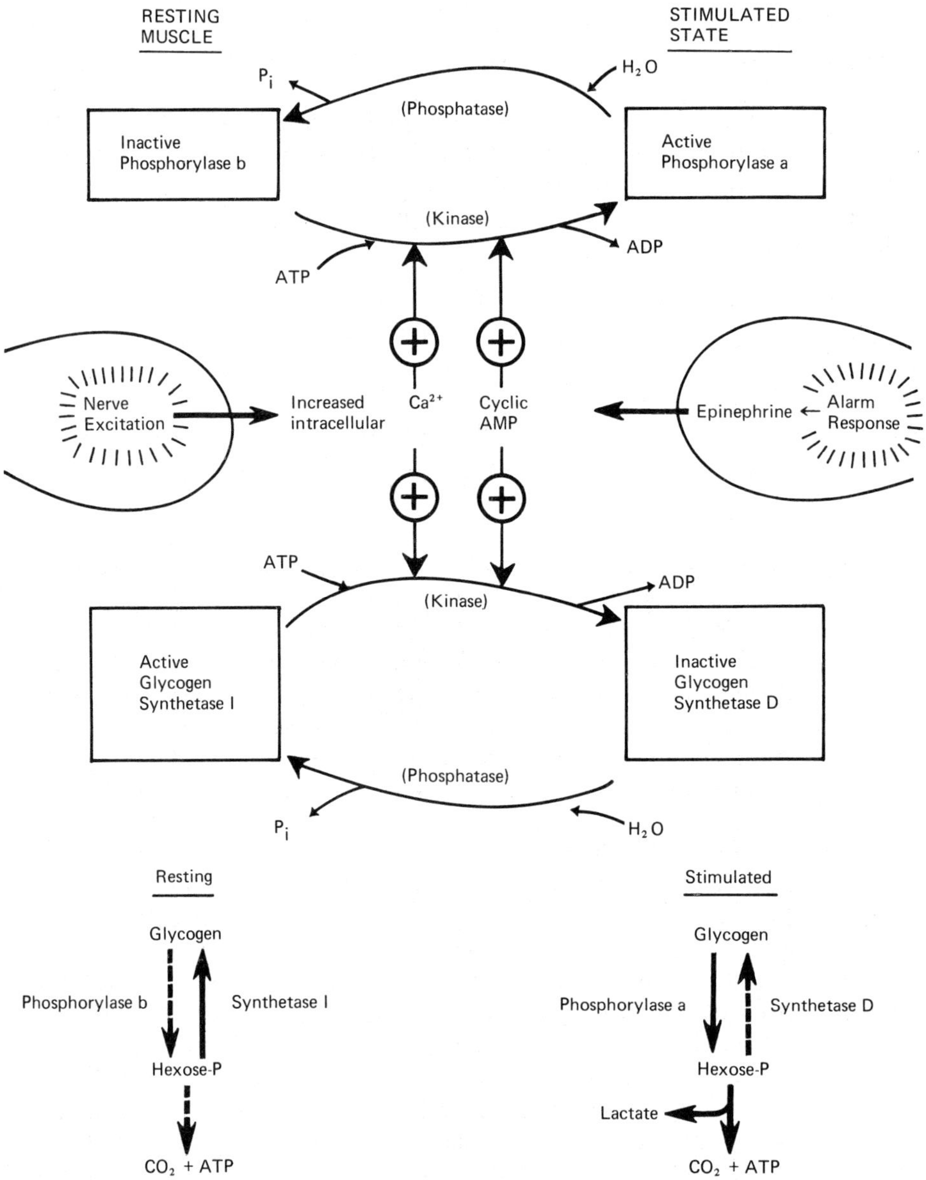

FIG. 4-15. Glycogen metabolism in muscle and its regulation. Either calcium ions (released in response to neural stimulation) or cyclic AMP (released in response to epinephrine) will convert phosphorylase to its active a form and the synthetase to its inactive D form.

Glycogen provides the most rapidly utilizable source of energy for muscle after its high-energy stores of ATP and phosphocreatine are exhausted (20 sec–30 sec maximal activity, as in a sprint). The utilization of glycogen then occurs at the maximal rates allowed by glycolysis (for 2 min–3 min) until the lactate accumulation becomes excessive and self-limiting. Thereafter, as in long-distance runs, the muscle is limited in its performance by the maximal rates of the reactions of its aerobic mechanisms for the oxidation of NADH and pyruvate. Muscle glycogen will be used preferentially as fuel at first, but it will gradually be replaced as the oxidizable substrate by fatty acids. Thus, the operation of muscle under conditions of long endurance comes to resemble once again that found in the resting condition. Relatively short periods of rest (several minutes) suffice to restore the resting levels of energy storage, even after several hours of activity at the slower rates.

Muscle Glycogenoses

Several disorders of glycogen metabolism have been described that variously involve the muscle tissues. Thus, the abnormal glycogen structures found in type III and IV glycogenoses of the liver have also been found in muscle. *Type II* glycogen-storage disease, or *Pompe's disease,* shows some liver involvement but its side effects are most severely manifested in muscle tissue, resulting in weakness of the skeletal muscle, enlargement of the heart, and death in infancy due to cardiac failure. An α-1,4-glycosidase is missing from these tissues, but the reason why this enzyme is required to prevent glycogen deposition is not clear.

Type V or *McArdle's disease* is a more common glycogenosis that affects only the skeletal muscles, which show a complete lack of phosphorylase. The liver phosphorylase is normal, and no hypoglycemia is observed. Apart from muscle cramps during severe exercise, the symptoms are generally mild, emphasizing the importance of fatty acids and other extramuscular fuels as energy sources for muscular contraction.

A few cases of type VII glycogen-storage disease in skeletal muscle have been attributed to very low levels of phosphofructokinase (1%–3% of normal). Like those with type-V disease, these patients also tolerate moderate exercise, which again emphasizes the point above about the alternatives to carbohydrate metabolism as a muscle energy source.

THE BRAIN

Among all the body organs, the brain is most dependent upon the metabolism of carbohydrates (Fig. 4-16). For example, hypoglycemic patients become severely confused when blood glucose concentrations fall below 20 mg/100 ml, and a deep coma ensues at values below 10 mg/100 ml. To

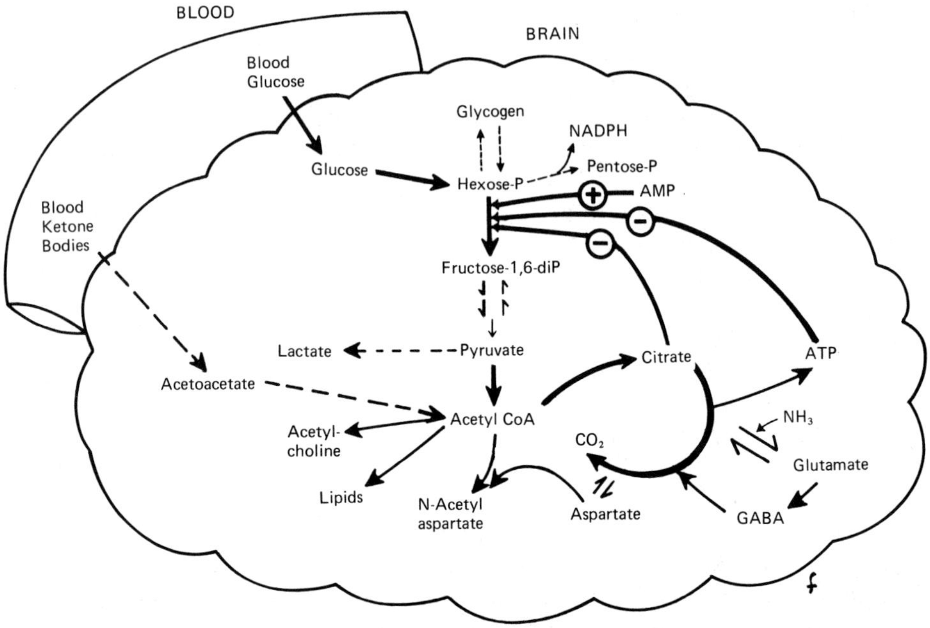

FIG. 4-16. Carbohydrate metabolism in the brain.

provide adequate energy, glucose catabolism must proceed by aerobic pathways, as attested by the even greater sensitivity of the brain to hypoxia (low blood oxygen) than to hypoglycemia (low blood glucose). In addition to providing the energy to maintain the electrochemical gradients required for nerve transmission, the metabolism of incoming glucose may sustain the syntheses of neurotransmitter substances, amino acids, lipids, and nucleic-acid components by providing the carbon precursors within the brain cells. The pentose–phosphate pathway operates to a minor extent to provide NADPH for some of these syntheses, but most of the glucose catabolism in the brain is by the glycolytic pathway.

GLUCOSE UTILIZATION IN THE BRAIN

Properties of the glucose uptake system of the brain are similar to those of the system in skeletal muscle. The brain hexokinase has a similar high affinity for glucose ($K_m \sim 10^{-6}\ M$), thus providing an effective trapping mechanism. Activity of the glycolytic enzymes is very high. Phosphofructokinase control is exerted by negative feedback from ATP and citrate and positive feedback is provided by AMP, which serve as the major regulatory influences.

The high activity of the mitochondrial citric-acid-cycle enzymes prevents lactate accumulation in the brain, so that most of the pyruvate undergoes oxidation to acetyl CoA. A minor portion of acetyl CoA is used for the crucial formation of the neurotransmitter, acetylcholine, and small amounts of the acetyl groups may be stored in brain as *N*-acetyl aspartate. The majority of the acetyl-CoA molecules, however, are oxidized for energy production by the citric-acid cycle.

Intermediates of the citric-acid cycle play important ancillary roles in the brain by providing the precursor keto acids for aspartate and glutamate synthesis, two amino acids of special significance in neural tissue function. The excessive utilization of citric-acid-cycle intermediates for these pathways may have deleterious effects upon normal carbohydrate metabolism; the toxic effects of ammonia upon the brain, for example, may be related to the depletion of the citric-acid-cycle α-keto acids that are used in NH_3 fixation. A bypass of the cycle—through glutamate decarboxylation to γ-aminobutyrate (GABA) and the conversion of the latter to succinate—diverts roughly 10% of the carbons from glucose that is catabolized in the brain.

OTHER SUBSTRATES IN BRAIN METABOLISM

As might be expected from its reliance upon incoming glucose, the brain has rather small intracellular energy stores. Despite its heavy dependence upon carbohydrate catabolism during prolonged fasting the brain can adapt to use blood ketone bodies, such as β-hydroxybutyrate, to provide substrates for the citric-acid cycle. Amino acids such as glutamate and aspartate will be catabolized in emergencies to provide oxidizable α-keto acids. The stimulation of brain tissue—for example, by application of electrical pulses—causes a rapid discharge of ATP and phosphocreatine stores. The glycogen content of brain is low (0.1% by weight) and is rapidly dissipated when the external supplies of glucose are inadequate for the demands of the tissue.

The mechanism for the control of the phosphorylase system in the brain is unknown. Although the brain contains an extremely active adenylate-cyclase system that is stimulated to produce cyclic AMP by neurohormones (e.g., norepinephrine and serotonin), its significance for carbohydrate metabolism is not clear, since there is no concomitant activation of phosphorylase as in the muscle system.

THE ERYTHROCYTE

Up to the stage of differentiation of the reticulocyte, human red blood cells possess the organelles typical of eukaryotic organisms. The nuclei, mitochondria, and so on become extruded during later development, however, with the consequence that the citric-acid-cycle and respiratory-chain activities are missing from the mature erythrocyte. It may seem ironic that this cell—whose major role is the transport of oxygen for the use of other tissues—is itself restricted to the anaerobic mechanisms of energy production.

GLUCOSE UTILIZATION IN THE ERYTHROCYTE

Glucose enters the erythrocyte from the surrounding plasma by means of an energy-requiring, facilitated transport system that is not influenced by insulin (Fig. 4-17). Under normal circumstances, over 90% of the glucose utilization by erythrocytes is by means of the glycolytic pathway, which provides the ATP required for energy-mediated transport reactions across the cell membrane. In addition to providing for glucose uptake, these transport mechanisms, which are coupled to ATP hydrolysis, promote the movement of Na^+ outward and K^+ inward to maintain the internal composition (high K^+, low Na^+) against the concentration gradients of blood plasma (low K^+, high Na^+). The triose-phosphate dehydrogenase reaction of glycolysis also generates NADH, which is necessary to reduce any excess of *methemoglobin* (an oxidized form of hemoglobin that does not combine with O_2) that may be formed in the erythrocyte by oxidizing agents.

A by-product of the glycolytic pathway that is found in uniquely high concentrations in erythrocytes is 2,3-diphosphoglycerate. The latter serves several functions: first, it is the major phosphate-containing compound in the cell and serves as an important anion to balance the intracellular cations as well as acting as a buffering agent; second, it may serve as emergency energy reserve for a cell that lacks phosphocreatine and glycogen stores; and third, it binds to hemoglobin to lower its affinity for O_2 and thus facilitates the unloading of oxygen to tissues. The pentose–phosphate pathway normally accounts for only a small percentage of total glucose catabolism, but under conditions where the NADPH requirements are increased, the activity of this "shunt" may accelerate many times. The major need for NADPH in the erythrocyte is to maintain the important intracellular reductant, *glutathione,* in its reduced–SH form; hence, the presence of agents that promote the oxidation of glutathione to the –S–S– form will activate the pentose–phosphate pathway to provide compensatory reducing equivalents in the form of NADPH.

HEMOLYTIC ANEMIAS

The importance of carbohydrate and energy metabolism for the survival of the erythrocyte is emphasized by several inborn errors of metabolism that lead to hemolysis and anemia. In the *type VII glycogen-storage disease* discussed under muscle metabolism, for example, the phosphofructokinase enzyme of the erythrocytes is partly deficient, which results in reduced efficiency of ATP production and mild hemolytic symptoms.

A more severe disorder is associated with another recessively inherited condition, *pyruvate-kinase deficiency,* in which the enzyme in the

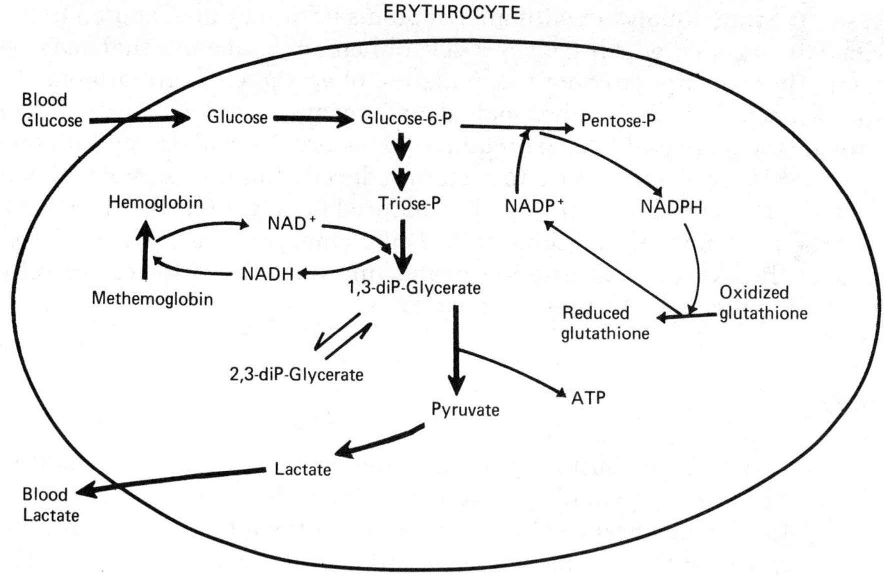

FIG. 4-17. Carbohydrate metabolism in the erythrocyte.

erythrocytes is specifically decreased. Since glycolysis is blocked before the last energy-producing step, there can be no net ATP generation from glucose catabolism. The affected cells also accumulate increased amounts of 2,3-diphosphoglycerate, but they cannot use the latter for ATP production because of the kinase deficit. The resulting lack of cellular energy sources causes an impaired maintenance of internal composition and disruption of membrane integrity. Deformation and other alterations in surface properties lead to the recognition by the spleen of such cells as abnormal, and they are consequently removed and destroyed by the reticuloendothelial system. Other similarly severe hemolytic anemias have been described with inherited lesions of the erythrocytic enzymes of the glycolytic pathway (triosephosphate isomerase, glucose-phosphate isomerase, phosphoglycerate kinase, and hexokinase), but these conditions seem to be very rare.

The significance of the pentose–phosphate pathway for the ability of the erythrocyte to withstand external stresses is demonstrated by a group of common, drug-induced, hemolytic anemias that are characterized by *glucose-6-phosphate dehydrogenase deficiency*. The diseases are transmitted as sex-linked recessive traits and are found most frequently among black people and inhabitants of the Mediterranean area. The black groups are very sensitive to antimalarial drugs such as primaquine, and this condition is often termed *primaquine sensitivity*. The Mediterranean groups show similar sensitivity to such drugs and may also react to the fava bean, which has thus led to the term *favism* to designate this disorder. In both cases, the individuals are without symptoms until they are exposed to the causative agents, which trigger a severe hemolytic anemia that may be fatal. These agents promote the oxidation of erythrocytic glutathione. In the absence of glucose-6-phosphate dehydrogenase, the cells are unable to provide the increased NADPH required to restore the oxidized glutathione to the –SH form. The failure to sustain reduced glutathione is associated with the appearance of deposits of denatured hemoglobin (Heinz bodies), and the resulting deformations of the cell membrane single out the defective erythrocytes for destruction in the sinusoids of the spleen, thereby producing the acute hemolytic episode.

BLOOD SUGAR

The composite contributions of each of the preceding processes result in the finely balanced pool of glucose that is found in the circulating blood (Fig. 4-18). In the absence of glucose intake in the diet, the normal individual will maintain a constant concentration of glucose in the blood of between 70 mg/100 ml and 90 mg/100 ml (4 mM–5 mM). Following a carbohydrate-containing meal, the concentration may rise transiently to about 150 mg/100 ml (8 mM), but it will usually return to normal within 2 h. The liver acts as the first buffer; it removes incoming monosaccharides in the portal circulation that are generated during digestion and converts them first to glucose, then to glycogen. Any overflow of glucose increases its concentration in the systemic circulation, which will in turn increase its uptake in the peripheral tissues (muscle, adipose tissue, and so forth). Upon reaching the pancreas, excess blood glucose activates the islet cells to release insulin, and the latter further enhances the facilitated uptake of glucose in the peripheral tissues.

Insulin, as noted previously, also increases glucose utilization for glycogen synthesis and depresses its production in gluconeogenesis; it is the major negative feedback agent involved in promoting a reduction of blood glucose to normal levels in response to hyperglycemia. In diabetes mellitus, the failure to produce insulin will result in the underutilization and overproduction of glucose and consequent chronic hyperglycemia. If a high-carbohydrate meal is eaten by a diabetic, the blood glucose concentration may even exceed the ability of the kidney to reabsorb it from the glomerular filtrate (about 220 mg/100 ml, 12 mM) and result in its excretion in the urine (glucosuria).

In the absence of carbohydrate intake, the blood glucose levels will remain within the limits noted because of the secretion of glucose into the circulation by the liver. During a prolonged fast, the kidney cells will also contribute significantly to the blood glucose. Other tissues may act only to remove glucose from the extracellular fluids, since they lack the necessary glucose-6-phosphatase for the conversion of other carbohydrate sources to free glucose. The lowering of the blood glucose concentration

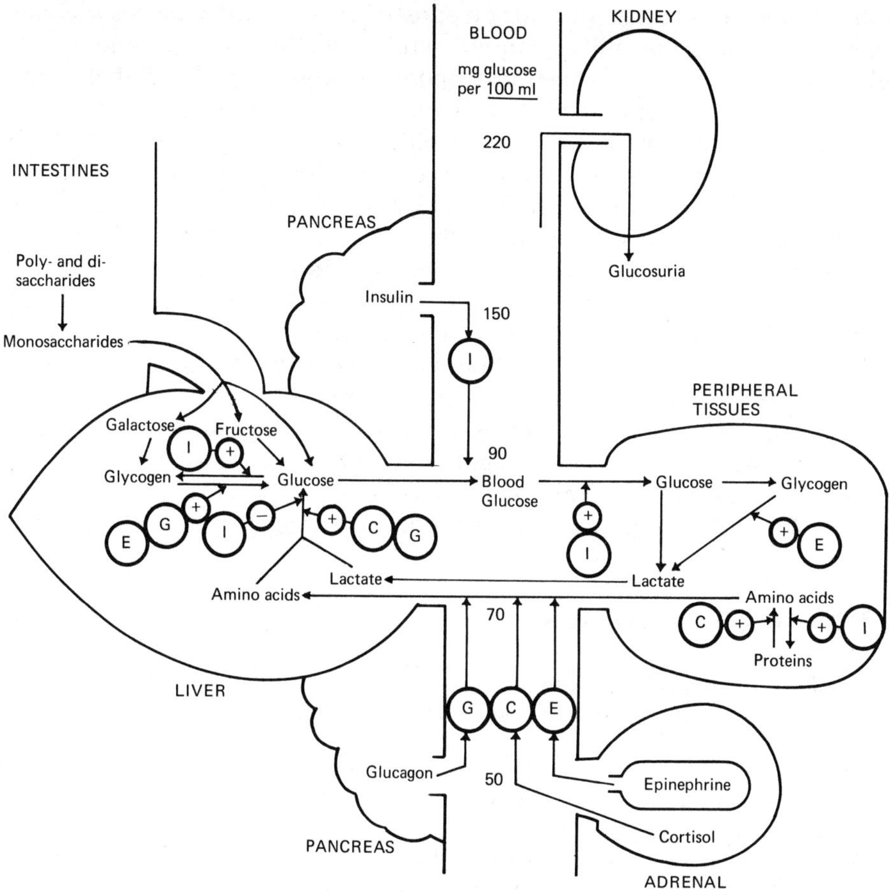

FIG. 4-18. Control of blood sugar levels. Counteracting factors serve to maintain the normal blood glucose level between 70 and 90 mg/100 ml. *Hyperglycemia* causes release of Insulin *(I)* from the pancreas to lower the blood sugar. Excessive elevation of blood glucose results in spillover into the urine (glucosuria). *Hypoglycemia* initiates release of agents to raise the blood sugar—glucagon *(G)* from the pancreas, cortisol *(C)* from the adrenal cortex, and epinephrine *(E)* from the adrenal medulla.

that reaches the pancreas stimulates the release of glucagon from the islet cells. This hormone has selective effects upon the liver to enhance glycogenolysis and redress the deficiency of glucose in the blood. This is the rapid and short-term response to the removal of food, but it will fail to maintain glucose levels for more than 24 h, at which time, the hepatic glycogen will be depleted. The second line of defense is provided by the adrenal cortex, which will take over in prolonged fasting and release glucocorticoid hormones. These hormones stimulate the slower synthesis of gluconeogenic enzymes in the liver and enhance protein breakdown in other tissues to provide the carbon substrates for gluconeogenesis. Since these actions are normally counteracted by insulin in response to hyperglycemia, and since this insulin response is absent in the diabetic, the excessive gluconeogenesis of diabetes may be viewed as the result of an unchecked action of the glucocorticoids. Conversely, diabetically high blood glucose levels may be found in conjunction with a normally functioning pancreas when adrenocortical hyperactivity (e.g., in Cushing's syndrome) produces an elevation of the glucocorticoids relative to insulin. Other agents that may cause an enhanced release of glucose, that is, *hyperglycemic agents,* include epinephrine, which enhances glycogenolysis in both the liver and muscle in response to stress, and growth hormone, which depresses glucose utilization while favoring a shift to the catabolism of fats in response to severe hypoglycemia. Obviously, it is the combination and balance among each of these interacting factors that determine how well the glucostatic regulation will be maintained in the blood and allows the carbohydrate requirements of the integrated organism to adjust to diet, growth, activity, and other physiologic states of the body.

SUGGESTED READINGS

Altszuler N, Finegold MJ: Glucose metabolism. In Becker FF (ed): The Liver: Normal and Abnormal Functions, Part A, pp 463–530. New York, Marcel Dekker, 1974

Beutler E: Glucose-6-phosphate dehydrogenase deficiency. In Stanbury JB, Wyngaarden JB, Fredrickson DS (eds): The Metabolic Basis of Inherited Disease, 4th ed, pp 1430–1451. New York, McGraw-Hill, 1978

Brewer GJ: General red cell metabolism. In Surgenor D MacN (ed): The Red Blood Cell, 2nd ed, Vol 1, pp 387–435. New York, Academic Press, 1974

Brewer GJ: Inherited erythrocyte metabolic and membrane disorders. Med Clin N Am 64:579–596, 1980

Carlson GM, Bechtel PJ, Graves DJ: Chemical and regulatory properties of phosphorylase kinase and cyclic AMP-dependent protein kinase. Adv Enzymol 50:41–115, 1979

Cohen P: The role of cyclic-AMP-dependent protein kinase in the regulation of glycogen metabolism in mammalian skeletal muscle. Curr Top Cell Regul 14:118–196, 1978

De Paola DP, Alfano MC: Diet and oral health. Nutr Today 12, No. 3: 6–11, 29–32, 1977

Denton RM, Halestrap AP: Regulation of pyruvate metabolism in mammalian tissues. Essays Biochem 15:37–77, 1979

Everse J, Kaplan NO: Lactate dehydrogenases: structure and function. Adv Enzymol 37:61–134, 1973

Felig P: Disorders of carbohydrate metabolism. In Bondy PK, Rosenberg LE (eds): Metabolic Control and Disease, 8th ed, pp 276–392. Philadelphia, WB Saunders, 1980

Fletterick RJ, Madsen NB: The structures and related functions of phosphorylase a. Annu Rev Biochem 49:31–61, 1980

Gitzelmann R: Newborn screening for inherited disorders of galactose metabolism. In Bickel H, Guthrie R, Hammersen G (eds): Neonatal Screening for Inborn Errors in Metabolism, pp 67–79. New York, Springer–Verlag, 1980

Goldhammer AR, Parodies HH: Phosphofructokinase: Structure and function. Curr Top Cell Regul 15:109–141, 1979

Greene HL, Slonim AE, Burr IM: Type I glycogen storage disease: A metabolic basis for advances in treatment. Adv Pediatr 26:63–92, 1979

Hems DA, Whitton PD: Control of hepatic glycogenolysis. Physiol Rev 60:1–50, 1980

Hers HG: The control of glycogen metabolism in the liver. Annu Rev Biochem 45:167–189, 1976

Howell RR: The glycogen storage diseases. In Stanbury JB, Wyngaarden JB, Fredrickson DS (eds): The Metabolic Basis of Inherited Disease, 4th ed, pp 137–159. New York, McGraw-Hill, 1978

Hue L: The role of futile cycles in the regulation of carbohydrate metabolism in the liver. Adv Enzymol 52:247–331, 1981

Lee EYC, Silberman SR, Ganapathi MK,et al: The phosphoprotein phosphatases: Properties of the enzymes involved in the regulation of glycogen metabolism. Adv Cyclic Nucleotide Res 13:95–131, 1980

Leung K, Munck A: Peripheral actions of glucocorticoids. Annu Rev Physiol 37:245–272, 1975

Levy, HR: Glucose-6-phosphate dehydrogenases. Adv Enzymol 48:98–192, 1979

Maker HS, Clarke DD, Lajtha AL: Intermediary metabolism of carbohydrates and amino acids. In Siegel GJ, Albers RW, Katzmann R, Agranoff BW (eds): Basic Neurochemistry, 2nd ed, pp 279–307. Boston, Little, Brown & Co, 1980

Purich DL, Fromm HJ, Rudolph FB: The hexokinases: Kinetic, physical, and regulatory properties. Adv Enzymol 39:250–326, 1973

Randle PJ: Phosphorylation–dephosphorylation cycles and the regulation of fuel selection in mammals. Curr Top Cell Regul 18:107–128, 1981

Randle PJ, Sugden PH, Kerbey AL, Radcliffe PM, Hutzon NJ: Regulation of pyruvate oxidation and the conservation of glucose. Biochem Soc Symp 43:47–67, 1978

Roach PJ: Glycogen synthase and glycogen synthase kinases. Curr Top Cell Regul 20:45–105, 1981

Segal S: Disorders of galactose metabolism. In Stanbury JB, Wyngaarden JB, Fredrickson DS (eds): The Metabolic Basis of Inherited Disease, 4th ed, pp 160–181. New York, McGraw–Hill, 1978

Utter MF, Barden RE, Taylor BL: Pyruvate carboxylase: An evaluation of the relationships between structure and mechanism and between structure and catalytic activity. Adv Enzymol 42:2–72, 1975

Uyeda K: Phosphofructokinase. Adv Enzymol 48:194–244, 1979

Valentine WN, Tanaka KR: Pyruvate kinase deficiency and other enzyme-deficiency hereditary hemolytic anemias. In Stanbury JB, Wyngaarden JB, Fredrickson DS (eds): The Metabolic Basis of Inherited Disease, 4th ed, pp 1410–1429. New York, McGraw-Hill, 1978

Wilson JE: Brain hexokinase, the prototype ambiquitous enzyme. Curr Top Cell Regul 16:2–54, 1980

Chapter Five Lipids

The ***lipids*** comprise a rather mixed group of substances that are not very soluble in water; instead, they are generally extracted from tissues with organic solvents such as chloroform and methanol, the so-called fat solvents. A common feature of many lipids is their content of ***fatty acids,*** which are long, straight-chain compounds with a carboxyl group at one end and varying numbers of double bonds along the carbon chains. The number of carbon atoms in the chains is generally 16, 18, or 20; the number of double bonds generally ranges from zero to four (Fig. 5-1A). Palmitic acid (16:0) is the most common saturated acid; oleic acid (18:1) is the most common monounsaturated acid; and linoleic acid (18:2), linolenic acid (18:3), and arachidonic acid (20:4) are the principal polyunsaturated acids.

The fatty acids rarely occur in the free form but exist most often as esters in which the carboxylic group of the acid is linked to various alcohols. The most important of these alcohols is glycerol (Fig. 5-1B), which has three hydroxyl functions. One, two, or all three of these hydroxyl groups may be esterified to fatty acids to produce *mono-, di-,* or *triacylglycerols,* respectively. When two of the hydroxyl groups of glycerol are linked to fatty acids and the third to a phosphate group, the resulting lipids are known as *phosphoglycerides.*

Another large class of lipids is based upon the condensed ring system of phenanthrene; this class comprises the ***steroids*** (Fig. 5-1C), which are substances derived from the parent structure, ***cholesterol.*** A large and heterogeneous collection of other "fat-soluble" substances—including the *fat-soluble vitamins* A, D, E, and K—may also be included in the general category of lipids.

Lipids play many roles in cells. Foremost is the structural function of phosphoglycerides, cholesterol and glycolipids in forming the characteristic membranes of the intracellular organelles and the cellular plasma

A

$CH_3(CH_2)_{14}COOH$ Palmitic acid (16:0)

$CH_3(CH_2)_{16}COOH$ Stearic acid (18:0)

$CH_3(CH_2)_7CH{=}CH(CH_2)_7COOH$ Oleic acid (18:1)

$CH_3(CH_2)_4CH{=}CHCH_2CH{=}CH(CH_2)_7COOH$ Linoleic acid (18:2)

$CH_3CH_2CH{=}CHCH_2CH{=}CHCH_2CH{=}CH(CH_2)_7COOH$ Linolenic acid (18:3)

$CH_3(CH_2)_4CH{=}CHCH_2CH{=}CHCH_2CH{=}CHCH_2CH{=}CH(CH_2)_3COOH$ Arachidonic acid (20:4)

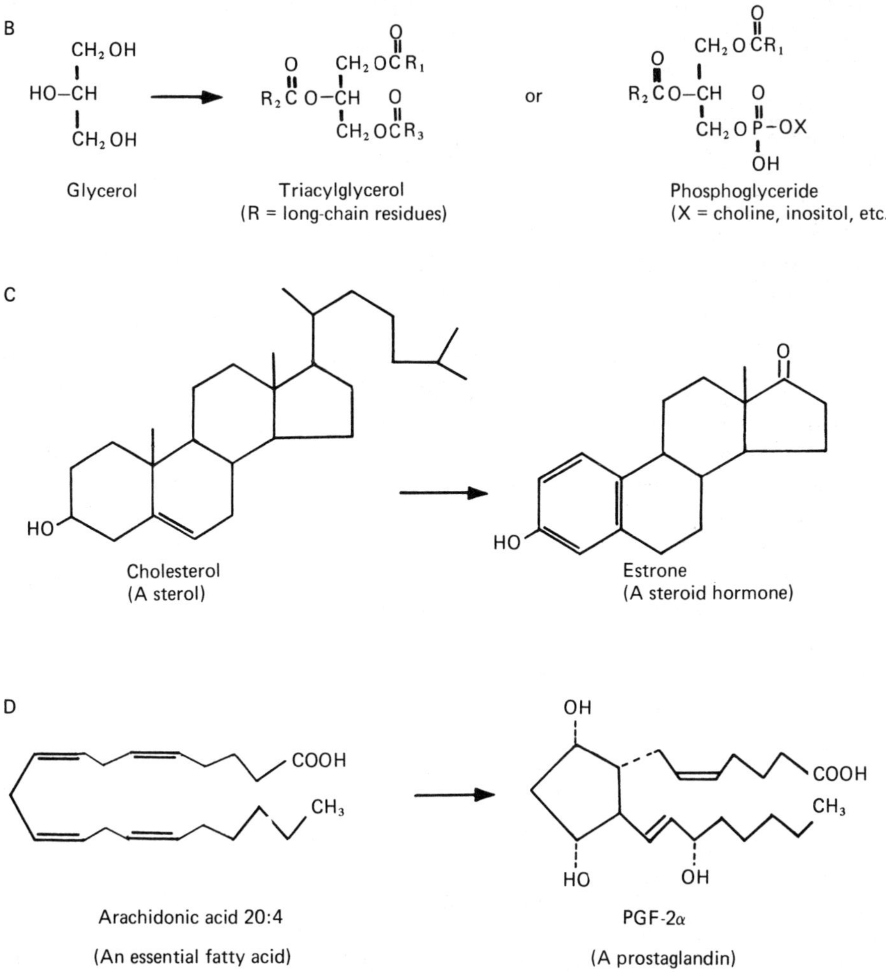

FIG. 5-1. Some important lipids. **A.** The common fatty acids. The first number in parentheses designates the carbon chain length; the second, the number of double bonds. Palmitic acid (16:0) has 16 carbons and no double bonds, whereas arachidonic acid (20:4) has 20 carbons and 4 double bonds. **B.** The glycerol-containing lipids. **C.** The steroid lipids and the relationship of cholesterol to the steroid hormones. **D.** The essential fatty acid, arachidonic acid, and its relationship to the prostaglandins.

membrane. In metabolism the fatty acids and their storage form, the fats, act as primary fuels for energy production in liver, heart, and other aerobic muscles. Fatty acids also serve as precursors for the key regulatory substances, the prostaglandins, whereas the sterols can give rise to a wide range of steroid hormones and other specialized products. Because of the connections between nutrition and diseases of lipid metabolism, especially the prevalent disorders that involve abnormal accumulations in the fat depots (obesity) or in the cardiovascular system (atherosclerosis), there has been a continuing preoccupation with lipids by the medical profession.

DIETARY FATS

As much as 40% of the body's caloric intake is derived from lipids in a normal diet, and all but a small proportion of this is contributed by the triacylglycerols, or fats, as they are commonly termed. The fatty-acid composition of the fats varies with their origin; animal fats usually contain more saturated and fewer polyunsaturated residues than plant fats do.

Although humans can synthesize their requirement of saturated and monounsaturated fatty acids, the polyunsaturated fatty acids must be obtained from the diet and hence are termed the ***essential fatty acids.*** The requirement for these substances in the tissues is probably related to three important functions: (1) they serve as precursors of the important hormone-like substances, ***prostaglandins*** (Fig. 5-1D); (2) they maintain the normal fluidity of the cell-membrane lipids; and (3) they prevent the deposition of cholesterol and other lipids on the blood vessel walls, which is a primary contributor to atherosclerosis. The quantity of cholesterol in the diet, normally only a few hundred milligrams per day, may play a secondary role in the development of atherosclerotic disease.

DIGESTION AND ABSORPTION

Since virtually 99% of the dietary lipids are fats, the major problem in lipid uptake from the intestinal tract concerns the ultimate fate of the constituent fatty acids and glycerol from the triacylglycerols. Four interacting factors are essential for the normal function of the intestines in fat absorption: (1) the secretion of hydrolytic enzymes from the pancreas to break the ester linkages of the fats; (2) the release of detergents in the bile to emulsify the fats and their breakdown products; (3) the uptake of the digestion products into the mucosal cells that line the villi of the small intestine; and (4) the conversion of these products of digestion into particles for transport from the mucosal cells into the lymph vessels and hence to the blood vessels.

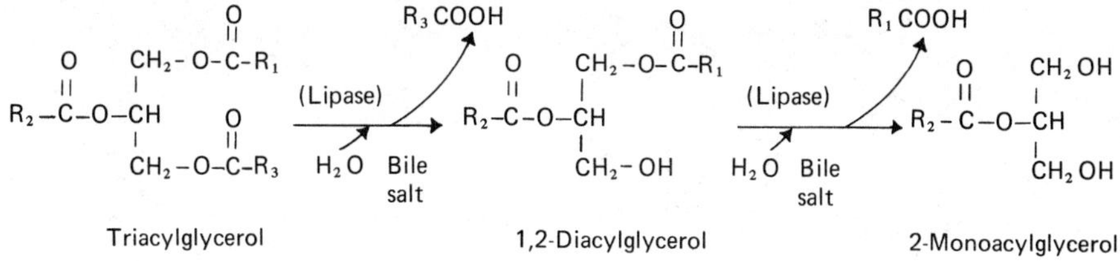

FIG. 5-2. Action of pancreatic lipase in the digestion of triacylglycerol.

A failure in any one of these processes will lead to an inadequate uptake of fats and a major derangement of intestinal function. The principal symptom in these disorders is the gross appearance of fat in the feces, a condition known clinically as *steatorrhea.* Typically, the accompaniments of steatorrhea are severe diarrhea with loss of water and electrolytes, inadequate absorption of other dietary components (especially the fat-soluble vitamins), and, if prolonged, eventual starvation due to insufficient caloric intake. Pancreatic diseases leading to an insufficiency of digestive enzymes, an obstruction of the bile duct leading to inadequate bile detergency, malabsorption defects of the intestinal cells, or a variety of disorders affecting the secretion and transport of lipids in the lymphatic vessels may all lead to defective fat absorption and steatorrhea.

LIPOLYSIS

This is a general term for the process of lipid hydrolysis by enzymes known collectively as the ***lipases.*** The process is important not only for the absorption of lipids in the intestinal tract, but also in the utilization of the absorbed fat by the body tissues after a meal, as well as in the mobilization of fat from storage depots in times of fasting. In the duodenum, small quantities of ingested fat are mixed with the pancreatic secretions, which contain lipase and buffers, and the bile, which contains detergent bile salts. The resulting alkaline emulsion of the triacylglycerols is attacked by the lipase, which rapidly releases fatty acids from the two ends of the fat molecule (Fig. 5-2). The products of lipolysis are mixtures of diacylglycerols, 2-monoacylglycerol, and the liberated fatty acids. These become converted to sodium soaps and add further to the detergency of the intestinal contents. Only a small amount of free glycerol is released, since the 2-monoacylglycerols are quite resistant to the action of pancreatic lipase.

BILE SALTS AND MICELLES

The detergent action of bile is due to the presence of ***bile salts,*** which are steroid derivatives produced by the oxidation of cholesterol in the liver. Although the steroid portion of bile salts has the characteristic hydropho-

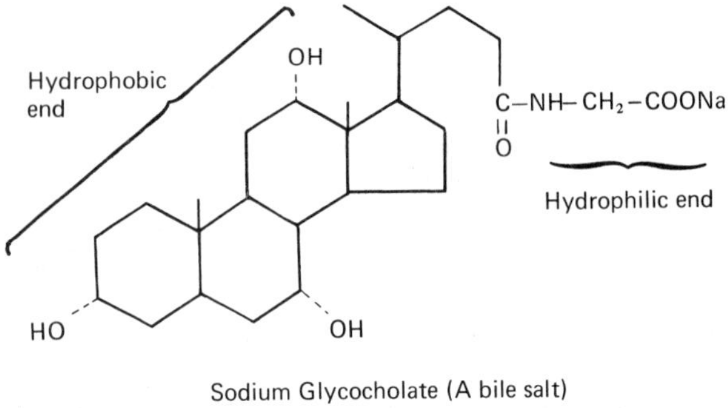

FIG. 5-3. The basis for the detergent action of the bile salts.

bic solubility of lipids, their oxidized side-chains, with the amino acids glycine or taurine attached, are hydrophilic, or quite compatible with water (Fig. 5-3). This dual miscibility is characteristic of detergents and soaps and is sometimes termed their ***amphipathic*** *property:* the hydrophobic end of the molecule mixes easily with other lipids, such as fats, in the intestinal contents; the hydrophilic end does not mix so readily with the lipids but favors contact with the bulk water phase of the gut. This forms the so-called lipid-in-water ***emulsion*** of small fat droplets. The emulsion is stabilized by the projecting hydrophilic ends of bile salts, which form a skin or interface in contact with the external water. The dispersion of fats as fine emulsions facilitates their interaction as substrates for the hydrolytic action of lipase, and thus the bile salts function as powerful activators of the digestive process (Fig. 5-4).

Another type of association that is promoted by the amphipathic nature of the bile salts is the formation of ***micelles.*** The principle involved is the same as that employed in the formation of emulsions, namely, the hydrophilic parts project out into the aqueous medium, and the hydrophobic parts remain associated together in the interior. In a micelle, however, there is no bulk lipid phase in the interior, the entire aggregate being only one molecule thick. Thus, the particles in a micelle are several orders of magnitude smaller than those in an emulsion, so that although the particles of the latter are large enough to be detectable by virtue of the characteristic milky turbidity that they cause, the micellar suspensions are optically clear. Emulsification is important for increasing the exposed surface to allow the reaction between lipase and the lipids, but the particles of an emulsion are too large to pass through cell membranes. On the other hand, micellar particles are readily absorbed by the mucosal cells of the small intestine. Since the monoglycerides and free fatty acids formed by lipolysis are also amphipathic by nature, they readily intersperse themselves among the bile-salt molecules, thus producing mixed micelles. Small quantities of cholesterol may also be sandwiched in among the hydrophobic regions of the micelles. A very similar structure is involved in the formation of cellular membranes from the amphipathic

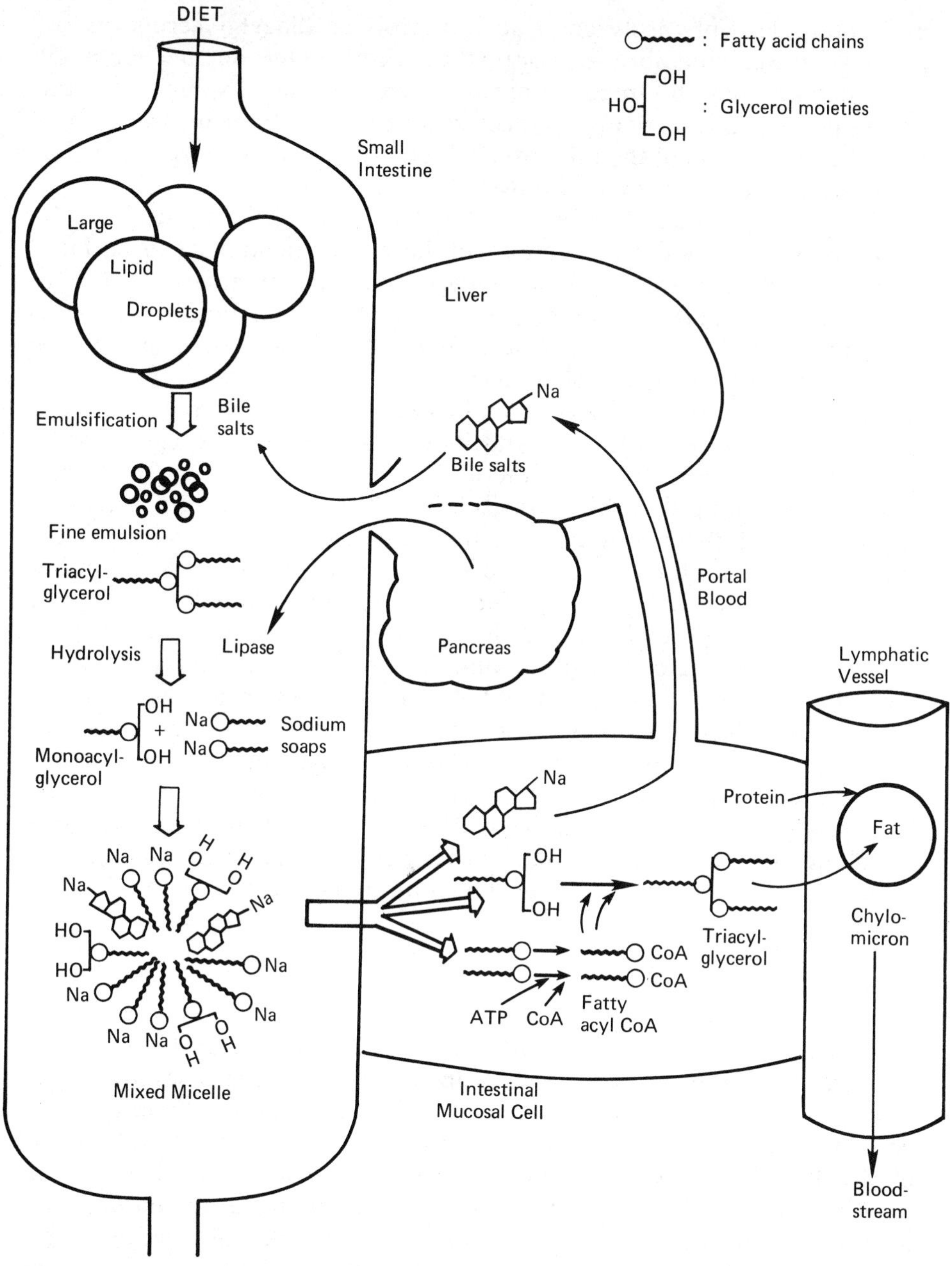

FIG. 5-4. Digestion and absorption of fat. Emulsification by bile salts precedes the hydrolysis by lipase in the intestine. The products of digestion with bile salts form micelles, much magnified here, but in reality much smaller than the particles in the emulsions. Micelles are taken up intact into mucosal cells, where the fats are re-formed and secreted with a thin protein coating as chylomicra into the lymph. Bile salts are recycled to the liver through the portal vein, which forms the enterohepatic circulation.

phospholipids. Unhydrolyzed triacylglycerols or diacylglycerols do not permeate the cell membranes, nor do they readily enter the micellar layer. It is thought that the micellar particles are the major vehicles for the uptake of the products of lipid hydrolysis during intestinal absorption and for the conservation of the bile salts, which are largely reabsorbed during the passage through the small bowel.

The final processes involved in the absorption of the lipids occur in the mucosal cells. With the exception of the small amounts of shorter fatty acids (*i.e.*, with a chain length of ten carbons or less, such as those occurring in milk fat) that pass directly into the portal blood and thence to the liver, the long-chain fatty acids and monoglycerides are reesterified together to form triacylglycerols. This is an energy-requiring process, since the fatty acids must first be activated to their acyl-CoA derivatives using ATP. Together with cholesterol and its esters, the resynthesized triacylglycerol is discharged from the lateral side of the intestinal mucosal cells as a fairly large lipid droplet, which is stabilized by a thin surface layer of proteins and phospholipids. This lipid particle is known as a ***chylomicron,*** and it enters the lymphatic vessels where it confers a typical milky appearance to the chyle or lymph fluid after a meal. The milky chyle with its fatty burden is slowly passed through the lymphatics into the bloodstream at the thoracic duct. Abundant chylomicra remain in the blood for several hours after the ingestion of fats, and they impart a persistent turbidity to the plasma. As they circulate, the chylomicra are acted upon in peripheral tissues by the enzyme, ***lipoprotein lipase,*** which hydrolyses the triacylglycerol to free fatty acids for use by the tissues and glycerol, which is taken up eventually by the liver. After removal of the triacylglycerol and some surface proteins the particles are converted into cholesterol-rich chylomicron remnants, which are taken up by liver cells.

THE LIVER

As in the metabolism of carbohydrates, the hepatic cells of the liver play a central role in the body's metabolism of lipids. In some cases, this function is direct, for example, in the biosynthesis of cholesterol and the conversion of the latter to bile salts or in the formation of the phospholipids of blood plasma, processes which are almost exclusively localized in the liver. In other instances, the liver may play a modifying or regulatory role, for example, in the storage of fats in the body. Unlike the abundant glycogen stores in hepatic cells, the amount of triacylglycerols deposited within the liver itself is normally quite small (less than 1% of its mass). Nonetheless, the balance among fat utilization, its synthesis in the body, and the consequent level of its deposition in other storage sites (or, pathologically, within the blood vessels and infiltrated organs) is controlled to a major extent by the metabolic events in the liver.

Some of the lipids from intestinal absorption reach the liver directly by the portal circulation. These include the bile salts and the short-chain fatty acids. The latter are of minor significance except in infants, in whom milk fats are a preponderant component of the diet. The short-chain acids travel in the blood as complexes with plasma albumin. Upon reaching the liver, they are catabolized by pathways of fatty-acid oxidation, and they appear not to be reutilized for incorporation into liver lipids to any extent. The bile salts, on the other hand, are taken up by the liver and enter the total pool along with newly synthesized bile salts. Thus, although the body contains several grams of bile salts in this pool within the cells and biliary tracts of the liver as well as in the concentrated storage receptacle of the gallbladder, only a few hundred milligrams of secreted bile salts escape from the reabsorptive process in the gastrointestinal (GI) tract and appear in the feces. The cyclic passage from the liver to the bile duct to the intestine to the portal vein and back to the liver is termed the *enterohepatic circulation* (Fig. 5-4). This circulatory route allows the conservation of most of the bile salts released in the bile and, incidentally, the minor recycling of some lipid-soluble waste products, such as the ***bile pigments*** (Chap. 6). Most of the latter, however, escape reabsorption and are excreted with the feces.

Other lipids enter the liver by its other major circulatory route through the hepatic artery. Dietary sterols, for example, reach the liver cells from the systemic circulation in the form of chylomicra remnants. These particles become attached to the membrane of the hepatic cell, probably by their external proteins. The major source of lipids entering the liver is the unesterified or free fatty acids that are released from the major fatty depots in adipose tissue and are transported in the systemic blood plasma complexed with albumin. This movement of fatty acids from the storage areas to actively catabolizing tissues constitutes the principal mechanism of energy mobilization in the body.

The liver sits in a key, apical position in relation to fat mobilization, processing, and biosynthesis. Imbalances of these counteracting systems may lead to metabolic disturbances of a very serious nature as well as to excessive deposition of fats in adipose tissue (obesity) or in the liver cells themselves (fatty liver).

The syndrome known as ***fatty liver*** may arise from a multitude of causes. Generally speaking, the excessive infiltration of the liver with fat signifies that the hepatic cells are inadequate to deal metabolically with the fats that are carried there by the circulation. This inadequacy may arise from toxic or degenerative damage within the liver in the face of normal fat intake and mobilization from the depots or, less frequently, in association with increased fat uptake or fat synthesis by liver cells. The most common causes of fatty liver are (1) the ingestion of *substances that are toxic* to hepatic cells (*e.g.*, ethanol, which is a factor in one third of all cases, and halocarbons, such as chloroform or carbon tetrachloride);

Fat oxidation: $-(CH_2)_n- \; + 3n[O] \rightarrow nCO_2 + nH_2O$

Carbohydrate oxidation: $-(CH_2O)_p- \; + 2p[O] \rightarrow pCO_2 + pH_2O$

FIG. 5-5. Explanatory scheme for the relatively high-energy yield for the oxidation of fat versus that of carbohydrates. Each carbon in a fatty-acid chain accounts for the consumption of 3/2 or 50% more oxygen than does each carbohydrate carbon. Moveover, the chain length is much greater in fatty acids, and thus $n >> p$ (by a factor of three on average).

(2) diets that are *chronically deficient in protein* (e.g., the syndrome known as *kwashiorkor,* which is one of the most serious nutritional problems in underdeveloped areas of the world); and (3) *infections or malignant disease* that affect the hepatic cells or biliary tract. All the above act by reducing the metabolic competence of the liver. The category of causes primarily related to the increased delivery of fat to the liver includes *starvation* and *diabetes mellitus,* which are associated with increased mobilization of fatty acids from the storage depots, and *obesity,* in which the accelerated lipogenesis exceeds the capacity of the liver to export the fat as lipoproteins.

FATTY-ACID OXIDATION

By virture of their chemical structure, the fatty acids constitute the most compact fuel for the body's energy requirements. On a molar basis, the complete oxidation of fatty acids yields several times the usable chemical energy that a carbohydrate does; for example, 130 mols of ATP derive from the oxidation of 1 mol of palmitic acid, versus 38 mols of ATP from the oxidation of 1 mol of glucose. On a weight basis, too, the energy yield is more than double (9 Cal/g fat vs. 4 Cal/g carbohydrate or protein). The reason for this higher energy yield is the same as that which renders gasoline, oil, and other petroleum products such effective fuels for heat and mechanical energy production, namely, the highly reduced state of carbon in the long alkyl chains. The bulk of the fatty-acid molecule consists of the repeated units $(CH_2)_n$, a structure that is maximally enriched in hydrogen.

We have seen from previous discussions (Chap. 3) that most of the energy conserved during biologic oxidations derives from the controlled transfer of electrons from hydrogen atoms in the respiratory chain, with coupled phosphorylation of ADP to ATP. Since the fatty acids consist essentially of carbon and hydrogen and thus contain much less combined oxygen than carbohydrates, their oxidation involves the consumption of proportionately more oxygen, and hence more total ATP is produced by oxidative phosphorylation (Fig. 5-5). More than half of the basal or resting energy requirement of most body tissues (including the liver but excluding

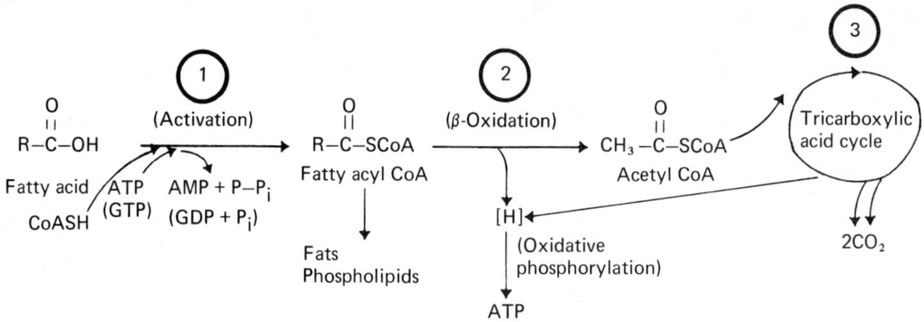

FIG. 5-6. The three stages of fatty-acid oxidation.

the brain) is produced by fat catabolism, and this preponderance is accentuated during fasting.

The pathways of fatty-acid oxidation are strictly confined to the mitochondria. There they are interlinked with the terminal catabolic reactions of the electron-transport system and the citric-acid cycle, which are also involved in the oxidation of carbohydrates (Chap. 4) and of amino-acid derivatives (Chap. 6). For convenience, the process may be considered to consist of three states (Fig. 5-6): (1) *activation* of the fatty acids by an energy-requiring reaction to the acyl-CoA thio esters (this step is also the prelude to the utilization of fatty acids for the biosynthesis of fats or phospholipids); (2) β *oxidation,* a cyclic series of reactions whereby hydrogens are sequentially removed from the alkyl chain and two-carbon units are split off as acetyl-CoA molecules (these are the reactions unique to fatty-acid oxidation); and (3) *citric-acid-cycle oxidation* of the acetyl-CoA products to yield CO_2 and more hydrogen for oxidative phosphorylation (this is the ultimate common pathway for the oxidation of carbohydrates, fats, and proteins).

The entry of the fatty acids into mitochondria requires the carrier molecule, *carnitine,* as noted previously in Chapter 3. The liver mitochondria have a limited capacity to utilize fatty acids directly without the intervention of carnitine-mediated transport. This is because the activation of some fatty acids may proceed in the intramitochondrial space by a reaction that utilizes the free energy of hydrolysis of guanosine triphosphate (GTP) to guanosine diphosphate (GDP). The reaction is analogous to the reversal of the conversion from succinyl CoA to succinate in the citric-acid cycle, so that in this case one molecule of fatty acids reacts with coenzyme A for each molecule of GTP hydrolyzed to GDP plus inorganic phosphate. The reaction can thus be coupled to the succinyl-CoA energy-yielding reaction:

1. Succinyl CoA + GDP + $P_i \rightarrow$ Succinate + GTP + CoASH
2. Fatty acid + CoASH + GTP $\rightarrow$ Fatty-acyl CoA + GDP + P_i

Thus,

3. Succinyl CoA + Fatty acid → Fatty-acyl CoA + Succinate

The reaction may be important chiefly for the shorter chain fatty acids (up to C8).

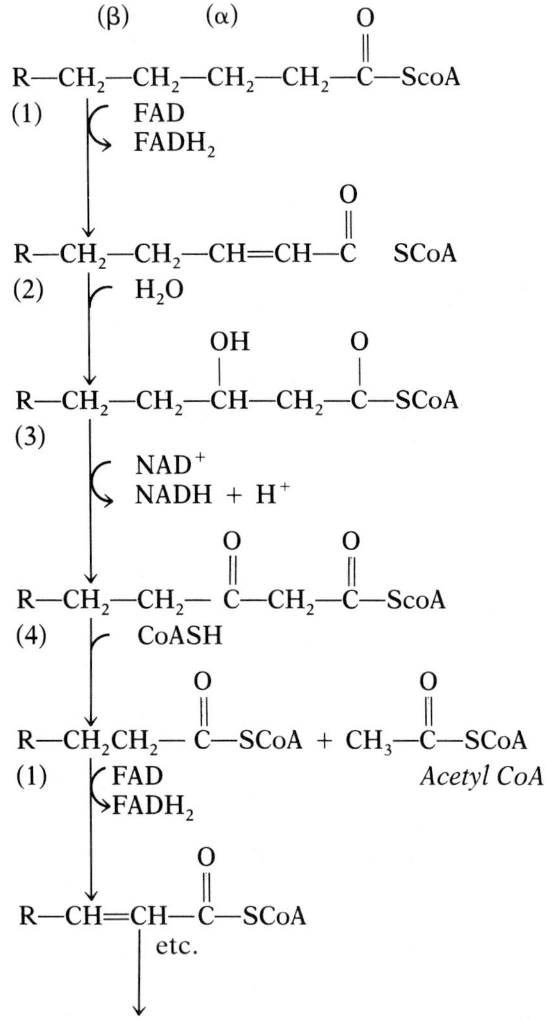

This initial energy-requiring activation step primes the fatty-acid molecule for all the subsequent energy-producing reactions in much the same way that phosphorylation primes the glucose molecule for catabolism in glycolysis.

The process of **β *oxidation*** is so named because the hydrogen and electrons of the fatty-acyl chains are removed and redistributed by oxidation-reduction reactions in such a way that the carbon atom in the β position of the chain becomes the most highly oxidized. The oxidized product is then cleaved so that the β carbon now becomes the carboxyl end of the shortened fatty-acyl chain after the removal of the terminal two carbons as acetyl CoA. The oxidation–reduction reactions are strictly analogous to the steps in the citric-acid cycle whereby succinate becomes converted to oxaloacetate: Reaction (1), which is analogous to that of succinate dehydrogenase, and reaction (3), which is analogous to that of malate dehydrogenase, involve the transfer of hydrogen, catalyzed by specific dehydrogenases, to the flavine and nicotinamide nucleotide coenzymes, respectively. (The transfer of the two electrons from NADH to the respiratory chain results in the formation of three molecules of ATP, whereas the transfer from $FADH_2$ taps into the chain through another flavoprotein at the same level as succinate dehydrogenase and thus yields two molecules of ATP.) Reaction (2), a hydration step, is analogous to that of fumarase. Reaction (4), which has no analogy in the citric-acid cycle, involves the cleavage of the β-keto-acyl compound of coenzyme A by a specific *thiolase* enzyme to yield acetyl CoA plus a fatty-acyl CoA that is two carbon atoms shorter and primed for the next round of β oxidation.

Thus the original fatty-acyl chain is sequentially catabolized; each round in the reaction adds 1 mol of H_2O and one of CoASH, extracts four reducing equivalents as $FADH_2$ and NADH + H^+, and produces 1 mol of CH_3CO–SCoA. The energy yield for the β-oxidation cycle is 5 mols of ATP for each round of reactions. Therefore, starting with a typical fatty-acyl group such as palmityl CoA (16:0), seven rounds will complete the β-oxidation process:

$$CH_3(CH_2)_{14}CO\text{—}SCoA + 7H_2O + 7\ CoASH + 7\ FAD + 7\ NAD^+ \\ \rightarrow 8CH_3CO\text{—}SCoA + 7\ FADH_2 + 7\ NADH + 7H^+$$

Considering the transfer of the reducing equivalents to the respiratory chain and subtracting the one molecule of ATP (which is equivalent to GTP) required for the initial activation of palmitic acid, we may express the stoichiometry of the reaction in energy terms, allowing for the 34 mols of H_2O derived in ATP formation:

$$CH_3(CH_2)_{14}COOH + 8\ CoASH + 7O_2 + 34\ ADP \\ + 34\ P_i \rightarrow 8CH_3CO\text{—}SCoA + 34\ ATP + 42H_2O$$

Thus, although β oxidation occurs in the presence of oxygen, in an energy sense it bears a resemblance to the anaerobic glycolytic sequence, since both processes lead to only incomplete oxidation of the starting substrates and to only a fraction of the total theoretic energy yield.

The completion of fatty-acid oxidation, as in carbohydrate catabolism, requires the operation of the citric-acid cycle. This is clearly seen from the stoichiometry of the citric-acid-cycle oxidation of the products from the β oxidation of palmitic acid:

$$8 \text{ Acetyl CoA} + 16O_2 + 96 \text{ ADP} + 96 \text{ P}_i \rightarrow 16CO_2 + 8 \text{ CoASH} + 96 \text{ ATP} + 104H_2O$$

Thus, the energy yield from this final common pathway of catabolism is nearly three times as great as that from β oxidation alone; the combined yield of ATP, as noted before, may be seen to total 130 mols for each mol of palmitic acid completely oxidized. The flux of electrons through the β-oxidation dehydrogenases will, of course, depend upon the coupling-site regulation of the flux through the respiratory chain; the citric-acid cycle will also play a cardinal role in maintaining the normal operation of the β-oxidation process by removing the product, acetyl CoA. As noted briefly in Chapter 3 and discussed more fully in the following pages under regulation of fatty-acid metabolism, a failure to provide adequate citric-acid-cycle activity in the liver results in damming up the flow of acetyl groups through oxidative channels and the diversion of the carbons into the production of ketone bodies. Although the latter may normally pass from the liver through the bloodstream to be metabolized as energy-yielding substrates, an excessive accumulation of these substances will produce serious pathologic effects.

FATTY-ACID SYNTHESIS

Up to this point, we have been considering the metabolic fate of the fatty acids that have reached the liver from dietary sources. Apart from the essential polyunsaturated fatty acids, the human is capable of synthesizing all the other fatty acids required either for structural lipids in membranes or for storage purposes. This total or *de novo* synthesis of fatty acids is referred to as ***lipogenesis,*** and the liver is one of the most active tissues in the performance of this function. Although any substance whose metabolism yields acetyl CoA as an end product is a potential precursor for lipogenesis, the most important source is the carbohydrates. Consider the example of the well-satiated diner who has excessive quantities of glucose accumulating in his hepatic cells. After replenishing any previous deficit in liver glycogen, any further synthesis of the polysaccharide will be limited by the feedback system discussed in Chapter 4. Now the excess glucose will be broken down as far as acetyl CoA and CO_2, which are both important prerequisites for fatty-acid formation. In general, it may be stated that whenever the dietary intake of carbohydrate exceeds the body's energy needs, the excess is converted into fat.

Lipogenesis is not simply a reversal of the β-oxidation system. As in glycogenesis and glycogenolysis, entirely different pathways are followed for the anabolism and the catabolism of fatty acids. One difference is that

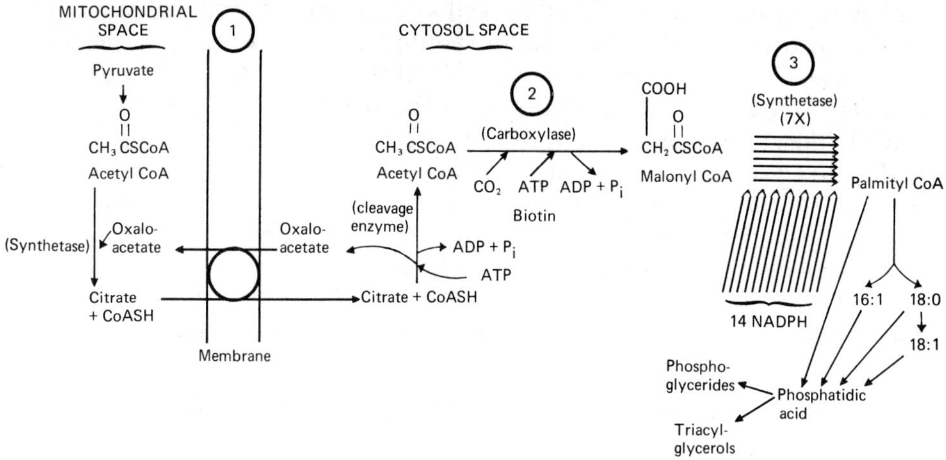

FIG. 5-7. The three stages of fatty-acid biosynthesis.

the synthetic process is found outside the mitochondria in the cytosol. Another is that although coenzyme A is important for lipogenesis, the intermediate compounds become attached to the sulfhydryl group of a protein known as the ***acyl carrier protein*** (ACP). Third, lipogenesis requires a carboxylation reaction involving CO_2, ATP, and the coenzyme biotin, none of which are needed for β oxidation. Finally, the reversal of the oxidation stages, which in β oxidation utilize the coenzymes FAD and NAD^+, shows an absolute specificity for NADPH to provide the reducing equivalents for lipogenesis. For convenience, let us consider fatty-acid biosynthesis in its three consituent phases (Fig. 5-7): (1) the provision of the starting material, acetyl CoA, which originates inside the mitochondrion and must be transported out to the cytosol; (2) the carboxylation of acetyl CoA to the true substrate of lipogenesis, malonyl CoA; and (3) the condensation of the two-carbon units and their reduction to long-chain, saturated acids, chiefly palmitic acid.

Although the reactions of lipogenesis occur in the cytosol compartment, the starting material, acetyl CoA, is generated within the permeability barrier of the mitochondrion. Since the inner mitochondrial membrane is impermeable to coenzyme A or its derivatives, the acetyl CoA must first be converted to a form that can cross the membrane. This may be accomplished by combining the acetyl CoA with oxaloacetate to form citrate; the mitochondrial membrane possesses a specific transport site or translocase system for tricarboxylic acids such as citrate and isocitrate. Once the citrate reaches the cytosol, it may be reconverted to acetyl CoA plus oxaloacetate by the *citrate-cleavage enzyme,* a reaction which is not simply the reverse of that of the condensing enzyme, but which requires energy input in the form of ATP as well as coenzyme A. The oxaloacetate may be transferred back into the intramitochondrial space by a specific translocase for this dicarboxylic acid. In this way, oxaloacetate plays a catalytic role as the handle that carries the acetyl group from the intramitochondrial compartment to the cytosol; the cell must sacrifice one mole of ATP to facilitate this transport cycle.

The second stage in lipogenesis—the carboxylation of acetyl CoA—is conducted in the cytosol by the enzyme known as *acetyl-CoA carboxylase.* The latter contains the vitamin *biotin* bound to its active center. In the presence of ATP, a molecule of carbon dioxide becomes attached to the biotin prosthetic group of the enzyme to form the highly reactive *carboxybiotin:*

1. CO_2 + Biotin—Enzyme + ATP → Carboxybiotin—Enzyme + ADP + P_i

In this activated form, the carboxyl group may be readily transferred to acetyl CoA to form the free carboxyl of the product, *malonyl CoA:*

2. Carboxybiotin—Enzyme + CH_3CO—SCoA → $CH_2(COOH)CO$—SCoA + Biotin—Enzyme

$$\text{Carboxybiotin—Enzyme} + CH_3CO\text{—SCoA} \rightarrow \underset{\substack{|\\ }}{\overset{COOH}{\overset{|}{CH_2}}}CO\text{—SCoA} + \text{Biotin—Enzyme}$$

This reaction is identical in form to the carboxylation of pyruvate by pyruvate carboxylase. Again, a molecule of ATP is spent by the cell in order to activate acetyl CoA for its condensation role in lipogenesis. As will be seen later, the carboxylation step is a prime regulator of fatty-acid synthesis inasmuch as it controls the entry of two-carbon units into the process.

All the subsequent steps in lipogenesis involve either the malonyl or the acetyl group, though their activated carboxyls are transferred from CoASH to ACP-SH (Fig. 5-8). All the carbons of the fatty acid produced, other than the two in the terminal CH_3CH_2–position that come from acetyl ACP, are derived from the malonyl ACP. Moreover, although the carboxylation process is an important priming step for lipogenesis, the –COOH that is added is stripped off the malonyl CoA as CO_2 during condensation reactions. Therefore, in reality, all the fatty-acid carbons ultimately derive from acetyl CoA, and none of the CO_2 taken up by carboxylation is incorporated into the fatty acids. As in the case of gluconeogenesis, we have a situation where the biosynthetic process is seemingly wasteful of cellular energy, utilizing a complex activation process rather than simply reversing the catabolic reactions. And as in the case of the gluconeogenic bypass from pyruvate to phosphoenolpyruvate, the metabolic rationale for producing malonyl CoA at the expense of ATP and then decarboxylating it again is to provide the driving force for an energetically unfeasible reaction, that is, the formation of the acetoacetyl group from the two acetyl

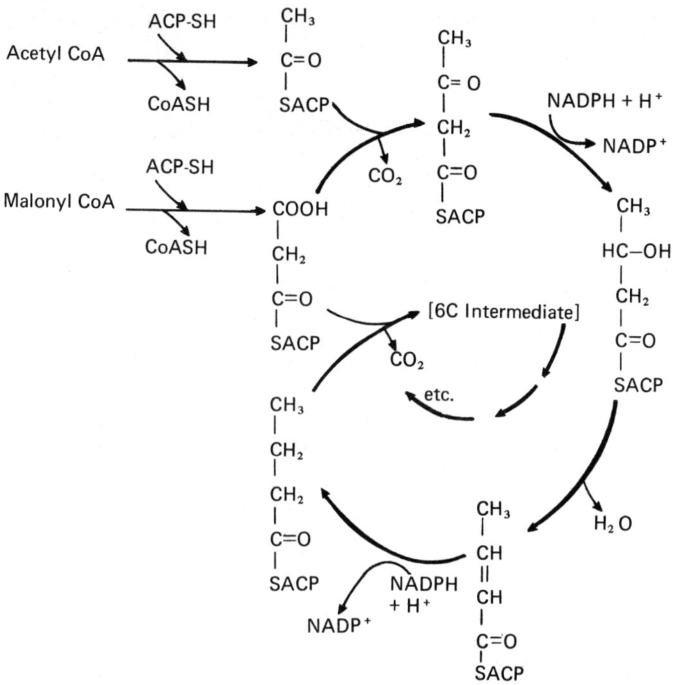

FIG. 5-8. Role of the acyl carrier protein (ACP) and of NADPH in the reductive biosynthesis of fatty acids.

groups. The energy decrease for the thiolase reaction of β oxidation that catalyzes the forward cleavage is so large that the reverse condensation by such a process normally leads to low yields of acetoacetyl CoA. This energy barrier is overcome by using the highly reactive malonyl group for the first and key condensation step, which leads to acetoacetyl–S–ACP and CO_2.

The subsequent steps in the cytoplasmic synthesis of fatty acids are similar to the reversal of the reactions in mitochondrial β oxidation. The keto group in the β position is first reduced by NADPH to a hydroxyl group; a double bond is introduced by a dehydration reaction; the double bond is reduced by NADPH to a saturated linkage; and the product, butyryl–S–ACP, then has a second malonyl group condensed with it to form the six-carbon β-keto acyl–S–ACP. The cycle of reductions and dehydrations is thus repeated until eventually a 16-carbon product, palmityl–S–ACP is formed (Fig. 5-8). Only when it reaches this endpoint does the fatty-acyl chain become free of the fatty-acid synthesizing complex of enzymes, which is displaced with CoASH to release palmityl CoA.

This entire process is performed with all the intermediate compounds tightly bound to a large multienzyme complex, ***fatty-acid synthetase.*** The overall stoichiometry of the fatty-acid synthetase process may be calculated

$$\text{Acetyl CoA} + 7\ \text{Malonyl CoA} + 14\ \text{NADPH} + 14H^+ \rightarrow \text{Palmityl CoA} + 7\ \text{CoA} + 7CO_2 + 14\ \text{NADP}^+ + 7H_2O$$

In order to determine the total energy input, we must allow for the ATP that is required (1) for the movement of acetyl CoA to the cytoplasm from inside the mitochondria (citrate-cleavage step) and (2) for the production of malonyl CoA from acetyl CoA (acetyl-CoA carboxylase step):

$$8\ \text{Acetyl CoA} + 15\ \text{ATP} + 14\ \text{NADPH} + 14H^+ \rightarrow \text{Palmityl CoA} + 15\ \text{ADP} + 15\ P_i + 7\ \text{CoA} + 14\ \text{NADP}^+ + 7H_2O$$

Many different metabolic fates may befall the palmityl CoA molecule (Fig. 5-7). It may be further elongated two carbons at a time in either the mitochondria or the endoplasmic reticulum. Palmitate or its elongation product stearic acid (18:0) may be desaturated in the endoplasmic reticulum to form the monounsaturated fatty acids, palmitoleic (16:1) and oleic (18:1) acids, respectively. It may be esterified to the hydroxyl groups of α-glycero-phosphate to form a ***phosphatidic acid*** which is the key intermediate in both fat and phospholipid biosynthesis. It may be esterified to the diacylglycerol that is formed by the dephosphorylation of phosphatidic acid to produce a triacylglycerol (in the liver, as in most body tissues, there is no pathway for the formation of monoacylglycerols nor for the esterification of the latter to diacylglycerols, as there are in the gastrointestinal mucosal cells). Finally, it may be transported back into the mitochondria, either as the free acid or as the carnitine ester, for β oxidation when the energy supplies in the cell are low.

REGULATION OF FATTY-ACID METABOLISM

The maintenance of the proper balance between fatty-acid catabolism and anabolism in the liver involves the influence of metabolites within the cell as well as the impinging effects of external forces, such as hormonal factors and dietary intake. These influences are obviously dependent upon the metabolic relations with other substances (carbohydrates and proteins), upon the energy supply of the liver cell, and upon the movement of lipids between the liver and other tissues of the body. In its relation to other tissues, the liver plays a role analogous to its part in blood glucose homeostasis, since it contributes significantly to the proteins, the phospholipids, cholesterol, and some of the fats in plasma lipoproteins, and it is a major consumer of the plasma free-fatty-acid pool.

In the control of β oxidation, a prime influence is the availability of substrate (Fig. 5-9). Fatty acids in the hepatic cell may originate from three sources: (1) mainly from the uptake of the unesterified fatty acids transported to the liver from adipose tissue (the regulation of the fatty-

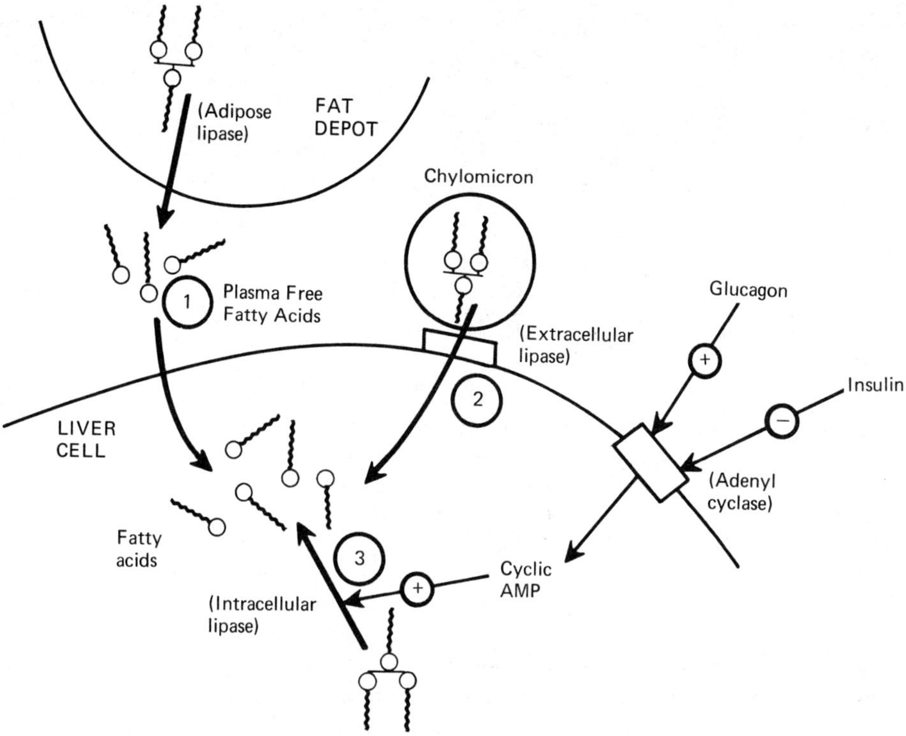

FIG. 5-9. The three sources of fatty acids in liver cells.

acid output from adipose cells is principally under hormonal control and is discussed in the next section); (2) secondarily from the uptake of fatty acids released from the attack of extracellular lipase upon chylomicra remnants (the major variable in substrate provision from this source will be the level of dietary fat); and (3) the fatty acids released by intracellular lipase acting upon the liver triglycerides. As in the case of the adipose-tissue lipase to be discussed in the next section, the action of the hepatic lipase is enhanced by those hormones that promote increased cyclic-AMP levels through stimulation of the adenylate cyclase system; in the liver, glucagon will thus be an important contributor to increased intracellular fatty-acid production.

A second controlling factor—as in the regulation of all oxidative reactions that are linked to the respiratory chain—is the energy content of the cell (Fig. 5-10). The β-oxidation cycle in mitochondria is dependent upon the flow of electrons through the coupling sites of oxidative phosphorylation; thus, the overall rate of fatty-acid utilization by oxidation will depend upon the relative concentrations of ADP and ATP. When the ADP levels are high (low cellular energy reserves), the coupling reactions may proceed to promote ATP synthesis and the flux of electrons from the

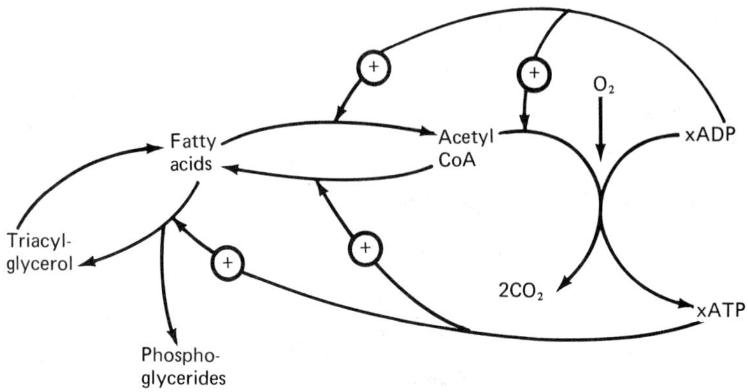

FIG. 5-10. Control of fatty-acid metabolism in the liver by energy levels. High ADP concentrations promote fatty-acid oxidation as well as oxidation of the hydrogens and acetyl-CoA products. High ATP concentrations inhibit oxidation of the end products of fatty-acid catabolism and favor the biosynthesis of fatty acids and esterified products.

flavoprotein-linked and NAD-linked dehydrogenases of β oxidation. When, on the other hand, the ATP levels are elevated, the above reactions will be slowed by respiratory control, accumulated fatty acids will be diverted to the biosynthesis of fats or phospholipids, and the net synthesis of palmitate or other elongated fatty acids will be favored in the liver.

A third level of control upon fatty-acid oxidation appears to occur at the point of entry into the mitochondria, normally mediated by carnitine and the membrane-spanning acyltransferase system. As noted below, when it is exposed to abundant amounts of carbohydrate the liver converts to a lipogenic mode. It would be inappropriate to allow β oxidation to proceed while fatty acids were being actively synthesized from glucose. A cross-over control between the catabolic and anabolic processes may be provided by the key metabolite, malonyl CoA, which acts as a potent reversible inhibitor of the mitochondrial carnitine acyltransferase. Thus, in the lipogenic state an increase of malonyl CoA, enhanced by the presence of high carbohydrate and insulin, blocks movement of fatty acids through the mitochondrial inner membrane and effectively shuts down β oxidation. Conversely, in the fasting or diabetic state, malonyl CoA levels drop, relieving the inhibition upon fatty-acyl-carnitine formation and the oxidation of fatty acids to generate energy, and ketone bodies proceed at maximal rates in the liver.

The ability of the citric-acid cycle to catabolize the acetyl CoA produced by β oxidation will be of tremendous importance in realizing the total energy potential from fatty-acid catabolism as well as in preventing the undesirable accumulation of ***ketone bodies.*** The latter comprise *ace-*

toacetic acid and its products of reduction or decarboxylation, β-*hydroxybutyrate* or *acetone,* respectively. These substances may arise inside the mitochondrial matrix from the reversal of the thiolase reaction, which normally favors the cleavage of the final acetoacetyl moiety of β oxidation but which can be reversed to condense two acetyl-CoA molecules when the concentrations of the latter become elevated. The acetoacetyl CoA in mitochondria gives rise to free acetoacetic acid; in the cytosol compartment, acetoacetyl CoA will be used to synthesize the sterol ring as will be discussed in the next section on cholesterol metabolism.

Under normal circumstances, the hepatic mitochondria produce significant amounts of these ketone bodies, which diffuse out of the liver into the bloodstream. Both skeletal and cardiac muscles, the brain, and other extrahepatic tissues have the means to reconvert acetoacetic acid to acetyl CoA, and they hence may derive a sizable proportion of their resting energy requirements from ketone-body catabolism. The liver, on the other hand, has no way of utilizing the free acetoacetic acid once the CoA group is removed, and therefore it is incapable of oxidizing ketone bodies itself. As discussed under the section in Chapter 3 on the regulation of the citric-acid cycle, the principal determinant of ketone-body production by the liver will be the ratio between acetyl-CoA production (principally from β oxidation) and the oxaloacetate levels (which is derived from the carboxylation of pyruvate and hence is dependent upon adequate carbohydrate utilization) (Fig. 5-11). Consequently, accelerated fatty-acid catabolism together with diminished carbohydrate utilization leads to the production of ketone bodies by the liver in excess of the capacity of the extrahepatic tissues to eliminate them. When this happens, the accumulation of these acidic substances will lower the *p*H of the body fluids, producing the condition known as *metabolic acidosis.* Large excesses of the ketone bodies will be excreted by the kidney, leading to *ketonuria.* In extreme cases, the acetone will be volatilized from the lungs and be detectable in the expired air. Starvation and diabetes—both of which lead to increased fatty-acid release from storage depots and to decreased carbohydrate metabolism in the liver—are the most common causes of ketosis.

The control of the *de novo* synthesis of fatty acids will also depend upon the availability of substrate and energy (Fig. 5-12). As in the case of β oxidation, the interrelations with carbohydrate metabolism in general and with the pentose–phosphate pathway and the citric-acid cycle in particular will be of utmost importance. The supply of NADPH for the reductive steps in fatty-acid biosynthesis will arise to a large extent from the pentose–phosphate pathway. The glucose-6-phosphate dehydrogenase reaction is inhibited by long-chain fatty acids, so that NADPH production by this pathway will be maximal only when the concentrations of fatty acids are low, signaling a demand for accelerated synthesis.

Another source of NADPH, the *malic enzyme,* is also of importance in the hepatic synthesis of fatty acids. This may be inferred from the fact

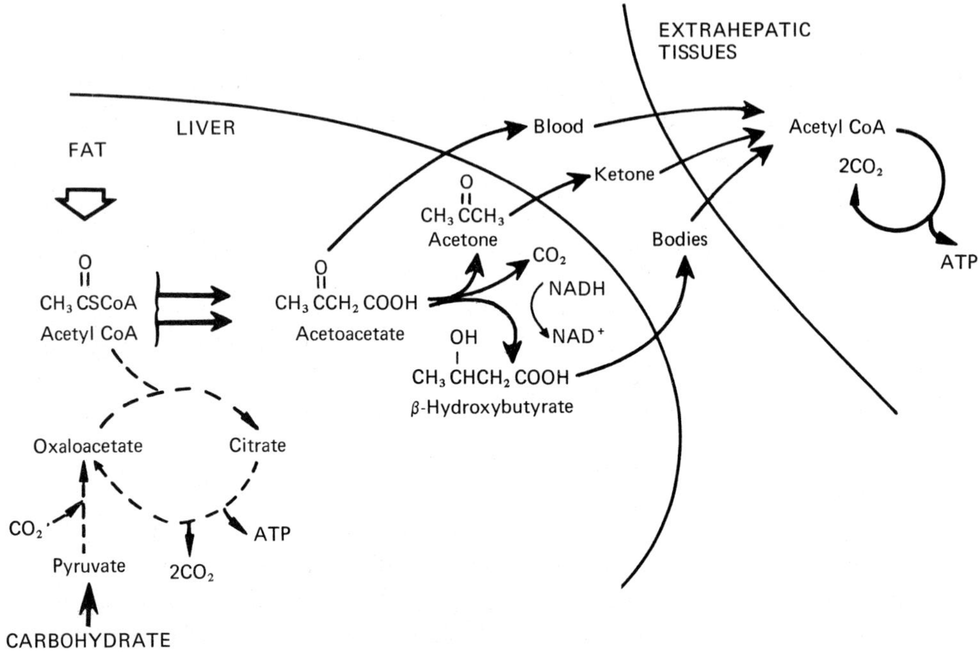

FIG. 5-11. Production of ketone bodies in the liver and their utilization by other tissues. Excess ketone-body formation (ketosis) results when fatty-acid catabolism predominates over carbohydrate catabolism. Note that acetoacetate is formed as shown in Fig. 5-14 (*i.e.*, by way of HMG CoA as an intermediate.)

that the amounts of the malic enzyme present in liver tissue are increased under dietary circumstances in which hepatic fatty-acid synthesis is increased (*e.g.*, in high-carbohydrate or low-fat diets). This is but one example of a general metabolic adaptation in both the liver and adipose tissue to the requirements for fat synthesis in relation to body nutrition.

The controls of the citric-acid cycle are of importance chiefly in the provision of citrate, which serves not only as the primary source of acetyl CoA, but also as an important allosteric regulator of fatty-acid synthesis. A high energy reserve within the mitochondrion is necessary before citrate may accumulate in sufficient amounts to pass out into the cytosol. Elevated concentrations of ATP will block the isocitrate dehydrogenase system by allosteric inhibition, thus causing a backup of tricarboxylic acids within the mitochondria. After migration to the extramitochondrial space, the citrate may be cleaved to acetyl CoA and oxaloacetate; the ATP-requiring citrate-cleavage enzyme is also subject to external control, since its activity is diminished in starvation and diabetes, whereas its synthesis is increased by high-carbohydrate or low-fat diets.

Another role of citrate and isocitrate is the powerful modulating effect of both of these tricarboxylic acids upon acetyl-CoA carboxylase, a key

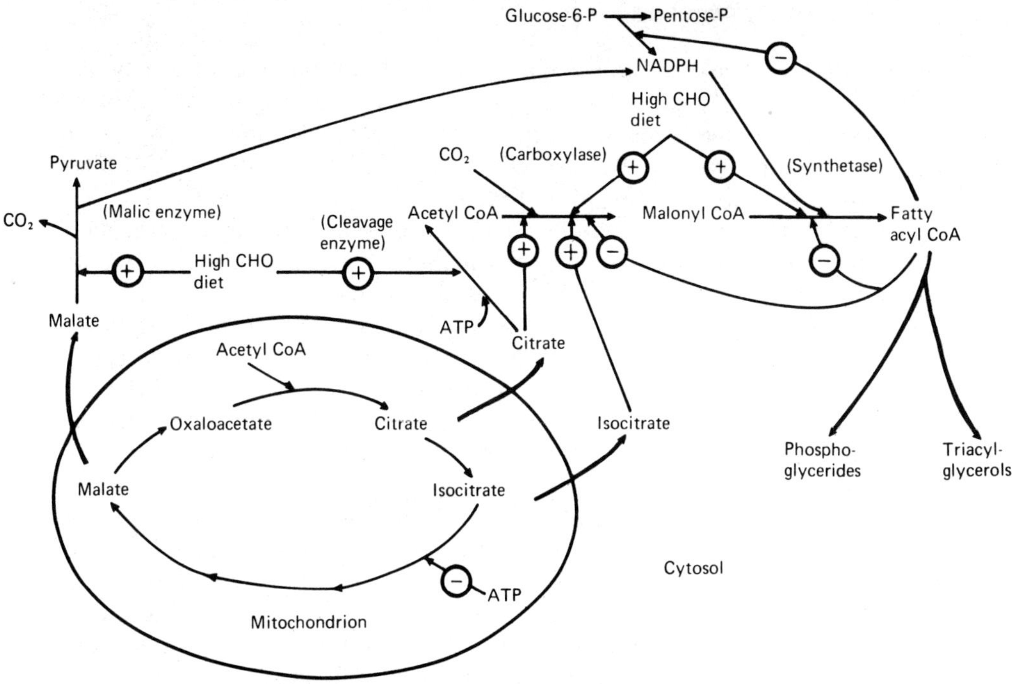

FIG. 5-12. Regulatory factors influencing fatty-acid biosynthesis in the liver. Stimulatory factors are depicted by circled plus signs, with the arrow indicating the reaction stimulated; negative modifiers are depicted by circled minus signs, with the arrow indicating the reaction inhibited. (*CHO*, carbohydrate.)

regulatory enzyme in fatty-acid synthesis. Elevated concentrations of citrate or isocitrate act to produce positive allosteric modification of the enzyme by increasing V_m, its maximal reaction velocity.

In these two ways, therefore, citrate from the mitochondria promotes malonyl-CoA production in the cytoplasm by providing the acetyl-CoA substrate in optimal amounts for carboxylation and by stimulating the activity of the carboxylase. The syntheses of the acetyl-CoA carboxylase and the fatty-acid synthetase may be under the same type of control as is the citrate-cleavage enzyme, since the activities of these lipogenic enzymes are enhanced in the liver following high-carbohydrate or low-fat dietary intakes. A final level of control is provided by the end-product inhibition of both acetyl-CoA carboxylase and the fatty-acid synthetase complex due to elevated levels of fatty-acyl CoA. Thus, an accumulation of the products of fatty-acid synthesis will automatically shut down their own formation; any excesses of long-chain acyl CoA in the liver under conditions of energy sufficiency will be diverted into fats and phospholipids for export to the fat depots and other extrahepatic tissues.

CHOLESTEROL METABOLISM

The ***sterols*** and their derivatives are essential components of the body. Unlike the triacylglycerols, which serve simply as energy stores, cholesterol and its fatty-acyl esters are important structural components of cell membranes and give rise to derivatives—such as bile salts, steroid hormones, and vitamin D—that play diverse and important functional roles in the body. The control of cholesterol metabolism is of key significance in the maintenance of a healthy cardiovascular system, and cholesterol imbalances contribute to the symptoms of atherosclerosis (as will be discussed further under blood lipids). Much of the biosynthetic and catabolic regulation that governs the turnover of the body pool of cholesterol takes place in the liver, the biliary system, the intestinal mucosa, and the interconnecting system known as the enterohepatic circulation. Accordingly, it is most convenient to consider cholesterol metabolism in this integrated context (Fig. 5-13).

As noted in the section dealing with lipid absorption, small amounts of cholesterol may enter the bile-salt micelles in the intestine and thence become absorbed by mucosal cells to enter the circulation in chylomicra. In the human gut, the amount of cholesterol that may be absorbed in this manner is severely limited (the maximum is about 0.5 g/day), and excess cholesterol in the diet is excreted in the feces. The cholesterol in chylomicra remnants is largely taken up in the liver, where it may either be esterified with fatty-acyl CoA to form the cholesterol esters of the liver cell membranes, or be converted to the bile salts. The latter pathway is probably the major catabolic route in terms of the amount of sterol turned over by the body. Moreover, the production of bile salts by the liver is under strict negative-feedback control. The bile salts returned from the enterohepatic circulation act as inhibitors of the enzymes in the liver that are involved in the oxidation of cholesterol to bile salts. It follows that an increase in the excretion of bile salts in the feces will, by preventing normal feedback inhibition, result in an increase in the catabolism of cholesterol by oxidation and hence to a decrease in the body's cholesterol pool. This excretion of bile salts may be accomplished, for example, by the ingestion of *cholestyramine,* an agent that forms a nonreabsorbable complex with the bile salts. The complex is eliminated, thus lowering the serum cholesterol concentration.

The liver and the intestinal mucosa are major sites for *de novo* biosynthesis of cholesterol from acetyl CoA (Fig. 5-14). The pathway occurs in the cytoplasmic compartment outside the mitochondria. The important reactions in sterol biogenesis are the first steps; these distinguish the system from other lipid metabolic sequences and play the chief role in regulatory mechanisms. As in the fatty-acid synthetic system, the first step is the formation of the acetoacetyl group, but this occurs by a reversal of the thiolase reaction. In the next stage, a third molecule of acetyl CoA is

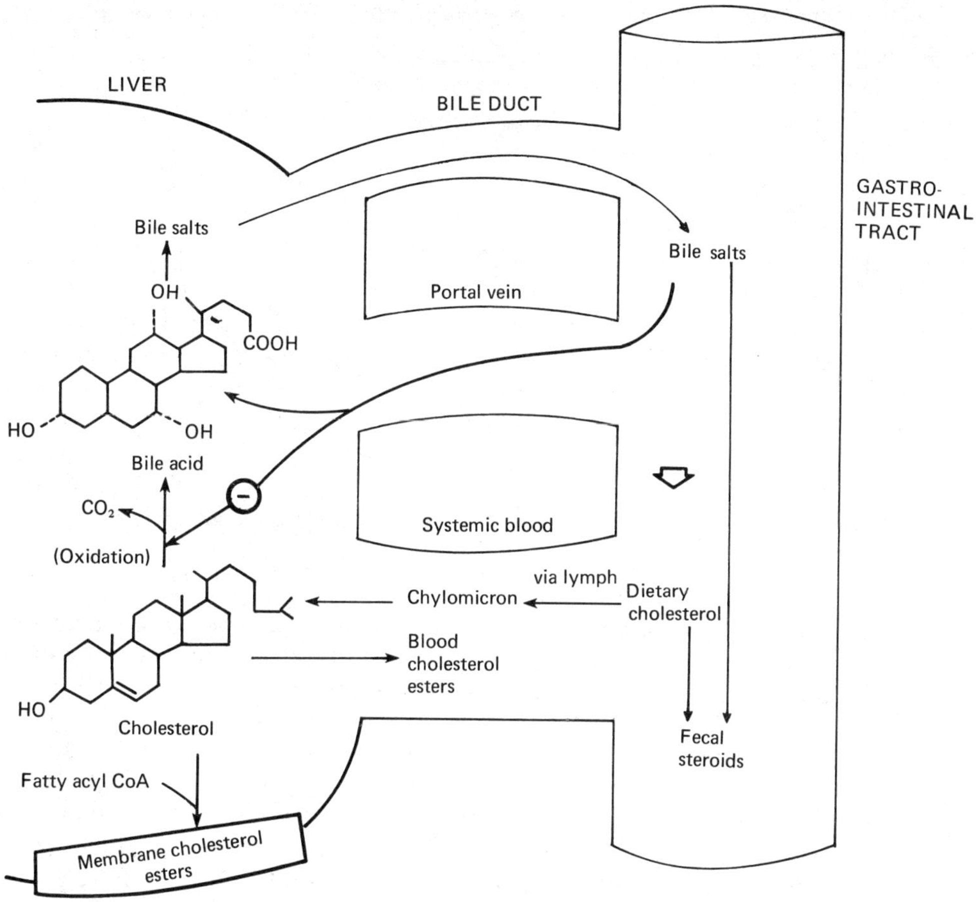

FIG. 5-13. Role of the liver in cholesterol and bile-salt metabolism.

condensed to form the important six-carbon intermediate, β-hydroxy-β-methylglutaryl CoA (HMG-CoA). In the liver, the latter molecule is also the precursor of free acetoacetic acid by the action of *HMG-CoA lyase* in the mitochondria. The key reaction in sterol synthesis appears to be the next one, in which HMG-CoA is reduced by two molecules of NADPH to mevalonic acid (3,5-dihydroxy-3-methylvaleric acid). The reaction catalyzed by the endoplasmic reticulum enzyme *HMG-CoA reductase* is irreversible and is the first committed step in the long series of reactions that lead to the building of the cholesterol ring structure. A number of energy-requiring steps lead to the condensation of six isoprenoid 5C intermediates, which are derived from the decarboxylation of mevalonic acid, to form the 30C compound squalene, which becomes cyclized to the sterol ring.

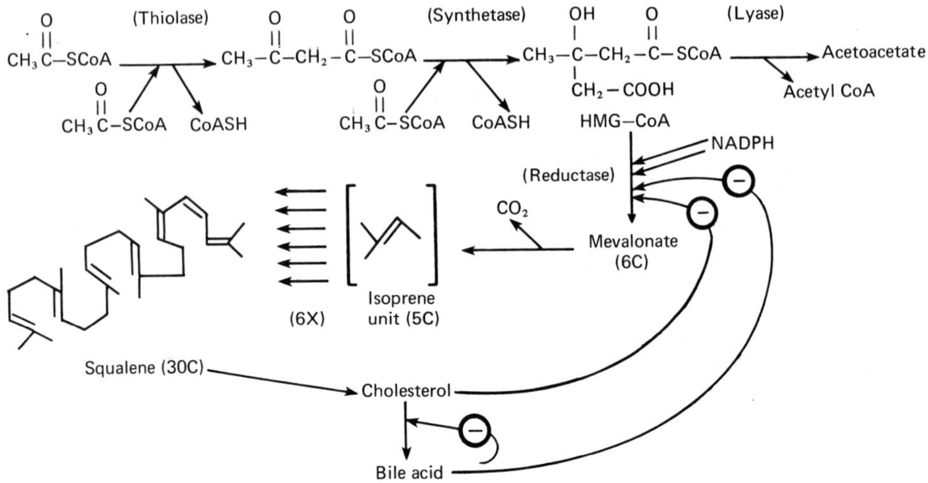

FIG. 5-14. Cholesterol biosynthesis. Negative feedback controls are indicated by arrows with circled minus signs.

The biosynthesis of cholesterol, as well as its oxidation to bile acids, is subject to negative feedback control by end-products. The main site of regulation seems to be at the committed reaction catalysed by the HMG-CoA reductase (Fig. 5-14). The feedback is mediated by uptake of serum cholesterol, chiefly as the lipoprotein fraction designated Low Density Lipoprotein (LDL), by endocytosis into cells. The released intracellular cholesterol then inactivates the reductase, shutting down sterol synthesis. Other sterols, such as the hydroxylated derivatives formed in bile-acid synthesis and other oxidation products of cholesterol, are even more potent in lowering the HMG-CoA reductase activity and may prove to be the actual intracellular modulators of feedback. We also know that HMG-CoA reductase is subject to inactivation following its phosphorylation by a kinase that is stimulated by cyclic AMP. The physiological significance of the phosphorylation–dephosphorylation process in long-term control of cholesterol synthesis, however, is not clear.

Recently there has been much interest in the development of safe cholesterol-lowering drugs that would selectively inhibit sterol synthesis in the body. One such hypocholesterolemic agent that has been used extensively in clinical investigations is p-chlorophenoxyisobutyrate (clofibrate). Clofibrate causes a lowering of serum cholesterol levels, and although the mechanism is unclear, it reduces the activity of HMG-CoA reductase, possibly either by inactivating the enzyme or enhancing its degradation in the liver. Unfortunately, clofibrate produces toxic reactions, including liver enlargement (hepatomegaly), and the mortality rate is reported to be high in patients treated with the drug for extended periods. Compactin (6-demethyl-mevinolin) is a potent competitive inhibitor of HMG-CoA reductase. Compactin, when administered in the diet, blocks cholesterol synthesis and decreases cholesterol levels in the serum. However, since mevalonate, the product of HMG-CoA reductase, is required for synthesis of coenzyme Q, dolichol, and other important isoprenoid compounds, the hypocholesterolemic effects of compactin may be vitiated by its harmful side reactions.

Cholesterol produced in the liver may be exported in two ways: (1) a portion can pass into the bloodstream, where it is esterified at the expense of phosphatidylcholine, and (2) a portion may be excreted together with bile-salt micelles in the bile fluid. An important potential consequence of this latter excretory pathway is the pathologic deposition of excess cholesterol accumulations in the form of bile stones that may lead to the obstruction of the bile duct.

PHOSPHOLIPID METABOLISM

The ***phospholipids,*** like the sterols, are essential components of the cellular membranes of the liver and other tissues. The basic structure of biologic membranes consists of phospholipid molecules arranged side by side in layers, similar to the layering of bile salts in a micelle (Chap. 1). Embedded at intervals throughout the thin sandwiches formed by these associations of amphipathic phospholipids are the occasional molecules of sterols and the functional protein molecules of the membranes. The proteins may exist only on one side of the membrane, either on the interior or exterior surface (as discussed in Chap. 3 in connection with the asymmetry of mitochondrial membranes), or they may penetrate the hydrophobic interior of the membrane, as in the case of the translocase proteins that are involved in moving hydrophilic molecules across the lipid-bilayer barrier from one aqueous compartment of the cell to another. The main function of the phospholipids, therefore, is to provide the basic organizing features of membranes as permeability barriers, as two-dimensional scaffolds for cellular and organellar forms, and as the hydrophobic milieu for membrane-bound enzymes, such as those involved in electron transfer, energy transduction, and the maintenance of chemical concentration gradients.

The biosynthesis of the phospholipids in the liver serves three main purposes: (1) to provide for the renewal and adjustment of the structural phospholipids in its own membrane structures, (2) to release the phospholipids that are circulated to other tissues via the plasma lipoproteins, and (3) to provide diacylglycerols for the biosynthesis of fats within the liver. In considering these different functions, it is simplest to discuss the last point first.

The liver has no source of monoacylglycerol for fat synthesis and no enzymes for the acylation of the ester that are analogous to those found in the intestinal mucosal cells. Rather, the diacylglycerol in the liver must

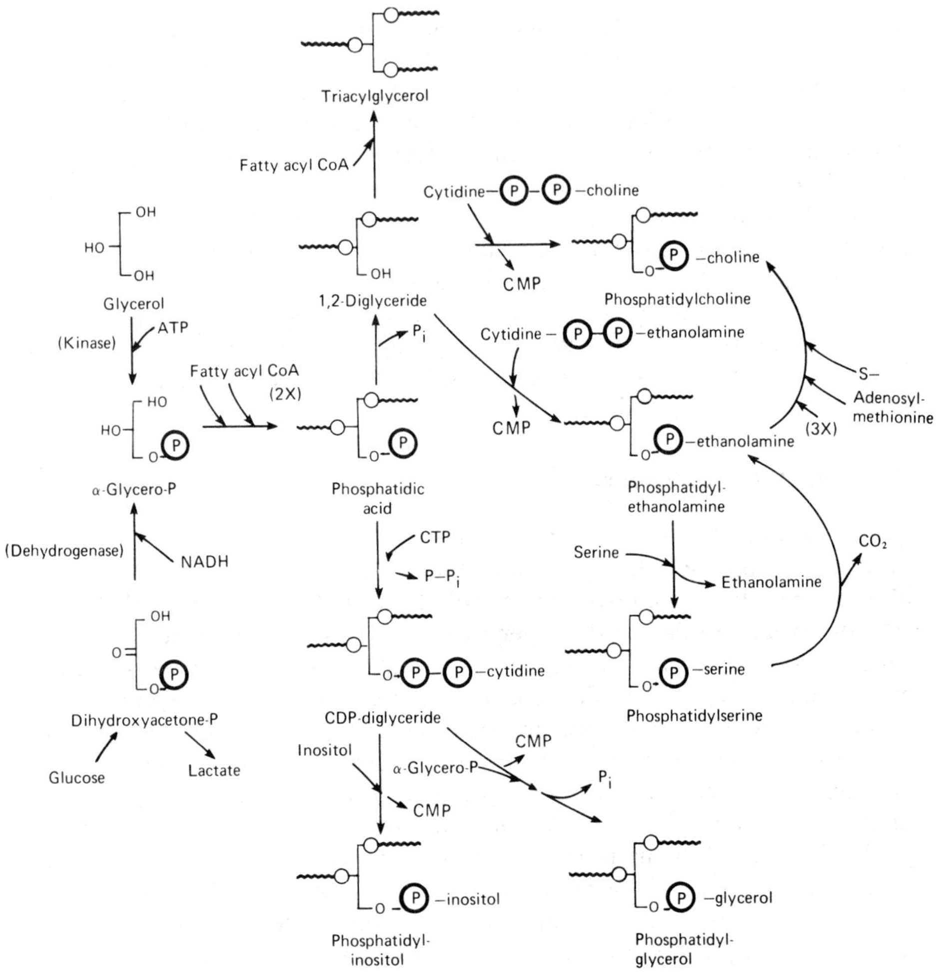

FIG. 5-15. Biosynthetic reactions of phospholipid metabolism.

originate from the phospholipid, *phosphatidic acid*. The latter, a central intermediate in both fat and phospholipid metabolism, may be derived from two sources: either from glycerol or from glycolysis (Fig. 5-15). Glycerol is presented to the liver following the actions of extracellular lipase upon the fats of chylomicra, and of intracellular lipase upon adipose tissue fats. In order to be reutilized, the glycerol must be phosphorylated by a specific kinase (at the expense of ATP) to produce α-glycerophosphate. The reactions of glycolysis can also produce α-glycerophosphate by the reduction of the triose-phosphate intermediate, dihydroxyacetone phosphate, by a specific dehydrogenase (at the expense of NADH). The α-glycerophosphate produced by either *glycerol kinase* or *α-glycerophosphate dehy-*

drogenase may then be esterified with two moles of fatty-acyl CoA to form phosphatidic acid, or diacylglycerophosphate.

Phosphatidic acid is itself only a minor component of cells, but it serves as a common intermediate of all the other phospholipids, as well as a precursor of fats, by undergoing dephosphorylation to produce diacylglycerols. In phosphatidylcholine formation, for example, the diacylglycerol moiety is combined with an activated phosphocholine group from the nucleotide, *cytidine diphosphate-choline* (CDP-choline). This reaction is analogous to the activation of glucose or galactose by combination with uridine diphosphate (UDP), and just as uridine forms the nucleotide coenzymes of carbohydrate metabolism, so does cytidine participate as the partner in lipid biosynthesis. An identical reaction transfers the phosphoethanolamine group from CDP-ethanolamine to diacylglycerol to form phosphatidylethanolamine. The latter may also be converted to phosphatidylcholine in the liver by the transfer of three activated methyl groups, as discussed further in Chapter 6. Together, the components containing choline and ethanolamine constitute over three fourths of all the phospholipids in the liver as well as in most other tissues.

The minor phospholipids include those with inositol, glycerol, or serine attached to the phosphatidyl group. In both phosphatidylinositol and phosphatidylglycerol, the phosphatidyl group is activated by reacting phosphatidic acid with CTP, and the product—CDP-diglyceride—then transfers the phosphatidyl group intact to the inositol or glycerol molecule. Parallel systems for the participation of either of the above cytidine-activated pathways in phosphatidylserine biosynthesis have not, however, been demonstrated in higher evolutionary forms. In the mammalian liver, the regeneration of the serine moiety in phospholipids is mostly accomplished by exchange displacement from phosphatidylethanolamine. Phosphatidylserine can also be decarboxylated to phosphatidylethanolamine, thus completing the sequence of conversions in the liver:

$$\text{Serine} \xrightarrow{\nearrow CO_2} \text{Ethanolamine} \xrightarrow{\searrow 3CH_3} \text{Choline}$$

where all the reactants are combined to the phosphatidyl group (see Chap. 6).

LIPOPROTEIN FORMATION

Virtually all the phospholipids in the liver or other tissues exist in the membranes, with very few of them occurring in the form of solutions or micelles. When they do occur in the soluble phase of cells or body fluids, phospholipids are invariably associated with special protein transporters known as ***lipoproteins.*** This phospholipid-carrier protein complex is an important vehicle for the transfer of phospholipids from their major sites of biosynthesis in the endoplasmic reticulum to the mitochondria, the Golgi apparatus, the plasma membrane, and other membranes and organelles within the cell. The lipoproteins are also involved in the extracellular movement of phospholipids from the liver to other tissues of the body. As noted earlier, the chylomicra contain cholesterol and triacylglycerols in addition to the phospholipids associated with proteins. The albumin of the blood plasma also forms lipid-protein complexes with free fatty acids and other lipid-soluble substances in the blood.

The formation of lipoproteins is a major synthetic function of the liver, and disturbances in the biogenesis of the protein portions will have severe imbalancing effects upon lipid metabolism in the whole body. An inability to carry normal amounts of phospholipids and cholesterol in the blood is seen in *Tangier disease,* which results from a congenital defect in the formation of a transport protein by the liver. A deficiency in the liver's capacity for albumin synthesis, which causes *analbuminemia,* leads to a decrease in the plasma's capacity to carry the fatty acids released from fat depots. Moreover, since the major mechanism for transferring the fats synthesized in hepatic cells from the liver requires their combination with a specific protein for transport in the blood, any interference with the production of this protein may lead to an intracellular accumulation of triacylglycerols. This is probably the cause of the fatty livers that are seen in patients with chronic dietary protein deficiency (as noted earlier) or in patients treated with excessive amounts of the protein-synthesis inhibiting antibiotic, tetracycline.

ADIPOSE TISSUE

For an average human in a healthy state of nutrition and fitness, fat accounts for about 15% of the body mass, with wide variations depending on age, physique, and sexual differences. The distribution of this lipid storage form is not as widespread in different cell types throughout the body as are glycogen stores. Rather, it is concentrated in particular loci: in parts of the peritoneum such as the omentum and mesentery, in layers under the skin, and in the bone marrow. These specific fat depots constitute a highly differentiated tissue, *adipose tissue,* which contains specialized cells of the reticuloendothelial system known as *adipocytes* or fat cells. The function of such cells is specialized for the purpose of fat retention. It should not be concluded from this, however, that the fat cells are metabolically inert; quite the opposite is true. The cells of adipose tissue have a capacity for fatty-acid synthesis that is comparable to that of the liver in times of plenty as well as active catabolic systems for the release of fatty acids to serve the needs of other tissues in times of want. Nor should it be assumed that fat deposition in the adipose tissues serves only as an energy reserve. The fatty deposits have other important roles to play in (1) insulating the deep tissues from excessive cold and heat; (2) cushioning the bones, tissues, and deep viscera from knocks and shocks; and (3) last but not least modulating the linear planes and angles of the body's frame into those gentle rotundities that we associate with esthetic and sexual appeal.

The adipose cell, when fully charged, consists of a thin rim of cytoplasm surrounding a large lipid droplet. As much as 90% of the adipose tissue mass may be constituted of lipids and of the latter, 99% is triacylglycerol. The fats in the depots are somewhat more highly saturated than those in the liver, since adipose tissue lacks the desaturating system to produce monounsaturated fatty acids. The product of adipose lipogenesis is therefore chiefly palmitic acid. Nonetheless, there is a high degree of unsaturation in the depot fat of humans on a random diet; over half of the fatty acids may be oleic (18:1) or linoleic (18:2). This proportion will modify slowly toward the composition of the dietary lipids, particularly with respect to the polyunsaturated fatty-acid content. The amounts of oleic acid (18:1), which is furnished principally by liver synthesis, will be sustained in the adipose tissue at high enough levels to keep the stored fat in the liquid state in order to compensate for alterations in the diet or the external temperature.

LIPOGENESIS

The formation of fatty acids *de novo* and their incorporation into the fats of adipose tissue utilize similar enzymes and pathways to those found in the liver. The fat cell, however, is not as abundantly supplied with cytoplasm or mitochondria as the hepatic cell, and thus the process of lipogenesis in adipose tissue relies much more heavily upon the provision of its precursors, intermediates, and coenzymes. In particular, the associated reactions of carbohydrate metabolism that lead to the generation of acetyl CoA, ATP, NADPH, and α-glycerophosphate will become rate-limiting for fat synthesis. This provides the rationale for the observation that it is not so much the amount of fat as the amount of carbohydrate in the average diet that determines the amount of lipids laid down in the depots.

The utilization of glucose in fat cells proceeds partly by the pentose–phosphate pathway to yield NADPH for the reductive reactions of lipogenesis (Fig. 5-16). The bulk of the glucose metabolized in adipose tissue (more than 80%), however, passes through the glycolytic pathway and the citric-acid cycle to furnish ATP and the mitochondrial citrate necessary for acetyl-CoA production in the cytosol. At least half of the NADPH required must be produced by the malic enzyme, which is highly active in adipose cells. In addition, the reactions of glycolysis are particularly important for fat synthesis in adipose tissue, since the latter, unlike the liver, has little capacity to utilize glycerol. Fat cells have very low glycerokinase activity, and they therefore must derive virtually all the α-glycerophosphate required for phosphatidic-acid synthesis by the reduction of dihydroxyacetone phosphate (produced during glycolysis) with α-glycerophosphate dehydrogenase.

In addition to the synthesis of fat from blood glucose, the adipose tissue can utilize circulating fatty acids. These reach the fat depots as triacylglycerols either in chylomicra or in lipoproteins synthesized by the liver. This is how the combined influences of dietary content and liver metabolism may modify the fatty-acid composition of depot fats. The fat cells contain an extracellular *lipoprotein lipase,* which releases free fatty acids from blood lipoproteins. The fatty acids are taken up, but the glycerol cannot be utilized effectively by adipose tissue and must return to the liver or other tissues for further metabolism. Thus, whether it uses endogenously generated fatty acids or those provided by the diet or liver synthesis, the fat cell requires a concomitant intake of glucose to provide the α-glycerophosphate for fat synthesis. This synthesis proceeds by the dephosphorylation of phosphatidic acid and the acylation of the resulting diacylglycerol, as in the liver. The fat cell, however, lacks the ability to form lipoproteins and hence cannot export its fats as such to the exterior and into the circulation; all the triacylglycerol formed in adipose tissue is secreted instead into the storage droplet at the interior surface of the globular cells.

ENZYME ADAPTATION AND DIET

Mention has been made of the interdependence of fat intake and fat deposition. Moreover, it should be apparent from the preceding discussions that the carbohydrate intake of the diet will grossly affect adipose fat synthesis. This is in part due simply to accelerated fat synthesis in the liver as well as because of the provision of more glucose substrate to drive lipogenesis in the fat cells themselves. The latter factor is exemplified by one of the actions of insulin, namely, the stimulation of glucose uptake and utilization by the fat cells, which thereby promotes a great acceleration of lipogenesis in the depots. High caloric intake and especially high-carbohydrate diets therefore promote fat synthesis by mass action (Fig. 5-16).

Another type of adaptation to the diet is parallel to that described previously for liver metabolism, namely, the adaptive changes in enzymatic activity due to allosteric modification and in the enzyme content by alterations of protein synthesis (Fig. 5-16). In the first category is the activation of acetyl-CoA carboxylase, which, as in the liver, is stimulated by elevated concentrations of citrate in the fat cells, thus ensuring that fat synthesis will proceed maximally only when there is abundant glucose in the cells for metabolism. In the second category is the regulation of the synthesis of the key enzymes of glucose metabolism and lipogenesis. In fasting, for example, the lipoprotein lipase of adipose tissue is decreased,

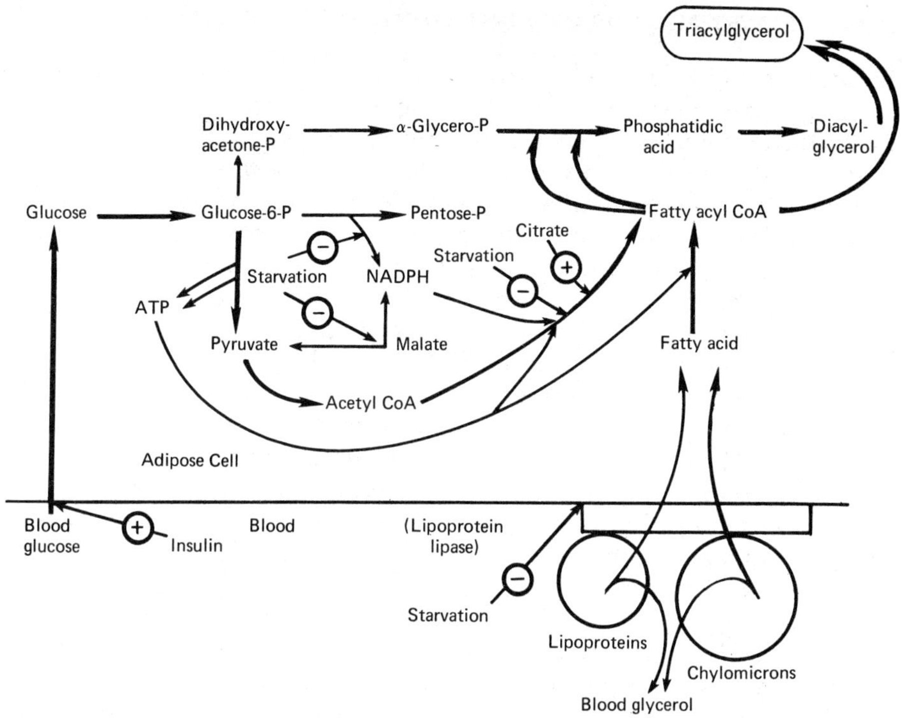

FIG. 5-16. Lipogenesis in adipose tissue and regulation of fatty-acid biosynthesis.

thereby preventing the utilization of circulating lipoprotein triacylglycerols for fat storage and ensuring that they may be used instead for the increased energy demands of other tissues. Additional controls exist in starvation to decrease the generation of NADPH by the pentose–phosphate pathway and malic enzyme, to reduce acetyl-CoA production by reducing the amounts of citrate-cleavage enzyme, and to reduce acetyl-CoA utilization by decreasing the levels of both acetyl-CoA carboxylase and fatty-acid synthetase in adipose tissue. These selective blockages of enzyme synthesis in adipose tissue conserve the glucose reserves of the body for energy-requiring processes but are reversed within days following the resumption of a normal caloric intake. High-carbohydrate or low-fat intakes may further enhance the synthesis of these lipogenic enzymes in order to provide accelerated fat production in the depots. In the fed state insulin will act to divert glucose for lipogenesis. The combination of elevated blood glucose and hyperinsulinism drive the utilization of carbohydrate for acetyl-CoA production. In addition to furnishing more substrate for pyruvate generation in fat cells, insulin will trigger the dephosphorylation of pyruvate dehydrogenase in adipocyte mitochondria. This converts the enzyme to its active state, thus stimulating oxidative decarboxylation of pyruvate to produce more acetyl CoA for fatty-acid synthesis.

LIPOLYTIC SYSTEMS AND HORMONAL CONTROL

In addition to its content of extracellular lipase, the fat cell contains an internal lipase system that acts solely upon the storage fat droplet (Fig. 5-17). Intracellular lipolysis requires two separate enzymes: (1) a lipase that acts slowly upon only triacylglycerols and is stimulated by cyclic AMP, and (2) an enzyme that hydrolyzes the diacylglycerols completely to glycerol as fast as they are formed by the first lipase. Since the first enzyme is rate-limiting in the overall lipolysis of triacylglycerol and since the cyclic AMP is liberated by adenylate cyclase in response to activation by certain hormones, the whole process leading to the release of free fatty acids from adipose tissue is said to be due to the action of *hormone-sensitive lipase.* This enzyme, normally present in an inactive state, is activated upon phosphorylation by a protein kinase that is in turn stimulated by cyclic AMP. The fatty acids can be reutilized by fat cells and be reincorporated into fats; more usually, however, the hormonal signal leading to their release is associated with an increased requirement for metabolic energy by other cells. In the latter case, the free fatty acids from the depots enter the bloodstream where they travel to other parts of the body in the form of a complex with the plasma albumin. The glycerol produced cannot be reutilized by the fat cells to any extent and its release into the bloodstream is a useful indication of the rate of lipolysis.

The membrane of the fat cell has two types of hormone receptors: (1) those that interact with hormones such as the catecholamines—epinephrine or norepinephrine—which increase the production of cyclic AMP by adenylate cyclase and hence activate lipolysis, and (2) the insulin receptors that counteract the adenylate cyclase activation by these stimulatory hormones and inhibit lipolysis. The effect of insulin is exemplified by the metabolism in adipose tissue in diabetes mellitus. In this disease, the absence of insulin results in the unchecked stimulation of the intracellular lipase by lipolytic hormones and a marked mobilization of free fatty acids from the fat cells occurs. A similar phenomenon may contribute to the accentuated lipolysis seen in starvation since the lowered blood-glucose levels will suppress insulin production and its release by the pancreas.

Exposure to cold, emotional stress, and accelerated physical activity may also result in greatly increased fatty-acid release from the depots as a result of either the secretion of increased amounts of epinephrine from the adrenal medulla into the general circulation, or the release of norepinephrine from stimulation of the sympathetic nerves that innervate the adipose tissue directly. Other hormones, such as glucagon, adrenocorticotropic hormone (ACTH), and growth hormone, also serve to accelerate adenyl-cyclase activation and hence free-fatty-acid release from the adipose tissue. In all these instances, the mobilization of fat responds to stress, increased muscular action, or increased energy needs by producing

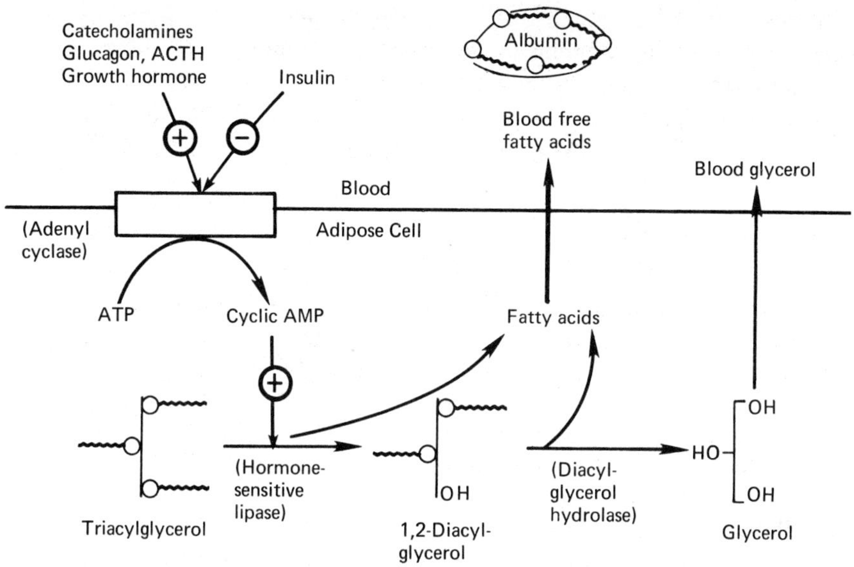

FIG. 5-17. Intracellular lipolysis in adipose tissue and its control by hormones.

large increases in the fatty-acid substances required for energy production in other tissues of the body.

The lipolytic response is of particular importance in a specialized type of fat depot referred to as *brown adipose tissue.* As opposed to normal, or white adipose tissue, which is deficient in cytoplasm, the brown adipose tissue cells are well-endowed with intracellular components, especially the mitochondia, the cytochromes of which confer the brown pigmentation. This fatty tissue acts as a heat-generating device and is particularly abundant in hibernating animals, in which it functions in the process of temperature elevation during arousal. The brown adipose tissue is also prevalent in newborn infants. In the newborn it invests and protects vital organs before other thermogenic systems in the body develop. The brown fat is well vascularized and supplied with sympathetic nerve fibers, and contains receptors that respond to circulating epinephrine or to neurogenic norepinephrine. Brown adipose tissue reacts to cold stress and catecholamines in two ways: (1) by enhanced lipolysis, and (2) by uncoupling of the oxidative phosphorylation that accompanies catabolism of the fatty acids released intracellularly by the hormone-sensitive lipase. The mitochondria of brown adipose tissue contain in their inner membranes a unique protein, thermogenin, which responds to elevated fatty-acid concentrations by equilibrating H^+ across the membrane, thus dissipating the proton gradient that would normally couple ATP formation to the β-oxidation of fatty acids. In this way the energy of respiration is released as heat instead of being captured as phosphorylation energy. This non-shivering form of thermogenesis, of greatest significance in the neonate, may also be important in later life in maintaining energy balance. There are indications that brown adipose tissue persists into adulthood, and could play a role in the development of obesity (see Chap. 7).

THE MUSCLES

It is a common misconception that the muscles act exclusively to catabolize carbohydrate for their energy needs. To be sure, there is some justification for this belief on the basis of the high glycogen content found in skeletal muscle and its rapid breakdown during exercise, as well as that provided by the capacity of muscle to function anaerobically in glycolytic reactions. Such metabolic events, however, are generally associated with vigorous contractions; as noted earlier the major source of resting muscle energy is from the oxidation of fatty acids (see Chap. 3). The cardiac muscle and the smooth muscles of blood-vessel walls probably utilize fats and their metabolites, the ketone bodies, to a greater degree than they do glucose, even during maximal activity. The sustained activity of skeletal muscles at moderate levels requires both carbohydrate and fat catabolism, with the latter predominating as the duration of exercise increases. This is notable in long-distance runners, whose leg muscles take up and oxidize progressively more fatty acids as their glycogen reserves become depleted. It should be recalled that the release of catecholamines, which will activate muscle glycogen utilization, will also promote lipolysis in the fat depots and hence will activate β oxidation in the muscle as a result of the consequent increase in fatty-acid substrates in the circulation.

In addition to utilizing fatty acids as such, the muscles have the enzymes necessary for the catabolism of the ketone bodies that arise from incomplete oxidation of fats in the liver (Fig. 5-18). The compound β-hydroxybutyrate, for example, may be oxidized by a specific dehydrogenase, which is highly active in the mitochondria of skeletal or heart muscle, to produce NADH and acetoacetate. The latter, together with the acetoacetate taken into the muscle from the circulation, may be activated by a mitochondrial enzyme that is absent from the liver but present in the muscles and other extrahepatic tissues:

$$\text{Acetoacetate} + \text{Succinyl CoA} \rightarrow \text{Acetoacetyl CoA} + \text{Succinate}$$

The acetoacetyl CoA may then be cleaved by thiolase to yield acetyl CoA for further energy production by oxidation in the citric-acid cycle. The heart muscle in particular derives a large amount of energy from the catabolism of acetoacetate by means of this pathway.

All the reactions described above are catabolic. This is characteristic of lipid metabolism in the muscles, which, as in the case of carbohydrate

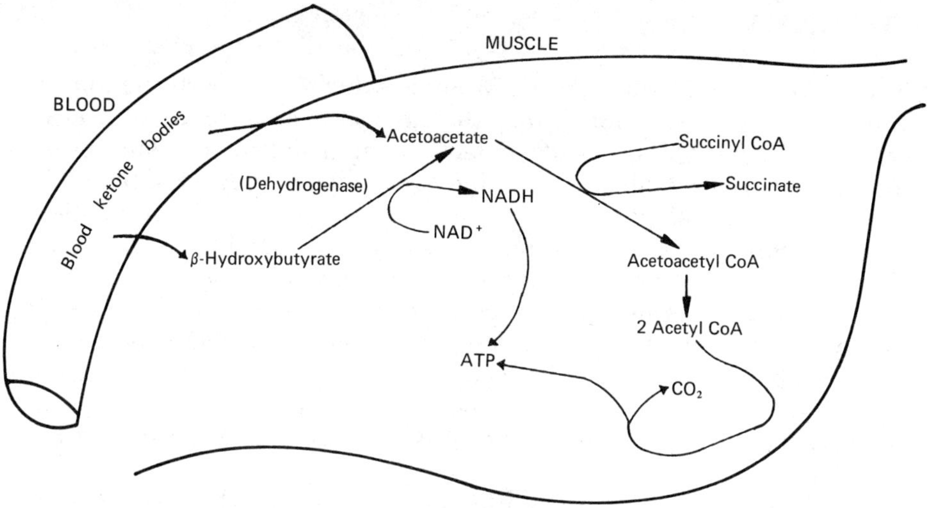

FIG. 5-18. Oxidation of ketone bodies in muscle.

metabolism, carry out little in the way of anabolic reactions of the lipids. Although fat storage may occur in the areas adjacent to muscle tissue, there is no capacity for fatty-acid or triacylglycerol syntheses in the muscle cells themselves.

ROLE OF CARNITINE

Although fatty-acid oxidation is stimulated by carnitine in most tissues, the muscle mitochondria appear to exhibit a much greater dependence upon the carnitine-mediated transfer reaction than do those of other tissues such as the liver. This is substantiated by two factors: (1) the concentration of carnitine in muscle tissue is very high, in keeping with its physiologic role; (2) in muscle, the mitochondrial GTP-coupled activation of fatty acids that is found in the liver is absent, and the production of fatty-acyl CoA occurs exclusively in the cytosol through an ATP-requiring reaction:

$$\text{Fatty acid} + \text{ATP} + \text{CoASH} \rightarrow \text{Fatty-acyl CoA} + \text{AMP} + \text{P–P}_i$$

Since the fatty-acyl CoA is generated outside the mitochondrion and since its inner membrane is impermeable to coenzyme A or any of its derivatives, no utilization of fatty acids by the muscle mitochondrion can occur unless carnitine is present. As discussed previously, fatty-acyl CoA reacts reversibly with carnitine to produce the freely permeable fatty-acyl carnitine compound, which may be transported into the mitochondrial interior where β oxidation can occur (see Chap. 3).

THE BRAIN

Neural tissue is particularly rich in lipids, which may constitute almost half of its total mass. While it may therefore seem appropriate to refer to certain individuals as fat-heads, in fact the brain and nerves contain very little triacylglycerol. Most of the lipids are complex lipids: the phospholipids; the lipids that contain the long-chain amino alcohol, sphingosine, or ***sphingolipids;*** and cholesterol, which is always found free rather than in the esterified state found in most other tissues. All these lipids may be synthesized in the brain from glucose and other small molecules provided in the bloodstream. Thus, the brain has a moderately high capacity for fatty-acid biosynthesis, but all the fatty acids synthesized are used to form phospholipids or sphingolipids rather than storage fats (Fig. 5-19). An active capacity for cholesterol biogenesis from acetyl CoA is also found in the brain.

Brain and neural mitochondria are virtually inert, however, insofar as the β oxidation of fatty acids is concerned, and all the acetyl CoA required for energy production in the citric-acid cycle, for sterol synthesis, and for other important synthetic reactions, such as the formation of the neurotransmitter acetylcholine, must be derived from glucose. The brain possesses some capacity to derive energy from the ketone bodies by the pathways discussed previously for muscle. In adult neural tissue, the rate of utilization of these substances is normally quite low relative to that of glucose catabolism, but during starvation and in the brain of newborn infants the oxidation of β-hydroxybutyrate in brain will be markedly increased and may yield an important alternative source of energy or of acetyl groups for lipogenesis during development when the blood glucose concentrations become too low.

LIPIDS AND MYELINATION

The lipids of neural tissue fall into two broad categories: (1) gray-matter lipids, which compose the neuronal membranes and (2) white-matter lipids, which make up the multilayered casing of the ***myelin*** sheath. Although most of the components of gray matter are similar to those of membranes in many other tissues, the myelin constituents are characteristic of this specific structure. The typical lipids of myelin include cholesterol, most of the sphingolipids, and some phospholipids. The brain of the fetus contains very little myelin; in the early neonatal period, however, there is a burst of synthesis of the myelin lipids that is coincident with the visible deposition of myelin around the nerves. Once the myelin has been laid down, these lipids remain quite stable throughout life. The adult brain, therefore, has a much lower rate of lipid biosynthesis than the infantile one.

In structure, the myelin sheath resembles many layers of cell membrane wrapped around one another. This is related to the method of pro-

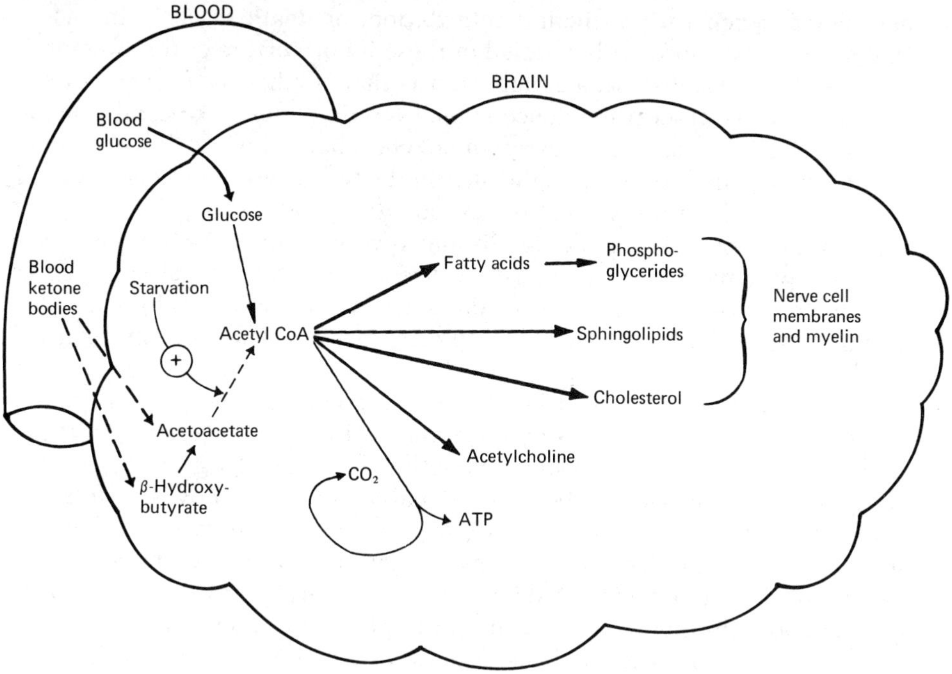

FIG. 5-19. Metabolism of lipids and lipid derivatives in the brain.

duction of the sheath, which results from a supporting cell, or glial cell, wrapping itself around the nerve axon to produce many turns of its own membrane in a roll. The unique lipids present in this membrane include the characteristic phospholipids known as *ethanolamine plasmalogens* (in which the fatty-acyl group of position 1 in phosphatidylethanolamine is replaced by an unsaturated, long-chain, ether-linked group) and the sphingolipid, *cerebroside* (in which galactose and fatty acid are combined with sphingosine).

The other sphingolipids that are present in neural tissues, as well as in other body cells, include *sphingomyelin* (which contains phosphorylcholine and fatty acids) and the *gangliosides* (which contain galactose, glucose, fatty acids, and the carbohydrate derivative, sialic acid). Sphingomyelin is found in both white and gray matter; it is a constituent of cell membranes in general as well as of myelin. The gangliosides are absent from myelin but appear to be specific components of the cell membranes in the neurons of gray matter.

LIPIDOSES AFFECTING THE NERVOUS SYSTEM

A number of genetically determined diseases have been described in which excessive accumulations of lipids occur in the nervous system. Generally, this pathologic deposition of lipid, or ***lipidosis*** as it is called, leads to neurologic degeneration, mental retardation, or death in early infancy. One of the most thoroughly studied of these inborn errors of lipid metabolism is *Tay-Sachs disease,* a gangliosidosis that is inherited as a recessive trait and has its highest incidence among Jewish people. The condition is characterized by the deficiency of an enzyme that is involved in the normal degradation of certain gangliosides in the brain, with the consequence that these lipids build up in the neurons and produce cerebral impairment, blindness, and early death. Similar involvement of the brain is seen in infant patients with *Niemann-Pick disease,* or sphingomyelinosis, and *Gaucher's disease,* or glucocerebrosidosis. In these disorders, sphingolipids accumulate in the brain and other tissues due to a deficiency of normal catabolic enzymes.

All these metabolic disorders have severe neuropathic sequelae, and no cures or preventive measures are known for the irreversible lesions that occur in the nervous system. Nonetheless, a knowledge of the metabolic defect has proved to be most valuable since it is now possible to detect, by chemical or enzymatic studies, the heterozygous state of those suspected to be carriers of the recessive genes. In this way, prospective parents or those having affected homozygous offspring may receive proper genetic counseling with respect to family planning. Alternatively, chemical and enzymatic studies of suspected homozygotes may be carried out by amniocentesis—that is, investigation of fetal cells from the amniotic fluid—in pregnant mothers with family histories of these inherited disorders. In these cases, the certain diagnosis *in utero* of such a destructive, debilitating disease provides compelling moral and medical arguments for the termination of the pregnancy by abortion.

BLOOD LIPIDS

As in the case of the blood sugar pool, the lipid balance in the circulation will be maintained by the competing processes of tissue uptake versus the input from either the diet or mobilization from the liver and fat depots. Blood sugar, however, consists of a single entity, soluble glucose, whereas the blood lipids are heterogeneous complexes—all bound to various plasma protein fractions—of almost equal amounts of triacylglycerols, phospholipids, and cholesterol, with lesser amounts of unesterified fatty acids. Moreover, in contrast to the blood glucose level, the normal fasting level of total blood lipids of about 500 mg/100 ml may be subject to much wider fluctuations (±200 mg/100 ml) and will return much more slowly to baseline levels after a meal.

Since probably no absolute tissue demands for blood lipids exist that are similar to those for blood glucose, the consequences of lowered lipid levels are insignificant in comparison with the effects of hypoglycemia. As will be noted in a later section, the pathology of the disorders in blood lipid metabolism tends to be related, rather, to excessive concentrations *(hyperlipemia).*

LIPOPROTEIN CLASSES AND COMPOSITION

The distinction between different classes of blood lipoproteins is based upon their relative densities. The *chylomicra,* for example, are lighter than all the lipoproteins and will float to the surface of serum upon standing. The composition of the lipoproteins is reflected in their densities; thus, the lightest fractions contain the most triacylglycerols, whereas the denser fractions have a progressively greater protein content relative to the lipids (Fig. 5-20). *Chylomicra* are composed mostly of fats, with a thin skin of protein. *Very-low-density lipoproteins (VLDLs)* have a high fat content but contain more cholesterol and phospholipids. *Low-density lipoproteins (LDLs)* have the highest content of cholesterol; *High-density lipoproteins (HDLs)* have the largest amounts of phospholipids. *Very-high density lipoproteins (VHDLs)* have a very low relative lipid content and include fatty acids complexed to albumin and some of the other lipoproteins after they have lost most of their bound lipids. With the exception of the albumin complexes, all the lipoproteins are categorized as ***globulins.*** When subjected to electrophoresis, the lipoproteins move at varying speeds toward the anode, except for the chylomicra which remain stationary. The albumin moves the farthest, followed by the α-globulins (HDLs) and the β-globulins (LDLs). The VLDLs migrate between the α and β regions and thus are sometimes denoted as the *pre-beta lipoproteins.* VHDLs other than albumin are not detected on electrophoresis.

TRANSPORT MECHANISMS

Each of the lipoprotein classes plays a distinct function in the transport of lipids in the blood. As noted in the previous discussion of lipid absorption, the chylomicra are produced in the gastrointestinal mucosa and carry the fats that are absorbed from the diet. Although most of this lipid is carried from the gut to the adipose tissue, any tissue may utilize the fatty acids from chylomicron fats if it possesses an active lipoprotein-lipase enzyme system.

The function of the VLDL fraction is also to carry triacylglycerols to the tissues, but in this instance, it transports exclusively those fats that are synthesized endogenously by the liver. The fatty acids from these triacylglycerols will therefore arise jointly from the dietary fats processed by the liver and from hepatic fatty-acid synthesis, chain elongation, and desaturation. The major recipient of these fats will be the adipose tissue. After removal of most of the triacylglycerol by lipoprotein lipase action the VLDLs are converted in blood to LDLs.

The LDL fraction carries the bulk of the plasma cholesterol to the

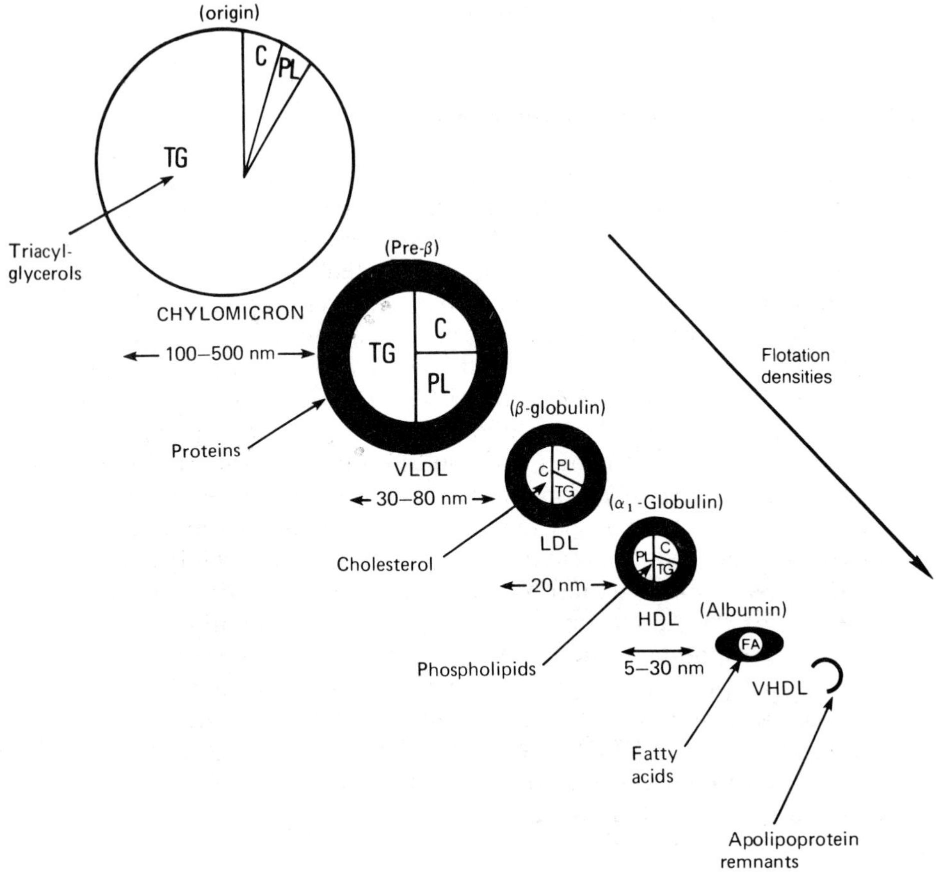

FIG. 5-20. Classes of blood lipoproteins. Different lipoprotein types are depicted according to their relative sizes, content of protein (dark outer rim), and the constituent lipids (*TG,* triacylglycerol; *C,* cholesterol; *PL,* phospholipids; *FA,* fatty acids) in the inner sectors. The electrophoretic designation relative to serum proteins is given in parentheses at the top for each fraction. (*VLDL,* very-low-density lipoproteins; *LDL,* low-density lipoproteins; *HDL,* high-density lipoproteins; *VHDL,* very-high-density lipoproteins.)

peripheral tissues, most of which is present as cholesterol esters. The lipoprotein complex formed in the liver contains free cholesterol, but it is esterified after its release into the circulation by a plasma enzyme—*lecithin-cholesterol acyl transferase*—that transfers an unsaturated fatty-acid group from position 2 of phosphatidylcholine (*i.e.,* lecithin) from the HDL fraction:

Cholesterol + Lecithin → Cholesterol ester + Lysolecithin

The other product, lysophosphatidylcholine or lysolecithin, has a single fatty acid esterified in position 1. Enzymes in the red blood cell and many other tissues have the ability to reconvert the lysolecithin to lecithin at the expense of fatty-acyl CoA.

The HDL fraction, in addition to facilitating the acylation of cholesterol, also serves to transport cholesterol back from peripheral tissues to the liver for excretion in the bile.

Both the cholesterol and the phospholipids of the LDL and HDL (α-lipoprotein) fractions are exchangeable with their counterparts in the membranes of the blood cells as well as other cell membranes that they may contact. The protein moieties are not exchangeable, however, and after the removal of their lipids, they are conveyed back to the liver in the VHDL fraction for reutilization.

The VHDL fraction contains the albumin-bound fatty acids. The latter represents the transport form of lipids mobilized from the fat depots by action of the hormone-sensitive lipase of adipose tissue. Although the total quantity of fatty acid in this fraction is small (10 mg/100ml–20 mg/100 ml) and represents only about 5% of the total fatty acids present in the plasma, the rapid utilization and liberation of the albumin-bound fatty acids result in an extremely fast turnover of this metabolically active pool. In the fasting state, as much as 70% of the total energy requirement of the body may derive from this fatty-acid fraction, with the total circulatory pool turning over completely every few minutes.

LIPID DEPOSITION IN ATHEROSCLEROSIS

This disease and its attendant complications of heart failure, cerebrovascular strokes, and hypertension are the principal causes of death in industrialized societies. Atherosclerosis is more prevalent among males than females and is associated with numerous genetic and environmental factors, such as hypertension, diabetes, smoking, lack of exercise, gross obesity, and elevated blood cholesterol. The last two features and the nature of the chemical lesion indicate that atherosclerosis may be viewed primarily as a disorder of lipid metabolism, and extensive investigation as well as curative and preventive measures have been based upon this rationale.

The lesion consists of the deposition of lipids, chiefly composed of cholesterol and its esters, in plaques or *atheromas* within the intimal lining of the arteries. The deposits are present even early in infancy in most American males, and the extent and severity of plaque formation is progressive with age. Eventually, the lumen may become occluded to the extent that the blood flow is impaired, which results in inadequate oxygen supplies to the brain or other vital organs, and backpressure on the major arteries that leads to high blood pressure and congestive heart failure. Alternatively, blood platelets and other cellular debris may accumulate around the damaged area, further obstructing key arteries such as those supplying the heart, kidneys, or brain, or the clotted material may break away to produce an embolism at some distant narrowing of the circulation.

Central to all these end results is the presence of lipid deposits in the early stages of plaque formation and the associated elevated concentrations of lipid, particularly cholesterol, in the bloodstream of persons in populations with a high incidence of atherosclerosis. In reducing the risk factors that may precipitate the high blood levels of cholesterol and associated lipids, it is reasonable to reduce the dietary intake of animal tissues that contain high quantities of this sterol, although the rate of cholesterol uptake from the human intestinal tract is quite limited. Additional factors in favor of utilizing plant rather than animal fat in the diet include the antagonism of cholesterol absorption by the plant sterol, β-sitosterol; other substances are also found in plants, including plant proteins and roughage, that appear to lower blood cholesterol levels. Among the more important factors are the much higher level of polyunsaturated fatty acids in plant tissue, and the abundance of saturated fatty acids in animal fats. The correlation of high blood cholesterol levels with a high ratio of saturated-to-polyunsaturated fatty acids in the diet and the higher incidence of cardiovascular disease among affluent meat eaters of Western civilization versus vegetarians have provided the rationale for the suggested control of dietary lipids and the promotion of polyunsaturated foods.

In addition to the dietary control of lipids, other measures that should not be overlooked include the administration of hormones or their analogs (both estrogens and thyroxine help to lower blood cholesterol levels), vitamins (nicotinic acid and pyridoxine have shown ameliorative effects), and drugs that interfere with cholesterol absorption (β-sitosterol), or with bile-salt reabsorption (cholestyramine). Drugs that inhibit enzymes of cholesterol biosynthesis in the liver (clofibrate, compactin) also lower serum cholesterol levels, but may have serious toxic side-effects. All these agents, either alone or in combination, as well as unidentified plant factors currently under investigation, have been used in the management of the atherosclerotic patient in attempts to redress the imbalance of blood cholesterol metabolism.

THE HYPERLIPIDEMIAS

Because of the broad range of normal values, it is difficult to define exactly an upper limit above which elevated blood lipids or *hyperlipidemia* would be said to be present. Several distinct disorders have been characterized in which individual lipids or lipoproteins are present in excess, and some of these, in which the hyperlipidemia is primary, are known to be inherited or inborn errors of metabolism. The first, designated as *type I hyper-*

lipoproteinemia, is associated with a congenital absence of lipoprotein lipase. The characteristic features include the presence of elevated levels of chylomicra, even during a fast, and the enormous increases of triacylglycerol concentrations noted in the blood after meals. The disease is frequently termed *hyperchylomicronemia* as a consequence. There is little increase in the blood cholesterol or phospholipid levels, and there appears to be no increased risk of atherosclerosis. Papular swellings and eruptions associated with fatty deposits in the skin *(xanthomas),* distension of the spleen and liver, and pancreatic involvement leading to acute abdominal pain are the most usual clinical findings. Successful management is generally obtained by restricting fat and caloric intake.

Type II hyperlipoproteinemia is clinically distinguished by an abundance of noneruptive xanthomas, not only in the skin, but extensively throughout the tendons and blood vessels. The vascular involvement will lead to a high incidence of atherosclerosis, even in early life. Chemically, the condition is characterized by normal triacylglycerol levels but gross elevations of serum cholesterol and a specific increase in the LDL fraction. Cells of these patients lack surface receptors for the LDL, with the result that uptake of cholesterol in peripheral tissues is impaired, and cholesterol synthesis is unchecked because the normal feedback inhibition is lacking. Therapy is directed toward lowering the cholesterol levels by the dietary and drug treatment methods described in the preceding section.

The condition classified as *type III hyperlipoproteinemia* exhibits variable increases of both serum cholesterol and triacylglycerol levels, as well as variable symptoms of xanthoma formation and atherosclerosis. It often appears with less severity and later in life than the type-II syndrome.

The increased triacylglycerol level in the blood of patients with *type IV hyperlipoproteinemia* is associated with the VLDL fraction and is enhanced by high carbohydrate intakes. It appears that in this disorder, the hepatic synthesis of fats from carbohydrates exceeds their rate of removal from the blood, and the condition is sometimes referred to as *endogenous* or *carbohydrate-induced hyperlipidemia.* The disorder is associated with obesity, vascular disease, and a family history of diabetes. Low-calorie, low-carbohydrate diets are the usual form of treatment.

In addition to the above inherited lipoprotein disorders, a number of secondary hyperlipidemias may arise from environmental factors or diseases. Excessive caloric intake alone may lead to exogenous or endogenous increases of blood triacylglycerol concentrations, depending on whether high-fat or high-carbohydrate diets are ingested. Ethanol, which serves as a good substrate for hepatic fatty-acid synthesis, may produce an endogenous increase of triacylglycerols in the blood when ingested in excessive amounts. Since the absence of insulin causes the increased mobilization of depot fat, diabetic patients may show an increased concentration of blood triacylglycerols and phospholipids, which are formed by the excess free fatty acids reaching the liver.

SUGGESTED READINGS

Bell RM, Coleman RA: Enzymes of glycerolipid synthesis in eukaryotes. Annu Rev Biochem 49:459–487, 1980

Bloch K, Vance D: Control mechanisms in the synthesis of saturated fatty acids. Annu Rev Biochem 46:263–298, 1977

Brady RO: Sphingolipidoses. Annu Rev Biochem 47:687–713, 1978

Danielsson HH, Sjövall, J: Bile acid metabolism. Annu Rev Biochem 44:233–253, 1975

Food and Nutrition Board, National Research Council, National Academy of Sciences: Toward healthful diets. Nutr Today 15, No. 3: 7–11, 1980

Fredrickson DS, Goldstein JL, Brown MS: The familial hyperlipoproteinemias. In Stanbury JB, Wyngaarden JB, Fredrickson DS (eds): The Metabolic Basis of Inherited Disease, 4th ed, pp 604–655. New York, McGraw-Hill, 1978

Goldstein JL, Brown MS: The LDL receptor locus and the genetics of familial hypercholesterolemia. Annu Rev Genet 13:259–289, 1979

Herbert PN, Gotto AM, Fredrickson DS: Familial lipoprotein deficiency (abetalipoproteinemia, hypobetalipoproteinemia, and Tangier disease). In Stanbury JB, Wyngaarden JB, Fredrickson DS (eds): The Metabolic Basis of Inherited Disease, pp 544–588. New York, McGraw-Hill, 1978

Hales CN, Luzio JP, Siddle K: Hormonal control of adipose-tissue lipolysis. Biochem Soc Symp 43:97–135, 1978

Havel RJ (ed): Symposium on Lipid Disorders. Med Clin N Am 66:317–553, 1982

Havel RJ, Kane JP: Therapy of hyperlipidemic states. Annu Rev Med 33:417–433, 1982

Jeffcoat R: The biosynthesis of unsaturated fatty acids and its control in mammalian liver. Essays Biochem 15:1–36, 1979

Lane MD, Moss J, Polakis SE: Acetyl coenzyme A carboxylase. Curr Top Cell Regul 8:139–195, 1974

Masoro EJ: Lipids and lipid metabolism. Annu Rev Physiol 39:301–321, 1977

McGarry JD, Foster DW: Regulation of hepatic fatty acid oxidation and ketone body production. Annu Rev Biochem 49:395–420, 1980

Miettinen TA: Effects of hypolipidemic drugs on bile acid metabolism in man. Adv Lipid Res 18:65–97, 1981

Nilsson–Ehle P, Garfinkel AS, Schotz MC: Lipolytic enzymes and plasma lipoprotein metabolism. Annu Rev Biochem 49:667–693, 1980

Oliver MF: Diet and coronary heart disease. Br Med Bull 37:49–58, 1981

Robinson AM, Williamson DH: Physiological roles of ketone bodies as substrates and signals in mammalian tissues. Physiol Rev 60:143–187, 1980

Romsos DR, Leveille GA: Effect of diet on activity of enzymes involved in fatty acid and cholesterol synthesis. Adv Lipid Res 12:97–146, 1974

Rumsby MG, Grang AJ: The myelin sheath—a structural examination. In Poste G, Nicolson GL (eds): The Synthesis, Assembly and Turnover of Cell Surface Components, pp 247–362. Amsterdam, North–Holland Publishers, 1977

Scanu AM, Landsberger FR (eds): Lipoprotein structure. Ann NY Acad Sci 348:1–436, 1980

Schroepfer GJ Jr: Sterol biosynthesis. Annu Rev Biochem 50:585–621, 1981; 51:555–585, 1982

Sémériva M, Desnuelle P: Pancreatic lipase and colipase. An example of heterogeneous biocatalysis. Adv Enzymol 48:319–378, 1979

Simons LA, Gibson JC: Lipids: A Clinician's Guide. Baltimore, University Park Press, 1980

Sodhi HA, Kudchodkar BJ, Mason DT: Cholesterol metabolism in clinical hyperlipidemias. Adv Lipid Res 17:107–153, 1980

Soutar AK, Myant NB: Plasma lipoproteins. In Offord RE (ed): Chemistry of Macromolecules IIB, Macromolecular Complexes, pp 55–119. Baltimore, University Park Press, 1979

Williamson DH, Whitelaw E: Physiological aspects of the regulation of ketogenesis. Biochem Soc Symp 43:137–161, 1978

Chapter Six Nitrogen Metabolism

Nitrogen-containing substances in the body are many and varied in their form and function. Although the carbohydrates and lipids act principally as the metabolic fuels for energy production and contribute secondarily to cell structure, it is the polymeric nitrogenous compounds that are responsible for the major attributes of living systems. The essential mechanisms for the operation and regulation of all our pathways of metabolism are provided by the structural proteins; the enzymes and other functional proteins such as myosin, hemoglobin, the cytochromes, and the lipoproteins; and the nucleic acids of the gene and the protein-synthesizing systems. Moreover, each of the monomers of these macromolecules has an independent metabolic life. Of no less importance than the nucleic acids are their constituent nucleotides—ATP, UTP, CTP, and GTP—which all act as key intermediates in metabolic reactions, and frequently as important regulatory signals either *per se,* or as their cyclic nucleotides. Each of the individual amino acids that compose the proteins may also contribute to carbohydrate synthesis via gluconeogenesis, to lipogenesis or energy production through catabolism to acetyl CoA, or, by special anabolic pathways, to the formation of other important nitrogen compounds, such as the catecholamines, the thyroid hormones, creatine, the heme porphyrin ring, and the nitrogen bases of the nucleotides and phospholipids. Many nitrogenous metabolites such as ammonia, amino acid or porphyrin derivatives can exert toxic influences if present in body fluids in excess, and hence there has been much interest in the inborn errors and other disorders of nitrogen metabolism. Recently attention has been focused upon the abilities of many small, seemingly simple oligopeptides, such as the endorphins and hormone-releasing peptides, to promote powerful controlling actions in the body.

DIETARY NITROGEN

The bulk of utilizable nitrogen in the diet comes from proteins. Several considerations are important in relation to dietary nitrogen: (1) The total amount of protein ingested, (2) the digestibility of the protein, (3) the amino-acid composition of the diet, and (4) the total caloric intake. The requirement for protein varies considerably with the individual's stage of development and state of health (*e.g.*, recuperation, pregnancy, lactation, and so forth). The average, normal adult, for example, maintains a healthy state on approximately 0.7 g protein/day/kg body weight, but the requirements of newborn infants will be over 2.0 g/day/kg, decreasing to about 1.0 g/day/kg by 5 years of age. Additional protein supplements of approximately 20% above the standard intake are recommended for pregnant or lactating females.

The *biologic value* of proteins in the diet refers to both their degree of utilization following digestion and their amino-acid composition. A number of factors—including pretreatment, heating, and so forth—will alter protein digestibility, but some proteins (*e.g.*, skin and hair keratins) are nonutilizable as such. In terms of the quality of proteins with respect to amino-acid content, the value of their source is the opposite of that for the lipids and their fatty-acid content: in general, plant proteins are less desirable than those from animals as sources of complete amino-acid mixtures. In the absence of an adequate intake of carbohydrates and lipids, the demands upon protein catabolism will increase to meet the energy requirements. Conversely, in the presence of a superabundance of protein with adequate amounts of other nutrients, the excess will be catabolized to contribute to the adipose-tissue stores and eventually obesity will result.

It should be noted that there is no requirement specifically for nucleic acids in the diet. The purine and pyrimidine bases are formed in the body from amino-acid precursors, and excesses from the dietary intake of these substances are broken down and excreted.

Protein deficiency is probably the major nutritional problem in the world today. Within our own society, families near the poverty line may have marginal protein intakes even though they may have adequate caloric intakes. In general, high-protein food has come to mean high-priced food, and this has injected an aura of socioeconomic discrimination into the study and remedy of protein deficiencies. Many recent surveys have documented evidence of the reduction in prenatal and perinatal complications of child development when protein supplements have been given to underprivileged pregnant and lactating mothers. Undoubtedly, many pediatric and geriatric health problems could be reduced by adequate programs to assist disadvantaged peoples at these two extremes within our culture.

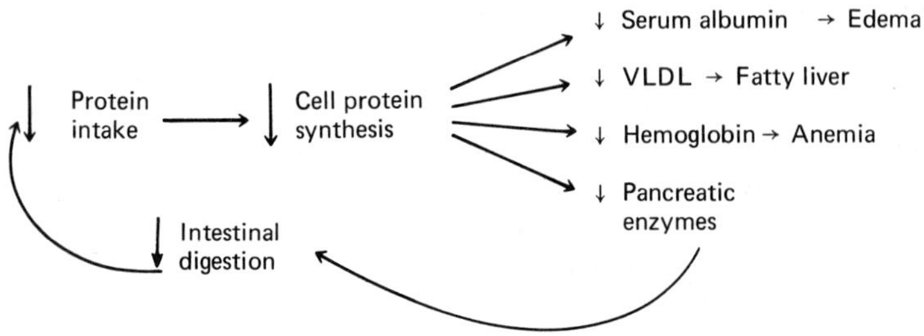

FIG. 6-1. The vicious cycle of kwashiorkor. Reduced protein intake results in impaired intestinal function and inefficient utilization of the small amounts of protein in the diet. *(VLDL,* very-low-density lipoproteins.)

In other lands, particularly the overworked agricultural regions of tropical Africa, Latin America, and Asia, the overall deprivation of food is particularly acute. This results in the syndrome of combined caloric and protein deficiency termed *kwashiorkor* (Fig. 6-1). The family members most affected are newly weaned infants, who cannot derive sufficient protein from their starchy diet to maintain normal growth. The failure of normal protein synthesis in the liver causes decreases in serum albumin with resulting edema, and decreases in the pre-beta lipoproteins or very-low-density lipoproteins (VLDL) result in the development of a fatty liver. Synthesis of hemoglobin is reduced, resulting in anemia. The entire intestinal function is severely affected, since there are inadequate amino-acid precursors for the synthesis of the pancreatic enzymes or for the renewal of intestinal mucosal cells. The resulting acute diarrhea, lowered resistance to infection, and tendency to chronic liver disease lead to a very high rate of mortality among such children. The supplementation of the diet with high-quality animal protein, such as powdered milk, or appropriately balanced vegetable protein mixtures, is essential to prevent the development of irreversible damage in these growing children.

ESSENTIAL AMINO ACIDS

Some amino acids may be synthesized in the body; others cannot and must be provided in the diet (Fig. 6-2). The latter are termed *indispensible* or ***essential amino acids,*** a misnomer, actually, since all are really essential for body functions. The so-called ***nonessential amino acids*** may replace one another in the diet since they are interconvertible; alternatively, nonessential amino acids may be formed from carbohydrate or lipid intermediates provided that an adequate source of total nitrogen is present. For the essential amino acids, however, no such metabolic routes exist, except in plants and bacteria. Consequently, humans require a continual,

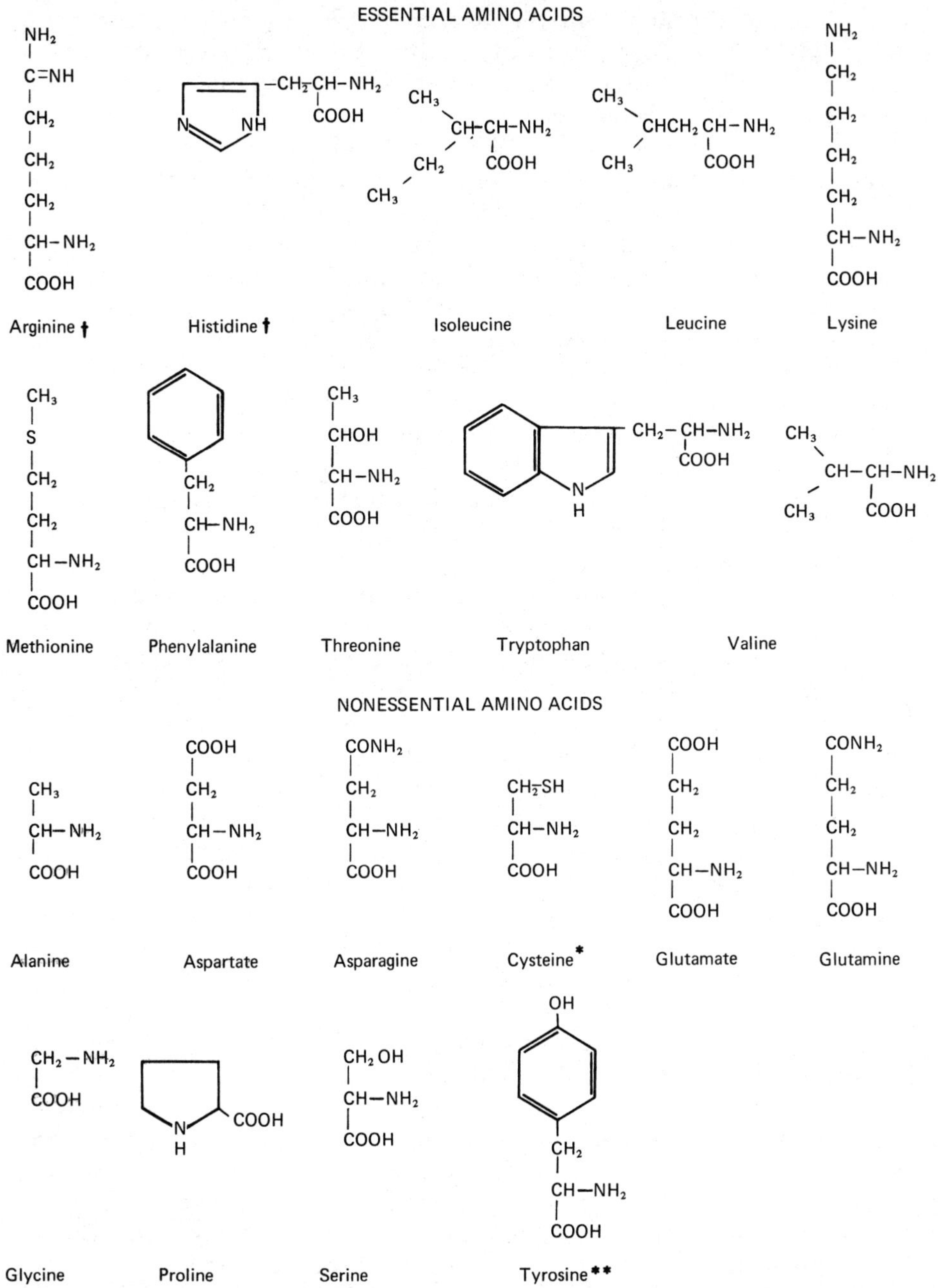

FIG. 6-2. Classification of the amino acids in proteins as essential or nonessential dietary components. (Arginine and histidine are synthesized slowly by the body and are required for normal function; cysteine is synthesized from methionine and tyrosine from phenylalanine.)

daily input of a balanced mixture of each of these amino acids to survive. An imbalance (*i.e.*, an increase in one amino acid relative to the others) or their ingestion at different meals rather than simultaneously can also lead to dietary insufficiencies.

Normal body growth demands the adequate ingestion of the branched-chain amino acids *leucine, isoleucine,* and *valine;* the aromatic amino acids *phenylalanine* and *tryptophan;* the aliphatic amino acids *threonine* and *lysine;* and the sulfur-containing amino acid *methionine.* Since phenylalanine is converted in the body into the nonessential, aromatic amino acid *tyrosine* and methionine is also metabolized to the nonessential, sulfur-containing amino acid *cysteine,* the inclusion of adequate quantities of these two "nonessential" components will greatly reduce or spare the requirements for their corresponding "essential" precursors. The basic amino acids, *arginine* and *histidine,* occupy a less definite niche in protein nutrition. Thus, low arginine diets, though adequate to sustain growth and body maintenance, may lead to defective spermatogenesis; low histidine diets promote eczema and reduced growth in infants and interfere with hemoglobin synthesis.

The reason that most plant proteins are inadequate is their lower content of certain essential amino acids in comparison with that of other sources, such as beef, whole milk, or whole egg. The last has often been considered as an ideal protein source, although this must be weighed against the high cholesterol content of the yolk and the inefficiency of egg production for mass nutrition. Milk and meat products obviously suffer the same limitations regarding efficiency of production in areas of limited agricultural productivity, and the associated increased ingestion of highly saturated animal fats must be weighed against their benefit as a protein source. Some animal proteins, particularly low-fat dairy products, will continue to provide good nutritional sources, but until the more affluent, producing countries develop more rational storage and distribution programs for dealing with surplus egg, milk, or meat products, rather than the current deliberate disposal by spoilage and wastage, the bulk of the world's people must depend, as they do now, upon cereal proteins supplemented by other vegetable sources.

It is not difficult to provide vegetarian diets that are adequate in protein, provided that no single grain or vegetable forms the sole or predominant intake. Thus, corn is severely deficient in both tryptophan and lysine, whereas beans are deficient in methionine; however, a balanced mixture of corn with soy meal or legume protein, if eaten simultaneously, will constitute a "good" protein source, since the interlocking deficiencies will offset one another. This is the rationale for the production and distribution of vegetable food mixtures such as Incaparina, a combination of corn, sorghum, and cottonseed, which has been used in deprived areas of Latin America. Another hopeful development is the application of plant genetics. The "opaque 2" mutants of corn have been introduced, which have a greater content of lysine, and the hybrid of rye and wheat, triticale, combines the ruggedness of the former with the high yield of the latter parent. Certain lines of triticale have a much higher total protein content and a lysine content equivalent to that of the high-lysine corn.

NITROGEN BALANCE

This measure is an indication of the state of protein anabolism or catabolism in the body. *Positive nitrogen balance* refers to the situation where the dietary intake exceeds the amount of nitrogen lost from the body by excretion, exfoliation, and so forth. Under these circumstances—which are seen during childhood growth, pregnancy, and recovery from illness—net synthesis of body protein and thus retention of nitrogen occur. In conditions of caloric inadequacy, as in prolonged fasting, the body proteins will be broken down to provide for energy demands and gluconeogenesis; hence, less nitrogen may be taken in than that which is excreted. This state is termed *negative nitrogen balance,* and it will also be observed in wasting diseases or when dietary protein is inadequate in quantity or quality. The omission of a single essential amino acid from the diet, for example, will result in negative nitrogen balance, since the presence of every natural amino acid is required to maintain the biosynthesis of proteins.

A healthy adult who ingests sufficient quantities of good quality protein will exist in a state of *nitrogen equilibrium,* where the amount of nitrogen intake is precisely balanced by the amount of nitrogen output. In a way, this equilibrium description satisfies our common-sense observation of nongrowing adults. The interpretation would be that the body uses only as much protein nitrogen from the diet as necessary to replace digestive enzymes, gastrointestinal (GI) cells lost in the feces, or any degenerated tissue components, such as skin cells or erythrocytes, that wear out during normal use. This "wear-and-tear" component of body protein replacement would be expected to be fairly small relative to the whole body mass. The bulk of the protein ingested is catabolized for energy production when the state of nitrogen equilibrium obtains.

Why, then, does the body have such high requirements specifically for protein as well as for essential amino acids? Why could these energy requirements not be met almost entirely by increased carbohydrate and lipid intakes? The answer to these questions requires a different concept of metabolism than is implied by the terms "equilibrium" or the replacement of components due to wear and tear.

THE DYNAMIC STATE OF BODY PROTEINS

Although the mature human body gives the appearance of being static in form during our day-to-day observations, this is in fact an illusion. Even when at rest, the cells and tissues are undergoing continuous exchanges of material and energy with their surroundings. We saw previously that "resting" nerve and muscle cells require energy expenditure to transport ions and metabolites and, thereby, to maintain the status quo of life. A substantial part of the material exchanged with the environment is related to this basal energy requirement. In addition, the actual machinery of metabolism—the enzymes, the formed elements of the cells, and the structural framework of tissues—all participate in an exchange of material with the world outside. This concept implies that all elements of the body cells are undergoing perpetual turnover—a process of degradation balanced by resynthesis—to produce a *steady-state* whereby the constant and equivalent comings and goings of molecules at any instant of stopped time produce the illusion that the process is at a standstill or *equilibrium* position. It is essential to make this distinction between the steady-state versus the equilibrium state. In fact, it is contradictory to speak of an equilibrium state in relation to life. The only time that a living organism approaches static equilibrium is when it dies; throughout its lifetime, the organism consumes materials and transforms energy states in order to avoid the inert state of equilibrium.

The steady-state of the body is exemplified by the daily turnover of our proteins and their constituent amino acids (Fig. 6-3). This has been termed the ***dynamic state*** of body proteins, and it applies as well to the state of the membrane phospholipids and other apparently static structural components of cells. When an amino acid, for example, is tagged by inserting an atom of heavy nitrogen (N^{15}), the labeled amino acid can be traced in its metabolic meanderings through the various tissues. Assume that you have eaten some of this tagged amino acid while you were in a state of nitrogen equilibrium: since your total N intake is balanced by your N output, you might anticipate that all the ^{15}N ingested would be excreted that day. Far from it, however; it would take many days for the labeled nitrogen to be eliminated from your body, and during that period, it would have traversed a metabolic odyssey of epic proportions.

Not surprisingly, you would find some of the label very rapidly incorporated into the cell structures and enzymes of the intestinal tract as well as into other expendable areas of the body. The plasma albumin and lipoproteins—which originate by syntheses in the liver—would also show a rapid uptake of the labeled amino acid from the intestines. The free amino-acid pool of the bloodstream would acquire the ^{15}N tag, and from thence it would permeate into all the tissue fluids. Newly formed members of the erythrocyte population would incorporate the ^{15}N amino acid into their hemoglobin and other proteins; the label would persist there for approximately 120 days, at which time that portion of the cells would age and die. Other organs in the body—the liver, kidneys, muscles, brain, and connective tissues, in that order—would gradually take up the amino-acid label, rapidly into some proteins and slowly into others. The liver en-

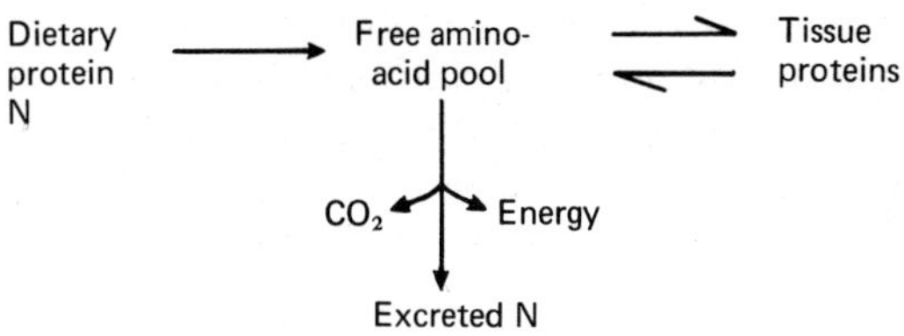

FIG. 6-3. The dynamic steady-state of body proteins. Although dietary nitrogen *(N)* is balanced by excreted nitrogen in the normal adult, a constant turnover and exchange of amino acids occurs within the body between tissue proteins and the free amino-acid pool of blood and other body fluids.

zymes, for example, that are inducible in response to physiologic changes are being rapidly resynthesized and degraded at all times, and thus they would quickly acquire the label and lose it rapidly, too. The proteins of collagen in the connective tissues, myosin in the muscles, or myelin in the mature brain would take up the amino acid very slightly, but the ^{15}N that was incorporated would persist for a year or more.

Mathematically, it is most convenient and precise to express these various rates of turnover in terms of the time required for one-half of all the molecules to be exchanged; this is termed the ***biologic half-life,*** or $T_{1/2}$, and its value will range from several minutes for inducible enzymes to several months for stable structural proteins. An average value for the body proteins as a whole yields a half-life of about three weeks; this implies that statistically, every three weeks one half of your protein components, on the average, will have been completely renewed by degradation and resynthesis. By calculation, it may be shown that the rate of total body protein synthesis in the state of nitrogen equilibrium is as high as 500 g/day, nearly five times the average dietary intake. Naturally, there must be efficient recycling of the amino-acid precursors and breakdown products of proteins to achieve this result, and it makes us appreciate, too, the nature of the truly immense exchange of body components with our environment.

DIGESTION AND ABSORPTION

Quantitatively, most of the nitrogen that enters the body of a person with a normal diet is in the form of protein. With few exceptions, the proteins do not directly enter the bloodstream from the gastrointestinal tract. Small amounts may be absorbed intact, and in some cases, these materials give rise to the allergic reactions of some individuals to certain foods. In nursing infants, the absorption of antibodies from the mother's milk is an important defense mechanism against infection, since the immune system is poorly developed in early life.

For the most part, however, the dietary proteins must be hydrolyzed to their constituent amino acids before absorption into the bloodstream may occur. Protein digestion is a complex, stepwise process that takes place in three stages: (1) in the stomach, (2) in the small-intestinal lumen, and (3) in the mucosal cells of the small intestine (Fig. 6-4). The first two stages involve the cleavage of the long polypeptide chains of proteins into short *oligopeptides.* The latter are absorbed into the mucosa, where they are cleaved into the free amino acids for passage into the portal bloodstream.

PROTEOLYSIS

This cleavage of proteins involves the hydrolysis of the peptide linkages by specific ***proteases,*** which act on large polypeptides, or ***peptidases,*** which act on short oligopeptides. As the proteins enter the digestive tract, they are first attacked by the protease, ***pepsin,*** in the stomach. This enzyme is produced by the cells of the gastric mucosa in an inactive zymogen form, ***pepsinogen;*** only when it reaches the acidic stomach contents is the zymogen converted to its active form, a process requiring the removal of part of the enzyme protein. Free pepsin acts autocatalytically to generate itself by hydrolyzing pepsinogen. Pepsin acts on dietary proteins selectively, attacking only at peptide carboxyl groups that bear aromatic-ring side-chains, and therefore may act at many points along the polypeptide chains.

The partly digested proteins next enter the neutralized small-intestinal contents, where they are attacked further by the pancreatic proteases, *trypsin* and *chymotrypsin.* As in the case of pepsin, both of these proteolytic enzymes are produced as inactive zymogens; this is to protect the secretory cells from being hydrolyzed themselves. *Trypsinogen* is activated by a protease produced in the intestinal tract, *enterokinase;* the trypsin produced acts autocatalytically upon trypsinogen and also acts as the trigger to activate other zymogens, *chymotrypsinogen* and *procarboxypeptidases.* Trypsin and chymotrypsin act in a complementary fashion by selectively cleaving peptide carboxyls bearing basic side-chains and aromatic side-chains, respectively. The combined actions of the proteases in the intestinal lumen produce progressively shorter residues (oligopeptides). Attack by the carboxypeptidases removes free amino acids from some of the carboxyl terminals. Hence, the products from the first two stages of protein digestion are small peptides, mostly tripeptides and dipeptides, and free amino acids. All these low-molecular-weight substances may be readily taken up in the intestinal mucosal cells by rapid transport systems.

The final or mucosal stage of protein digestion involves the action of a number of intracellular peptidases upon the absorbed oligopeptides. This constitutes an effective system for the uptake of protein amino acids from the intestine, since the small peptides are removed by hydrolysis as

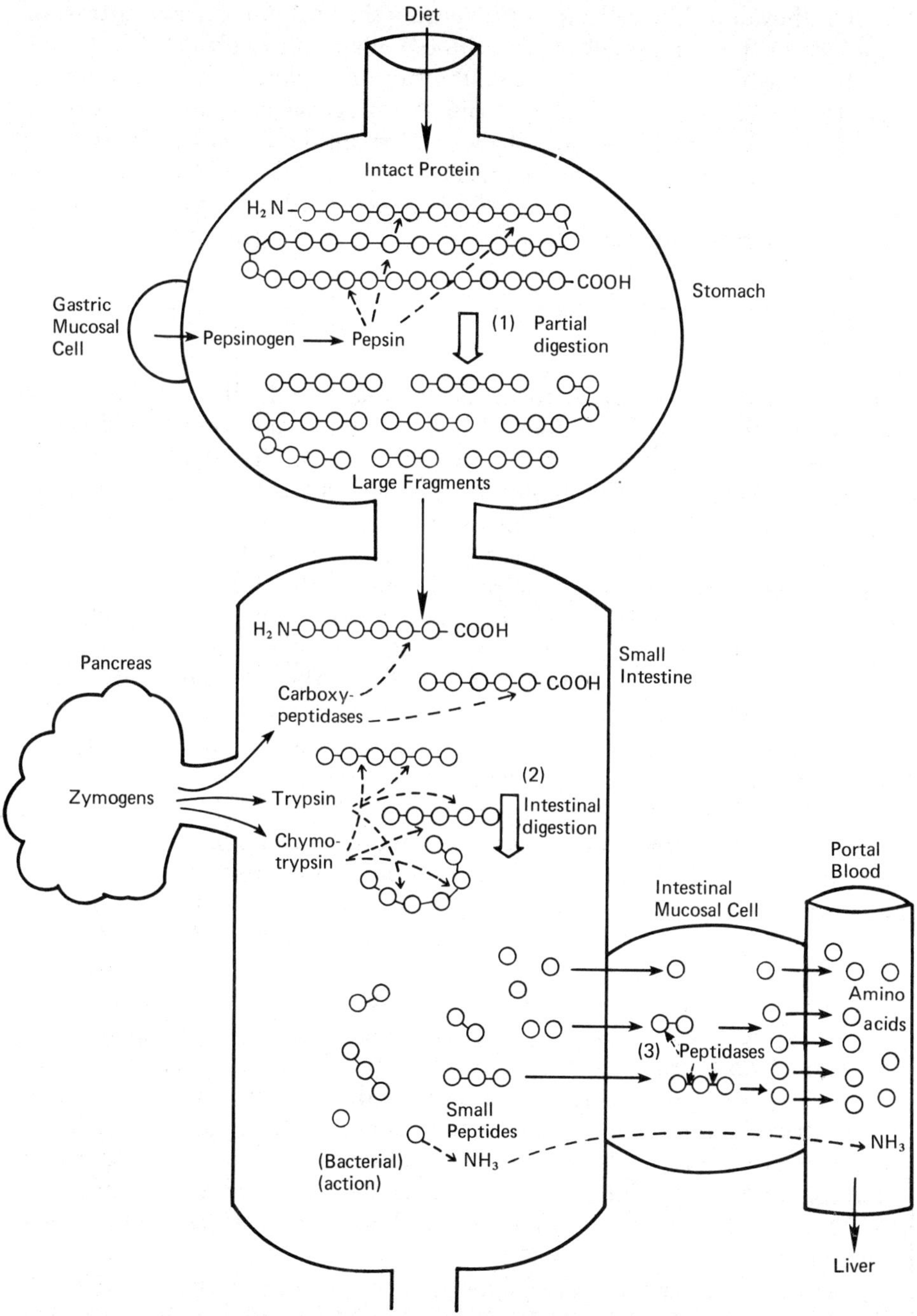

FIG. 6-4. Digestion and absorption of dietary proteins.

soon as they enter the cells, thus preventing the buildup of a concentration gradient in the mucosa. The free amino acids that are absorbed into the cell by energy-coupled transport systems or are produced by the action of cellular peptidases then diffuse from the mucosal cells into the portal bloodstream. A small fraction of the dietary amino acids will be metabolized to toxic ammonia by bacteria in the gastrointestinal tract. The ammonia is normally absorbed, and it passes to the liver by way of the portal circulation for detoxification.

THE LIVER

As is the case with so many forms of metabolism, the liver is the hub of activity in the interconversions of body proteins and amino acids (Fig. 6-5). The reasons for this preeminence of the liver in the nitrogen economy of the body are many. The first and most obvious is provided by its anatomic position. After a protein meal, the liver cells will get first crack at the intake of dietary amino acids and other products of digestion that are transferred from the gut by the portal vein. In this way, large surges of nitrogenous compounds, which may be deleterious to other organs, will be modulated by the liver's capacity to intercept the absorbed substances before they reach the systemic circulation. Another anatomic advantage is provided by the connection of the liver to the biliary tract, which allows it to excrete certain noxious, nitrogenous, waste metabolites (notably bile pigments) directly into the gastrointestinal tract.

A second reason for the key position of the liver is provided by the full expression of the enzymes involved in amino-acid metabolism within the hepatic cells, in contrast to those in other body tissues. This enables the liver to have a wide-ranging capability for the breakdown, modification, and synthesis of the nitrogenous substances presented to it. The metabolic primacy of the liver with respect to the free amino acids pertains to three broad functional processes: (1) the degradation of amino-acid carbon skeletons to provide for cellular energy and gluconeogenesis, (2) the formation of nonessential amino acids and derivatives such as nucleic-acid bases from simple precursors, and (3) the detoxification of ammonia and other catabolic end products such as urea, uric acid, bile pigments, and so forth. In many instances, the metabolic versatility and responsiveness of the liver to changing body needs are enhanced by the ability of the enzyme systems to be induced by changes in the diet or hormonal factors.

Third is the very rapid turnover, in terms of both synthesis and breakdown, of proteins by the liver. The net balance between protein catabolism and anabolism in the liver is of particular importance for the rest of the body because of three considerations: (1) the liver synthesizes many pro-

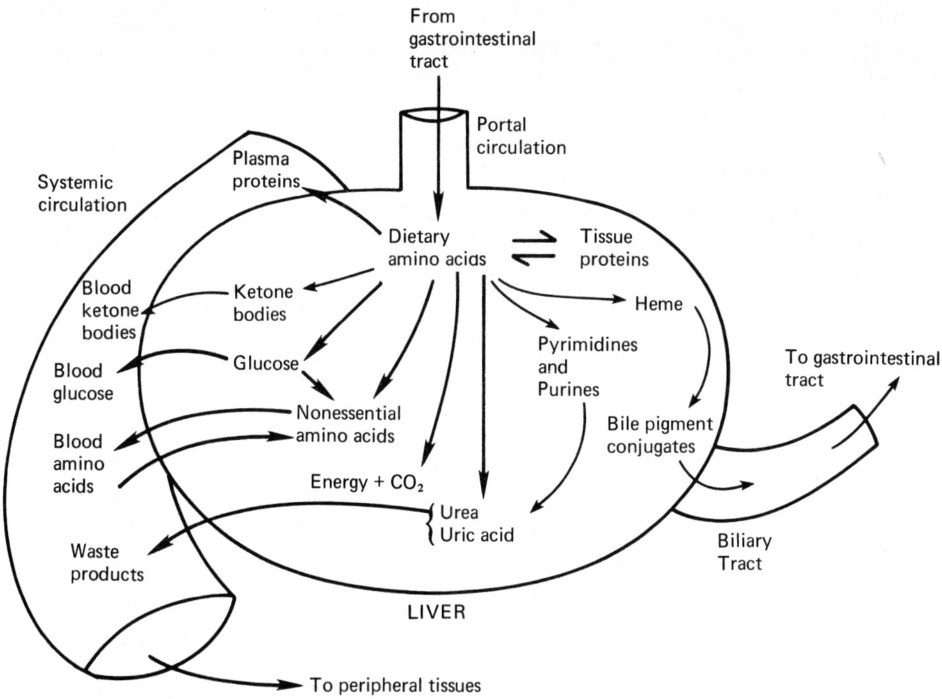

FIG. 6-5. Role of the liver in nitrogen metabolism.

teins for export in the plasma; (2) the formation of intracellular enzymes in the liver influences metabolism throughout the body; and (3) some liver proteins may be rapidly broken down to provide a labile reserve of amino acids in times of dietary deprivation. The rate of turnover of proteins in the liver is faster than in any organ other than the pancreas. An additional feature that distinguishes protein synthesis in liver is the fact that it is selectively enhanced by hormones such as the glucocorticoids, which cause protein catabolism in other tissues. In starvation, for example, muscle protein will be broken down to provide energy, while in the liver, the conversion of amino acids into proteins—such as the enzymes for gluconeogenesis (from amino-acid carbon) or for urea formation (from amino-acid nitrogen)—will be greatly increased.

Thus, the liver acts as an *aminostat* to monitor the input from the diet and to meter the output to the periphery. Both plasma proteins and free amino acids are maintained at quite constant levels in spite of diurnal fluctuations of supply and demand. Peripheral tissues may utilize these sources of nitrogen from the blood for biosynthesis, or, on the other hand, they may degrade their constituent proteins to release amino acids into the bloodstream under catabolic conditions. Upon recirculation to the liver, the organ will employ its unique complement of enzymes to remove excess amino acids from the blood and convert them to glucose, ketone bodies, CO_2, and urea.

AMINOTRANSFERASES

These enzyme reactions are of great importance in the overall economy of nitrogen metabolism because they act to interconvert a wide range of amino acids and are involved in the conservation of organic nitrogen as amino groups. The ***aminotransferases*** are a family of enzymes with differing specificities for the amino acids that act as their substrates, with differing distributions among the various tissues of the body, but with a common coenzyme and a common mode of action. The coenzyme is vitamin B_6 in the form of *pyridoxal phosphate.* The mode of action is simply the transfer of an amino group ($-NH_2$) from one α-amino acid to produce a second α-amino acid, and the pyridoxal group is bound to the enzyme that acts as the intermediary in the amino transfer:

$$\alpha\text{-Amino acid}_1 + \alpha\text{-Keto acid}_2 \xrightarrow[\text{(Pyridoxal-P)}]{\text{Aminotransferase}} \alpha\text{-Keto acid}_1 + \alpha\text{-Amino acid}_2$$

There is very little energy change in such a transfer, and the reaction will proceed in either direction according to the supplies and demands of reactants. Thus, if any amino acid_2 is in short supply, and amino acid_1 as well as keto acid_2 are present in abundance, the amino transfer will, in the presence of the enzyme, go from left to right, leading to the anabolism of amino acid_2. Conversely, the enzyme may lead to the catabolism of the same amino acid when the latter is presented—either through ingestion or by mobilization from peripheral organs—to the liver in excess.

Virtually all amino acids in the body, with the exception of lysine and threonine, are acted upon by specific aminotransferases. The liver has a particularly wide range of such enzymes, but it lacks the aminotransferases for the branched-chain amino acids, leucine, isoleucine, and valine. These amino acids are subject to aminotransfer reactions only in peripheral tissues such as the kidney and muscle.

Among the different aminotransferases, glutamate is the reactant most commonly involved. The aminotransferase enzymes are named according to the pairs of amino acids with which they react (Fig. 6-6). *Glutamate-alanine aminotransferase* and *glutamate-tyrosine aminotransferase* are the most prominent members of this enzyme family found in the hepatic cells. Their activities are greatly increased in response to the administration of glucocorticoids. These increases are due to the induction of synthesis of the enzyme protein in the liver by the hormones and are associated with the induction of a number of enzymes of carbohydrate metabolism that are also involved in gluconeogenesis utilizing these

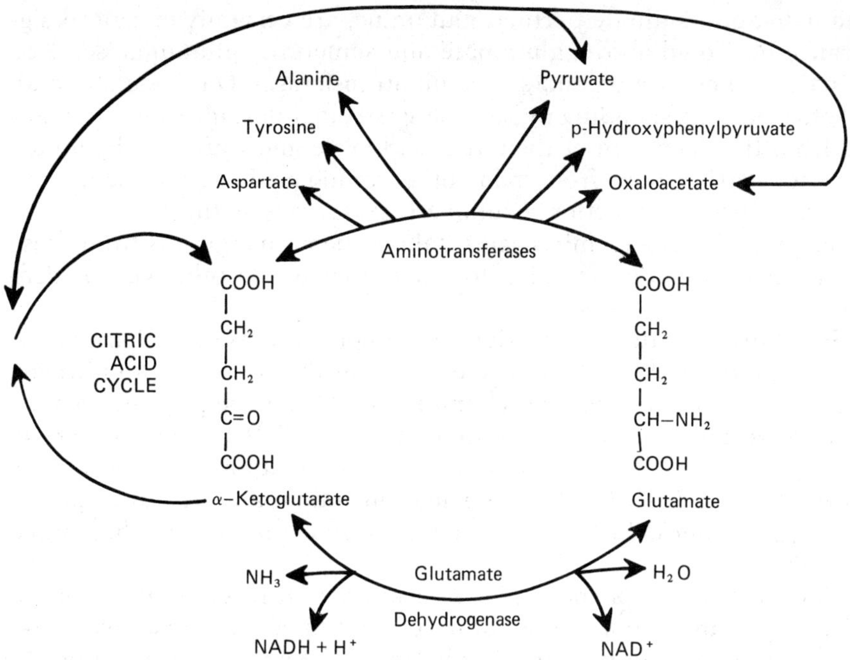

FIG. 6-6. Major amino group transfer reactions (transaminations), with α-ketoglutarate and glutamate as the common coreactants, showing the linkages to the citric-acid cycle via the α-keto acids and to the glutamate dehydrogenase reaction.

amino-acid reactants and the keto acids which result from aminotransferase reactions.

Damage to the liver cells is characterized by the release of aminotransferase enzymes into the bloodstream. Taken together with clinical signs, assays for bile pigments and other serum enzymes such as alkaline phosphatase, tracing the activities of serum aminotransferases provides valuable information in the diagnosis and management of liver diseases. Thus, in acute hepatitis, the activity of glutamate-alanine aminotransferase in serum may be elevated up to 100 times its normal value. In chronic liver disease, the level of *glutamate-aspartate aminotransferase* in the serum will also be elevated; it is also greatly increased in the serum in cases of damage to the heart muscle, which is also associated with elevations in serum creatine phosphokinase and lactate dehydrogenase isoenzymes.

GLUTAMIC-ACID METABOLISM

Although there is no single amino acid that plays a unique role in nitrogen metabolism analogous to that of glucose in carbohydrate metabolism, glutamate and its amide partner, glutamine, are certainly of central significance. As noted above, glutamate and sometimes glutamine serve as the major amino-group donors for aminotransferases. On the other hand, the keto acid corresponding to glutamate, α-ketoglutarate, is readily available from the operation of the citric-acid cycle and thus may be an acceptor for $-NH_2$ groups from many other amino acids within the liver in the reverse transfer reaction. Glutamate also serves as the pivot point in the balance of various amino acids, taking the amino groups from those that are in excess and donating to others that would otherwise be deficient.

In addition to its key function in aminotransferase reactions, glutamate participates in another type of reaction that is crucial to nitrogen metabolism, namely, ***oxidative deamination.*** This is the process whereby the α-amino group is released as free ammonia (NH_3) while the carbon chain is oxidized to α-ketoglutarate. The reaction requires a nicotinamide coenzyme (NAD or NADP) for the oxidation and is catalyzed by a specific enzyme, *glutamate dehydrogenase,* that does not act upon any other amino acids (Fig. 6-6).

Several features of the oxidative-deamination reaction are of physiologic importance: first is the ability of an aminotransferase reaction—that is, (1)—to be coupled with the glutamate-dehydrogenase reaction (2):

1. Alanine + α-Ketoglutarate → Pyruvate + Glutamate
2. Glutamate + NAD^+ → α-Ketoglutarate + NH_3 + NADH + H^+

Thus,

3. Alanine + NAD^+ (or $NADP^+$) → Pyruvate + NH_3 + NADH (or NADPH) + H^+

In this way, a wide range of amino acids that enter the liver, either from ingestion or from peripheral tissues, may be catabolized to α-keto acids and reduced nicotinamide coenzymes (for energy or glucose production) and NH_3 (for urea formation). Second, there is the fact that when supplied with reduced nicotinamide coenzyme, the dehydrogenase may act in reverse to synthesize glutamate and other amino acids from NH_3, as will be discussed in the next section. Third is the location of the dehydrogenase inside the mitochondria, where it is accessible to the keto acids and the nicotinamide coenzyme reactants of the citric-acid cycle. The final feature is the regulatory nature of the enzyme: it is allosterically inhibited by ATP and GTP but activated by ADP and GDP, so that maximal rates of oxidative deamination of amino acids in the liver will be obtained when the energy reserves of the tissue are lowest.

NONESSENTIAL AMINO ACIDS

With the exception of tyrosine (which is formed directly from the essential amino acid, phenylalanine) and cysteine (which is formed from methio-

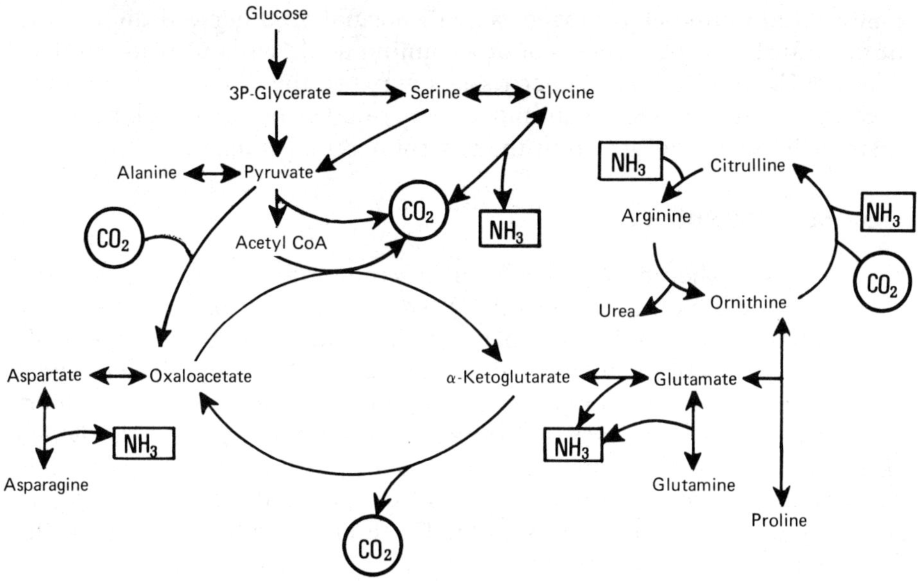

FIG. 6-7. Formation of nonessential amino acids.

nine), the nonessential amino acids are produced mainly from glutamic acid in the liver (Fig. 6-7). *Proline, ornithine,* and *arginine* obtain their carbon and amino nitrogen by the metabolism of the whole glutamate molecule. *Alanine, serine,* and *glycine* derive their carbon from the three-carbon intermediates of glycolysis and their amino nitrogen from glutamate; *aspartate* derives its carbon from the four-carbon intermediate of the citric-acid cycle, oxaloacetate, and again the nitrogen is obtained by amino transfer from glutamate.

In fact, a combination of the glutamate-dehydrogenase reaction operating in the reverse sense

2′. α-Ketoglutarate + NH_3 + NADH (or NADPH) → Glutamate + NAD^+ (or $NADP^+$)

with the reversal of the appropriate aminotransferase reaction

1′. Glutamate + α-Keto acid → α-Ketoglutarate + Amino acid

may lead to the biosynthesis of any amino acid, provided that the body is furnished with a supply of the corresponding α-keto acid and a source of NH_3:

3′. α-Keto acid + NH_3 + NADH (or NADPH) → Amino acid + NAD^+ (or $NADP^+$)

Since this process is the reversal of oxidative deamination and since it requires an input of reducing energy in the form of NADH or NADPH, it is termed ***reductive amination.*** Although most of the amino nitrogen for nonessential-amino-acid biogenesis will normally be derived directly by amino transfer from excesses of other amino acids without prior conversion to NH_3, the liver and other tissues may use the reductive amination to conserve and recycle the ammonia generated either by cellular metabolism or by bacterial action in the gastrointestinal tract.

AMMONIA METABOLISM

The processes whereby the liver can fix ammonia into organic linkages fall into three general categories: (1) *reductive amination,* which as discussed above, leads to the biogenesis of glutamate and other amino acids and the retention of the nitrogen of ammonia; (2) the *production of amides* such as glutamine, which may act as a temporary reservoir and transport form of ammonia; and (3) the *formation of carbamyl phosphate,* which gives rise to biosynthetic products such as the pyrimidine bases of nucleic acids or to urea for eventual excretion from the body. Each of these reaction systems requires energy expenditure on the part of the cell—the first, in the form of reduced nicotinamide coenzymes; the last two, in the form of ATP—for the conversion of inorganic NH_3 to organic N. All are of cardinal importance for survival because of the highly toxic effects exerted by free ammonia if it is allowed to build up in body fluids.

Glutamine

This compound is produced from glutamic acid by the addition of ammonia to its terminal carboxyl group; hence, it is an *amide,* not an amine as the name falsely implies (Fig. 6-8). Its production is by an energy-requiring reaction, promoted at the expense of the hydrolysis of a molecule of ATP, which is catalyzed by an enzyme known as *glutamine synthetase.* The liver also possesses the capacity to reconvert glutamine to glutamate and ammonia, but this reaction cannot be catalyzed by glutamine synthetase, for the action of this enzyme is irreversible. Instead, a separate enzyme, *glutaminase,* is responsible for the breakdown of glutamine by hydrolysis to glutamate and ammonia; as with other hydrolytic reactions, there is a substantial loss of energy, so this enzyme, too, functions irreversibly and the energy released cannot be recaptured as ATP.

Obviously, the actions of glutamine synthetase and glutaminase must be regulated in concert to prevent an energy-dissipating cycle of the type discussed earlier regarding simultaneous kinase and phosphatase action. Under normal circumstances, glutamine synthetase will be active in the liver to maintain the intracellular ammonia concentration below toxic levels. The glutamine produced can safely accumulate in the cells without toxic effects, and some will pass out into the bloodstream, where it constitutes the largest single component of the free amino-acid pool (Fig. 6-9). Under conditions of active gluconeogenesis, the protein and amino-acid catabolism in peripheral tissues such as the muscles will generate

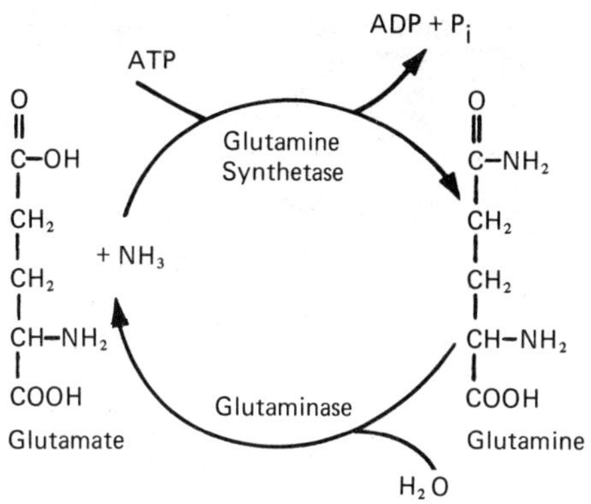

FIG. 6-8. Reactions in glutamine metabolism.

ammonia, which may be converted to glutamine in those tissues for safe transport to the liver through the bloodstream. Glutaminase will then be activated in the liver to yield glutamate and free ammonia: the former may be catabolized by glutamate dehydrogenase to provide reduced nicotinamide coenzymes and carbon sources for glucose synthesis plus additional ammonia; the latter, thus generated by the combined actions of glutaminase and glutamate dehydrogenase, will then be eliminated from further metabolic errantry by transformation to urea, the final end product of nitrogen metabolism.

The Urea Cycle

Urea (NH_2CONH_2) was historically the first and is certainly one of the simplest of the organic substances to be synthesized by chemical means. Formally, it is the diamide of carbonic acid, and it is formed by the fixation of NH_3 to another end product of metabolism, CO_2. This deceptively easy chemical change, in analogy with the apparent simplicity of the catabolism of the two-carbon acetyl group to CO_2, involves a very complex interaction of enzymes and intermediates. In another parallel with the final pathway of carbon metabolism, the terminus of nitrogen metabolism also utilizes a catalytic cycle of reactions. In fact, the parallelisms between the two cycles are based upon firmer connections than mere similarities in their mechanisms: first, both the citric-acid cycle and the key initial reactions of the urea cycle are localized within the liver mitochondria; second is the provision of the CO_2 and ATP that are required for urea production by means of the operation of the citric-acid cycle; third is the sharing of common four-carbon intermediates so that the two cycles actually mesh together. Thus, the interlocking urea and citric-acid cyclic pathways could be referred to, after the biochemist who elucidated them both, as the "Krebs bi-cycle."

The urea cycle involves two amino acids that do not enter into the peptide linkages of body proteins—*ornithine* and *citrulline—and two that*

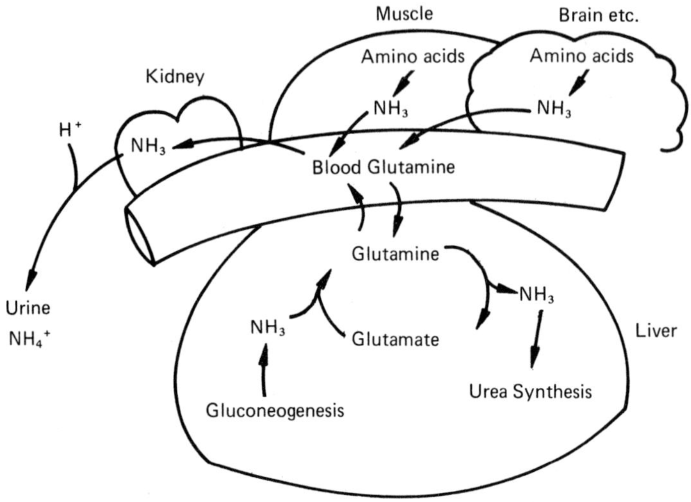

FIG. 6-9. Role of the liver in glutamine metabolism and the transport of nitrogen originating from ammonia to and from peripheral tissues.

do—arginine and *aspartate.* Ornithine plays an analogous role to that of oxaloacetate in the citric-acid cycle, that is, it acts as a catalytic carrier for the groups being modified in the cycle but is not itself changed by the process. The ornithine molecule has a terminal NH_2 group that serves as the handle to receive molecules of CO_2 and NH_3 during their transformations to the organic precursors of the urea molecule, which are shown in outline in Figure 6-10.

The first portion of the urea cycle involves the activation of CO_2 and one molecule of NH_3 by the formation of the compound mentioned earlier, *carbamyl phosphate.* This reaction requires the energy input from two molecules of ATP and the action of the specific enzyme *carbamyl-phosphate synthetase* (Fig. 6-11). The enzyme is allosterically activated by the cofactor, *N-acetyl glutamate,* and probably exerts a controlling influence over the whole cycle. In addition to its role as precursor in urea formation, carbamyl phosphate also contributes to the biogenesis of a portion of the pyrimidine ring structure by combining with a molecule of aspartate. In the path for ureogenesis, the carbamyl group (NH_2–CO) is transferred intact to the terminal NH_2 group of ornithine, using the high energy generated by cleavage of the activated phosphate to drive the reaction. The enzyme catalysing this process, *ornithine-carbamyl transferase,* is a unique component of liver cells, and its presence in elevated amounts in the blood plasma is a sensitive diagnostic indicator of hepatic disease. The product of the reaction, citrulline, now has the NH_2CONH– the *ureido* group, attached to its terminus. In order to undergo further change, the citrulline

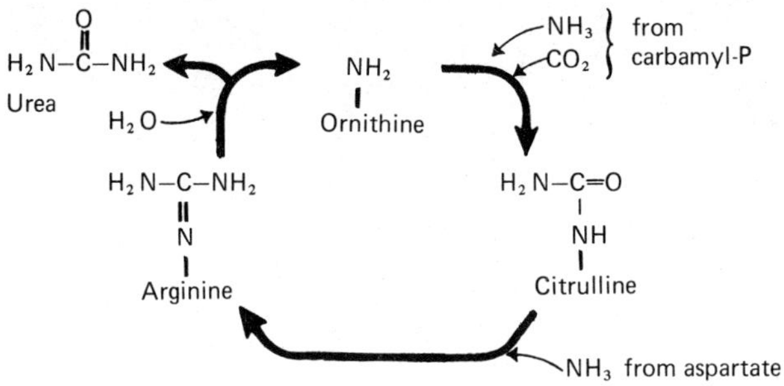

FIG. 6-10. Outline of the urea cycle, showing how the terminal NH_2 of ornithine acts to carry the carbon and two nitrogen atoms that eventually form the urea molecule.

must move out of the mitochondria, together with aspartate, into the hepatic cytoplasm, where they combine with each other in the presence of the appropriate synthetase enzyme and ATP to form an intermediate with the complex name, *argininosuccinic acid.* The latter is broken down immediately by a cleavage enzyme in such a way that the NH_2 group from the original aspartate is now attached to produce the $NH_2C(=NH)NH$–, the *guanidino* group, of arginine, and the remainder of the aspartate molecule is removed as fumarate.

The reactions to this point have resulted in the net conversion of ornithine to arginine. Since ornithine itself may be synthesized from glutamate, we can comprehend how the liver is able to satisfy most of the body's needs for arginine by these reactions of the urea cycle. Part of the arginine will be required for protein synthesis here and in other tissues, part will be used for the production of creatine (which is particularly plentiful in muscle), but most will be channeled into the final reaction of the cycle, which is catalyzed by the specific enzyme *arginase.* This enzyme is found exclusively in the cytoplasm of hepatic cells and only in the livers of those organisms, like man and other mammals, that convert their waste nitrogen to urea. Other animals, such as fish or birds, that excrete only ammonia or uric acid instead may carry out the reactions leading to arginine, but their hepatic cells lack arginase activity. This enzyme catalyzes the hydrolysis of the $NH_2C(=NH)$–, or *guanido* portion, from the terminus of arginine to release urea and to regenerate free ornithine. The latter may return inside the mitochondria to resume its catalytic function by initiating another turn of the cycle.

The overall conversion process requires some additional balancing and tying up of the tag ends. This is where the citric-acid cycle contributes by renewing the four-carbon intermediates involved (Fig. 6-12). Recall that the cleavage of argininosuccinate produces arginine and fumarate; the latter must be reconverted to aspartate so that the urea cycle may

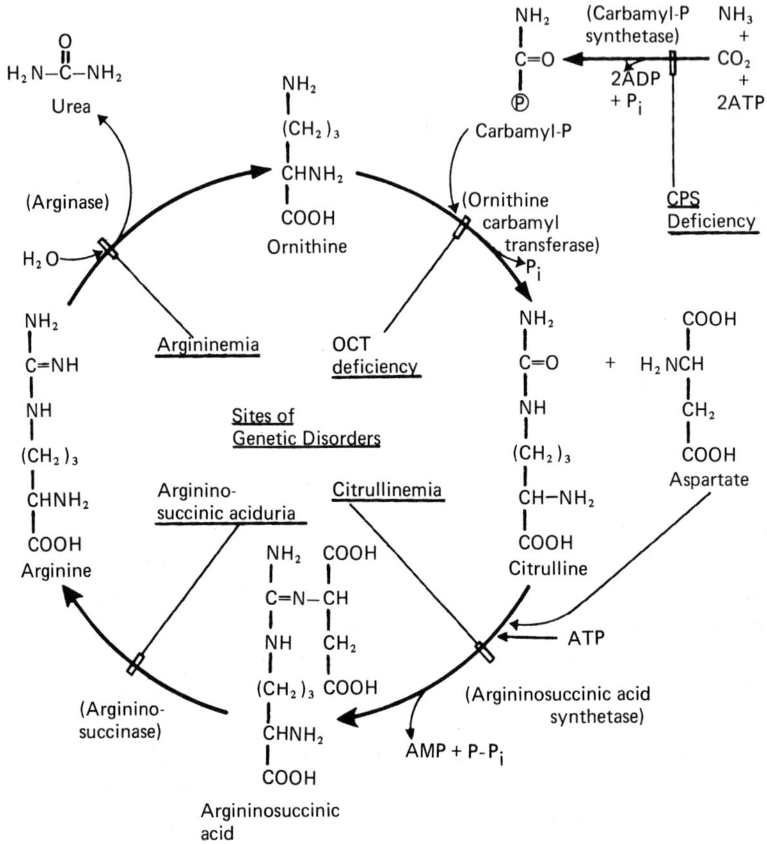

FIG. 6-11. Details of the urea cycle. *(CPS,* carbamyl phosphate synthetase; *OCT,* ornithine carbamyl transferase.)

function in a catalytic manner. This is accomplished by the transport of fumarate back into the mitochondria and its oxidation to oxaloacetate (by way of malate) by the enzymes in that segment of the citric-acid cycle. Meanwhile, oxidations in the citric-acid cycle coupled to phosphorylations through the respiratory chain will provide the ATP-energy requirements for urea synthesis. A total of 3 mols of ATP are used, but because argininosuccinic acid synthesis involves cleavage of ATP to AMP, a total of 4 high-energy equivalents are consumed in one turn of the urea cycle. Finally, the oxaloacetate may be converted to aspartate by amino transfer from glutamate through the action of intramitochondrial glutamate-aspartate aminotransferase. Additional revolutions of the citric-acid cycle are necessary, of course, to produce the carbon dioxide that supplies the eventual carbon atom of urea.

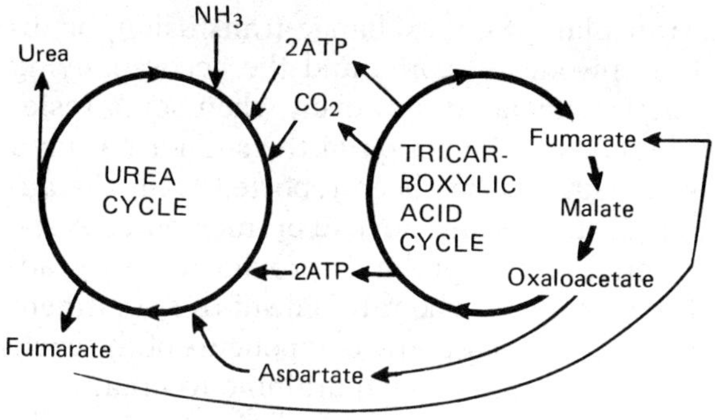

FIG. 6-12. Interrelations of the urea cycle and the citric-acid cycle, the "Krebs bi-cycle."

DISORDERS OF AMMONIA METABOLISM

The foregoing reactions are confined predominantly to the cells of the liver. Extrahepatic tissues possess some of the individual enzymes involved—for example, those to synthesize carbamyl phosphate or arginine in these other regions of the body—but, like the livers of fish and birds, none has significant quantities of the enzyme arginase. It follows that the other tissues lack the capacity for urea biosynthesis, and hence in cases where liver function fails, the overall ability of the body to remove toxic ammonia in the form of nontoxic urea will be diminished.

In many cases of liver disease, damage within the hepatic tissue obstructs the blood flow through the organ, which normally accounts for one quarter of the output from the heart. As a result of the obstruction, the portal vein develops collateral veins that bypass the liver to empty directly into the vena cava and thence to the systemic circulation. This *portacaval shunt* permits the delivery of absorption products from the gastrointestinal tract directly to body tissues without processing by the liver, and hence this condition greatly exacerbates the already serious consequences, of impaired detoxification of ammonia and other noxious substances by the diseased liver cells.

Hepatic Coma

Among the symptoms of severe liver disease is the development of bouts of unconsciousness at irregular intervals, which become more frequent near the terminal stages. This comatose condition is referred to as *hepatic coma,* and it is thought to be related to the elevations of blood ammonia along with other toxic substances that are accumulated as the result of inadequate liver metabolism. Since many of these products result from the action of intestinal bacteria upon the ingested protein and amino acids, therapy of the comatose patient is generally directed to decreasing dietary protein intake and the reduction of bacterial activity by the administration of nonabsorbable antibiotics.

In extreme cases, liver transplantation, exchange transfusion, or dialysis may be attempted. The episodes of coma and the accompanying hyperammonemia may be partly alleviated, however, when some residual, functioning liver tissue remains if the action of the subnormal urea cycle is accelerated. Thus, some success has been reported from the administration of arginine, either as the free amino acid or, more effectively, as basic proteins such as protamine, which probably provide a more gradual and prolonged release of free arginine. The rationale of this treatment is to beef up the concentrations of all the catalytic components of the cycle and thus to enhance the flux of ammonia through arginine to urea.

Another similar approach to this problem is to administer *N*-carbamyl glutamate, which, like *N*-acetyl glutamate, acts as a cofactor for the carbamyl-phosphate synthetase reaction in the liver. Unlike the *N*-acetyl compound, however, which is broken down rapidly upon ingestion, the *N*-carbamyl derivative is not subject to hydrolysis and therefore effectively persists to boost the activity of the first step of the cycle where ammonia enters. Of course, such treatments, which are designed only to provide temporary relief from elevated ammonia accumulations, do not act upon the basic causes of liver disease and must be considered strictly as palliative measures for the management of the comatose state.

Genetic Disease

Inherited deficiencies of each of the five enzymes of the urea cycle may occur (Fig. 6-11). In each case, the acute symptoms of ammonia toxicity—with episodic vomiting, seizures, lethargy, or other neuropathic signs—may be seen in the neonatal period. In some instances, the conditions may prove lethal in the early months of life, but even if the infants survive, they generally display severe mental retardation.

In *citrullinemia* and *argininosuccinicaciduria,* the respective urea-cycle intermediates accumulate in the blood and other body fluids as the result of a deficiency of the appropriate liver enzymes for the synthesis or breakdown of argininosuccinic acid. Upon reaching the kidneys, these high concentrations of urea-cycle intermediates will exceed the capacity for reabsorption from the glomerular filtrate, and consequently up to several grams per day of citrulline or argininosuccinic acid will spill over into the urine. The specific aminoaciduria may thus be readily detected by urinalysis. *Argininemia* with the excretion of arginine in the urine due to a deficiency of arginase has also observed. In each of these disorders of urea metabolism, there is, in addition to the accumulation of urea-cycle intermediates, a dramatic increase in the blood ammonia levels, particularly following the ingestion of proteins.

In two other rare genetic diseases—*carbamyl-phosphate synthetase deficiency* and *ornithine-carbamyl transferase deficiency*—marked hyperammonemia without the accumulation of urea-cycle intermediates is a common feature. Although the enzymes may not be totally absent in these conditions, it appears that even a partial impairment of any reaction of the urea cycle will reduce the capacity of the liver to eliminate ammonia effectively, especially when amino-acid catabolism is accelerated after a meal. The acute intoxication and the irreversible brain damage in long-term cases are more likely to be related to the hyperammonemia than to the accumulation of the cycle intermediates. No therapy is known for these conditions, but careful management by restricting the dietary protein intake during infancy to the minimum amounts compatible with growth and development has shown promise in reducing the severity of the acute neurologic disturbances and possibly in lowering the risk of irreversible brain damage as well.

METABOLISM OF AROMATIC AMINO ACIDS

Amino acids with aromatic side-chains include *phenylalanine, tyrosine,* and *tryptophan*. The first two, which contain a simple benzene ring, are closely interrelated since tyrosine is derived from phenylalanine metabolically. Tryptophan, which contains a conjugated indole ring, gives rise through metabolism to the B vitamin, niacin. These metabolic connections lead to complex nutritional linkages. Thus, although phenylalanine is essential and tyrosine is nonessential in the diet, an ample supply of tyrosine will exert a sparing action upon the amount of phenylalanine required. As we shall see, in situations where the conversion of phenylalanine to tyrosine is impaired, tyrosine will become an "essential" amino acid.

Pellagra, the vitamin-deficiency syndrome associated with inadequate supplies of niacin in the diet, may also be encountered in individuals who lack normal activity for the conversion of tryptophan to niacin. The absence of vitamin C (which is required for tyrosine metabolism) or of vitamin B_6 (which is required for tryptophan metabolism) also leads to disturbances in the utilization of the aromatic amino acids.

Catabolic Pathways

Aromatic rings are produced in nature by the agency of plant and bacterial enzymes; the higher organisms, such as man, lack these enzymes and only possess the destructive capacity required to break the rings down eventually to CO_2. Many other tissues in the body convert the benzene and indole rings of the aromatic amino acids into biologically important derivatives such as pigments and hormones, but, in quantitative terms, it is the liver that carries on the bulk of the catabolic reactions that account for the degradation of these compounds and its attendant energy production.

Phenylalanine catabolism is readily described: it simply involves the addition of a hydroxyl group to the benzene ring to produce tyrosine (Fig. 6-13). This reaction is catalyzed by an enzyme that is logically termed

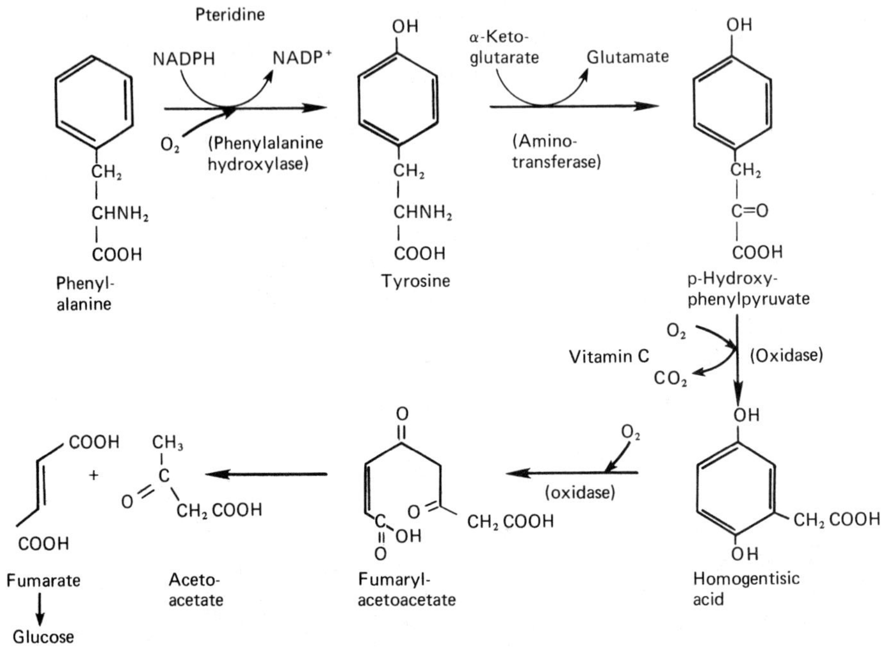

FIG. 6-13. Catabolism of phenylalanine and tyrosine.

phenylalanine hydroxylase, which resides only in the liver cells. The hydroxylation process requires molecular oxygen, which provides the O of the OH group, and two coenzymes, NADPH and a pteridine-containing cofactor, which act in concert to furnish the H. It is important to note that the reaction is completely irreversible; tyrosine can supplement some of the needs for phenylalanine, but it cannot be employed to replace phenylalanine itself by a reversal of the hydroxylation process.

Tyrosine catabolism also occurs in liver cells. The first step, as with many amino acids, involves an amino-group transfer, which is catalyzed by the tyrosine-glutamate aminotransferase referred to previously. The production of glucocorticoid hormones or the ingestion of high concentrations of tyrosine will induce the synthesis of more tyrosine-glutamate aminotransferase in the liver, which in turn will speed the breakdown of the amino acid and produce more glucose. An oxidase that requires O_2 and ascorbic acid rearranges the keto acid to homogentisic acid. In vitamin-C deficiency, particularly during the newborn's early days of life, the oxidase will only be partially active. The precursors will then back up in the blood and tyrosine and its keto acid *p*-hydroxyphenylpyruvate, will be excreted in the urine. The condition in infants is termed *neonatal tyrosinemia,* and it is quite prevalent in premature babies whose liver enzymes have not reached full maturity. This harmless and common anomaly is distinguished from the severe hereditary tyrosinemia discussed in a following section by its transient nature and its ready reversibility; the latter may be brought about by reduced protein intake or increased vitamin-C intake.

The opening of the aromatic ring of the homogentisic acid is achieved in the presence of O_2 by the action of *homogentisic-acid oxidase,* which produces an open-chain, eight-carbon product that is subsequently cleaved in half to yield acetoacetate and fumarate as the end products of tyrosine catabolism. The acetoacetate will be exported from the liver in the bloodstream to serve as an energy-yielding substrate for extrahepatic tissues, principally the muscles. The fumarate, on the other hand, may be directly oxidized in the liver by the citric-acid cycle, or, following conversion to oxaloacetate, it may enter the path of gluconeogenesis. Thus, half of each tyrosine molecule that is oxidized may give rise to glucose, whereas the other carbons must be utilized outside the liver in pathways described previously for ketone-body metabolism.

Tryptophan is also catabolized mainly in the liver (Fig. 6-14). The first reaction of the rather involved sequence is also the most interesting, because it responds to cortisol in a fashion analogous to tyrosine-glutamate aminotransferase activity. The activity of the enzyme responsible, named *tryptophan oxygenase,* is enhanced many times either by tryptophan itself or by the glucocorticoid hormones. The latter act as true inducers of the synthesis of new enzyme, but the substrate, tryptophan, interacts directly with the enzyme that has already been synthesized to maintain it in a stabilized form that resists breakdown. These two influences of hormone and substrate ensure the efficient utilization of excess tryptophan to prevent its accumulation in the liver as well as to provide for the energy needs of the body. The enzyme splits open the pyrrole-ring structure of the indole nucleus to yield the substituted benzene derivative known as *kynurenine.* The latter is further broken down in the liver by the removal of three carbons of its side-chain as alanine, which will go on to form glucose; the balance of the molecule, *3-hydroxyanthranilate,* may be oxidized to CO_2 and acetoacetate (the latter passing to extrahepatic tissues for utilization) or rearranged in several steps to form the pyridine ring of NAD and NADP.

It is estimated that under normal conditions, over 95% of tryptophan metabolism proceeds by the former oxidative path. Thus it is not surprising that niacin, the vitamin that supplies the ring for nicotinamide coenzymes by direct conversion, may be replaced in the diet by tryptophan, with approximately 60 mg of dietary tryptophan being required to replace 1 mg of dietary niacin. It is estimated that roughly half of the usual niacin requirement (20 mg/day) is provided from tryptophan in the diet.

The symptoms of dermatitis, diarrhea, and dementia—the so-called 3Ds characteristic of *pellagra* or niacin deficiency—may be seen when the amount of tryptophan is low in the diet or its utilization is impaired. An example of the latter situation is encountered when vitamin B_6, pyridox-

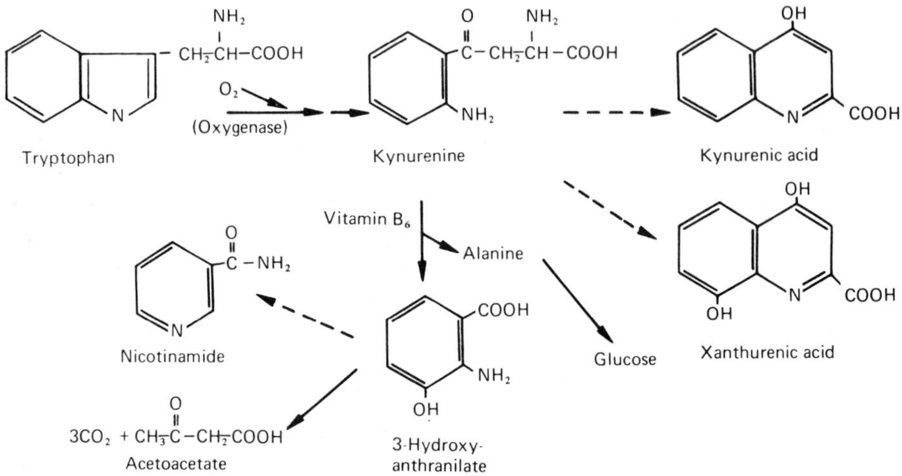

FIG. 6-14. Tryptophan catabolism.

ine, is in short supply. The reason for this anomaly is that pyridoxal phosphate is necessary for the metabolism of kynurenine derivatives to 3-hydroxyanthranilate. In the absence of pyridoxine, tryptophan metabolism is interrupted, and the kynurenine intermediates become diverted to other end products, *kynurenic acid* and *xanthurenic acid.* The appearance of these abnormal metabolites in the urine after the ingestion of a tryptophan load is diagnostic of abnormal tryptophan metabolism and, more particularly, of pyridoxine deficiency. The detection of high levels of xanthurenic acid in infant urine in cases of neonatal convulsive disorders, for example, has led to successful therapy of the disorder through correction of the pyridoxine inadequacies of the babies' feeding formulas.

Biosynthesis of Special Products

An enumeration of the many biologically active substances produced in the body from the aromatic amino acids emphasizes their importance, in addition to their obvious roles in protein synthesis and energy production (Fig. 6-15). Phenylalanine, by its metabolite tyrosine, is the source of melanin, the brown-colored component of pigmented cells in the skin, brain, and elsewhere. The same two amino acids may also give rise to the humoral agents, epinephrine and norepinephrine, of the adrenal glands and sympathetic nervous system, respectively. By reaction with iodine, the thyroid glands convert tyrosine into the hormone, thyroxine. In the gastrointestinal tract, regions of the nervous system, and mast cells, tryptophan may be converted to the neurohumoral agent, serotonin. Disturbances in any of these metabolic conversions may produce severe pathologic sequelae, as noted below.

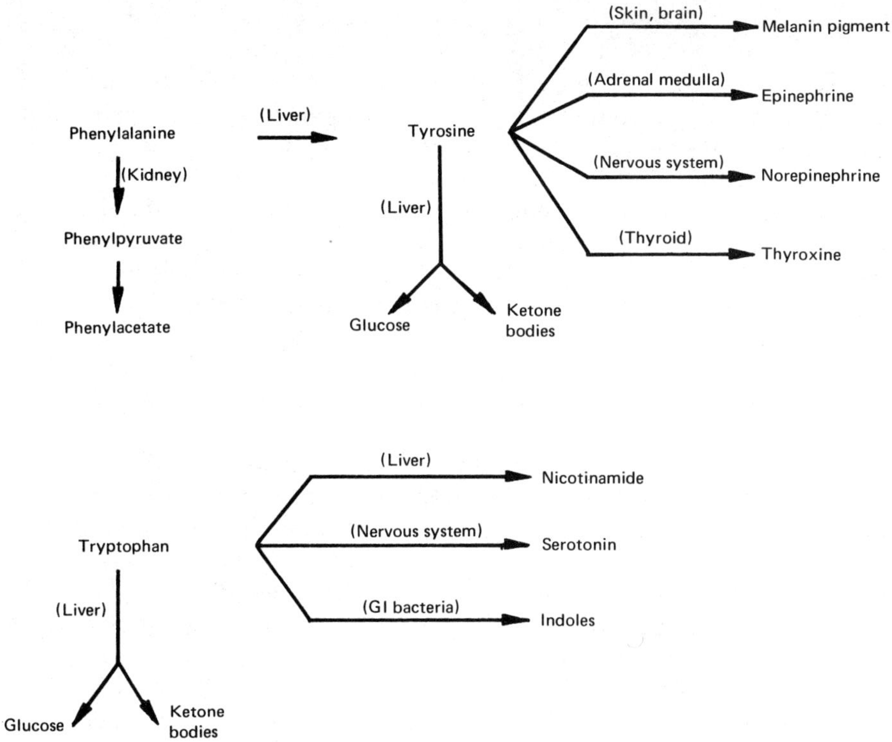

FIG. 6-15. Biosynthesis of special products from aromatic amino acids in various body tissues.

Genetic Diseases

The inborn errors in the metabolism of aromatic amino acids span the range from quite benign to severely debilitating (see Chap. 2). In the former category are the two syndromes that were first characterized as metabolic disorders of genetic origin, *alkaptonuria* and *albinism.* Alkaptonuria, which is inherited as a Mendelian recessive trait, is caused by the absence of one of the enzymes in the main pathway of tyrosine catabolism, homogentisic-acid oxidase. The pathway intermediate, homogentisic acid, builds up in the liver and passes into the bloodstream. The kidney, which does not normally encounter this metabolite in the renal tubules, treats it as a foreign substance and actively secretes it into the urine. As a result, homogentisic acid is rapidly cleared from the blood, and it does not accumulate as do the metabolites above an enzymatic block in most metabolic disorders. Upon standing in contact with air, the homogentisic acid in the urine turns black with time. The condition may also be detected when the patient's urine gives a positive test for reducing substances but negative tests for specific sugars, such as glucose or galactose. Otherwise, the disorder is symptomless in early life; with age, the gradual accumulation of homogentisic acid in tissues and, presumably, its oxidation by processes analogous to those that occur in urine lead to a generalized pigmentation of the connective tissues, termed *ochronosis*. This is associated with characteristic arthritic changes that affect the large joints and spine in later life.

In the albino, a side reaction of tyrosine metabolism in the pathway leading to the formation of melanin fails to occur. The enzyme, tyrosinase, which is normally abundant in the pigmented cells or melanocytes of the body, is lacking in such cases, with the result that the hair, skin, and eye color is pale or absent. Apart from hypersensitivity to sunlight and some impairment of vision, the albino does not have serious symptoms due to the disorder, and this condition is compatible with a normal and productive life.

Among the more serious inborn errors are those that affect phenylalanine and tyrosine metabolism: *phenylketonuria* and *hereditary tyrosinemia*, respectively. The biochemical defect in the latter condition resembles that seen in neonatal tyrosinemia, with increased excretion of tyrosine and its keto-acid analog. In the hereditary disorder, however, there is no amelioration with age, since the enzyme *p*-hydroxphenylpyruvate oxidase does not develop. Patients will fail to thrive, and some may succumb to the disease in infancy; those who survive show chronic degenerative changes of the liver and kidneys. In such individuals, high concentrations of aromatic amino acids in the diet have deleterious effects and should be avoided. The intake of tyrosine and its precursor, phenylalanine, should be controlled at levels just adequate to maintain growth, and this regimen should be instituted in early life for infants diagnosed as having the acute hereditary disorder.

Dietary control has produced the most remarkable results in cases of phenylketonuria. In this disease, phenylalanine intake must be strictly curtailed, but tyrosine becomes an essential dietary component and should be provided in abundance. The reason for this is due to the fact that the liver of phenylketonuric individuals lacks the enzyme phenylalanine hydroxylase. Thus, the normal conversion of dietary phenylalanine to tyrosine is impaired, so that the only tyrosine available for the body's requirements must come from the diet. On the other hand, incoming phenylalanine has no avenue of utilization other than protein synthesis, and the excess will accumulate in liver cells and spill out into the bloodstream, where its concentrations may be 50 to 100 times greater than normal. When the phenylalanine reaches the kidney, its high level may exceed the capacity for renal reabsorption, so the excess will overflow into the urine. This specific aminoaciduria, when accompanied by an elevation of the amino acid in the blood, is typical of the metabolic disorder, distinguishing it from a kidney defect, and it is readily detected by urinalysis. Continuous increases in the amount of phenylalanine reaching the kidney will induce the synthesis of an aminotransferase to convert some of the excess phenylalanine to its keto acid, *phenylpyruvic acid* (Fig. 6-15).

It was the presence of the latter phenylketone body in the urine that led to the name used to describe the condition as well as to the diagnostic test used for many years, which is based upon the typical green color resulting from the reaction between ferric chloride and phenylpyruvic acid. Other abnormal metabolites are also produced and excreted by the kidney, such as phenyllactic acid and phenylacetic acid; the latter compound has such a powerful, mousy aroma that, as in the first cases reported, the disease may be detected on the basis of a "sniff" test.

In the absence of intervention, the consequences of phenylketonuria are dire indeed. Although they may apparently be normal at birth, afflicted infants fail to develop mentally so that in later life their IQs generally fall in the idiot range (<20). Happily, this tragic outcome may be averted if the disease is detected in the newborn period, preferably by confirming that very high phenylalanine levels persist in the blood. This may now be accomplished on the basis of a mass screening of newborn infants for phenylketonuria, as well as for a number of other genetic disorders that result in abnormal increases of amino acids in the bloodstream.

If phenylketonuria is detected in the first few weeks of life and dietary management is rapidly undertaken, the brain damage will be prevented. It is important to regulate the amount of phenylalanine ingested to provide just enough for growth and development, but to avoid its excessive buildup in body fluids. It is likely that excess phenylalanine or one of its abnormal metabolites exerts toxic effects upon the developing nervous system, because children maintained upon low phenylalanine diets in early life will attain normal intelligence and brain function. After they are beyond about 8 to 10 years of age, they will tolerate ordinary diets quite well and show few symptoms of phenylalanine toxicity. Conversely, children who have been detected with the disease beyond the early postnatal period, that is, at a stage when impaired mental function has already become evident, will show no reversal nor will they gain normal intelligence if dietary control is instituted then. It is obviously imperative to diagnose phenylketonuria and treat it as early as possible.

Tryptophan utilization is disturbed in the hereditary condition, *Hartnup disease,* so named after the first detected family. Unlike the disorders mentioned thus far, this condition is related primarily to the defective absorption of the amino acid rather than to its impaired metabolism. Tryptophan malabsorption is manifested in two general ways. First, and most serious, is the defective uptake of tryptophan in the gastrointestinal tract, which leads to deficient niacin formation by the liver and hence to the symptoms of pellagra, while delayed retention of tryptophan in the gut leads to its conversion to abnormal indole metabolites by bacterial action in the colon. Second is the inability of the kidney to reabsorb tryptophan and other neutral amino acids from the glomerular filtrate, with the result that a generalized aminoaciduria, without elevations in the blood concentrations, is observed. The affected individuals display the skin lesions characteristic of pellagra, a sensitivity to sunlight, and the motor ataxia characteristic of cerebellar dysfunction. The treatment of patients with Hartnup disease has included the supplementation of the diet with additional niacin to alleviate the dermatologic and neurologic lesions, together with protection from sunlight and the periodic sterilization of the gastrointestinal tract with antibiotics to reduce the formation of the bacterial breakdown products of tryptophan that may exert toxic effects.

METABOLISM OF SULFUR-CONTAINING AMINO ACIDS

As in the case of the aromatic amino acids, the sulfur-containing amino acids are equally as important for their role in providing essential metabolites as they are for the part they play in protein structure and function. In the latter context, the disulfide or oxidized form, *cystine,* provides the covalent bridges that act as cross-linkages for many proteins; the reduced or sulfhydryl form, *cysteine,* serves often as a participant in the catalytic process at the active center of many enzymes.

The other sulfur-containing amino acid, *methionine,* bears a similar nutritional relation to cysteine as that between phenylalanine and tyrosine discussed earlier. Methionine, an essential amino acid, may be spared in the diet by provision of the nonessential amino acid, cysteine, which is derived chiefly in the liver from the irreversible breakdown of methionine. The latter, as its name implies, carries a methyl group attached to its sulfur atom, and this serves as the source of a number of key methyl compounds in various organs and tissues.

Active Methyl Transfers

Before it is capable of being transferred to other compounds, the methyl group of methionine must first undergo activation by a reaction of the sulfur atom with ATP (Fig. 6-16). This activation process is unique in that all three of the phosphate groups of ATP are released, two as inorganic pyrophosphate and the third as inorganic phosphate. Obviously, a great deal of energy will be released by such a hydrolytic process, and this is used to drive the formation of a highly energized, coenzymatic form, *S-adenosylmethionine.* The latter product contains the adenosine part of the ATP, which is joined to the sulfur atom by the carbon in position 5 of its ribose. The trisubstituted sulfur atom, or *sulfonium,* confers great lability upon its attached methyl group, with the result that the cleavage of the S^+–CH_3 linkage will release an amount of energy equivalent to that of the

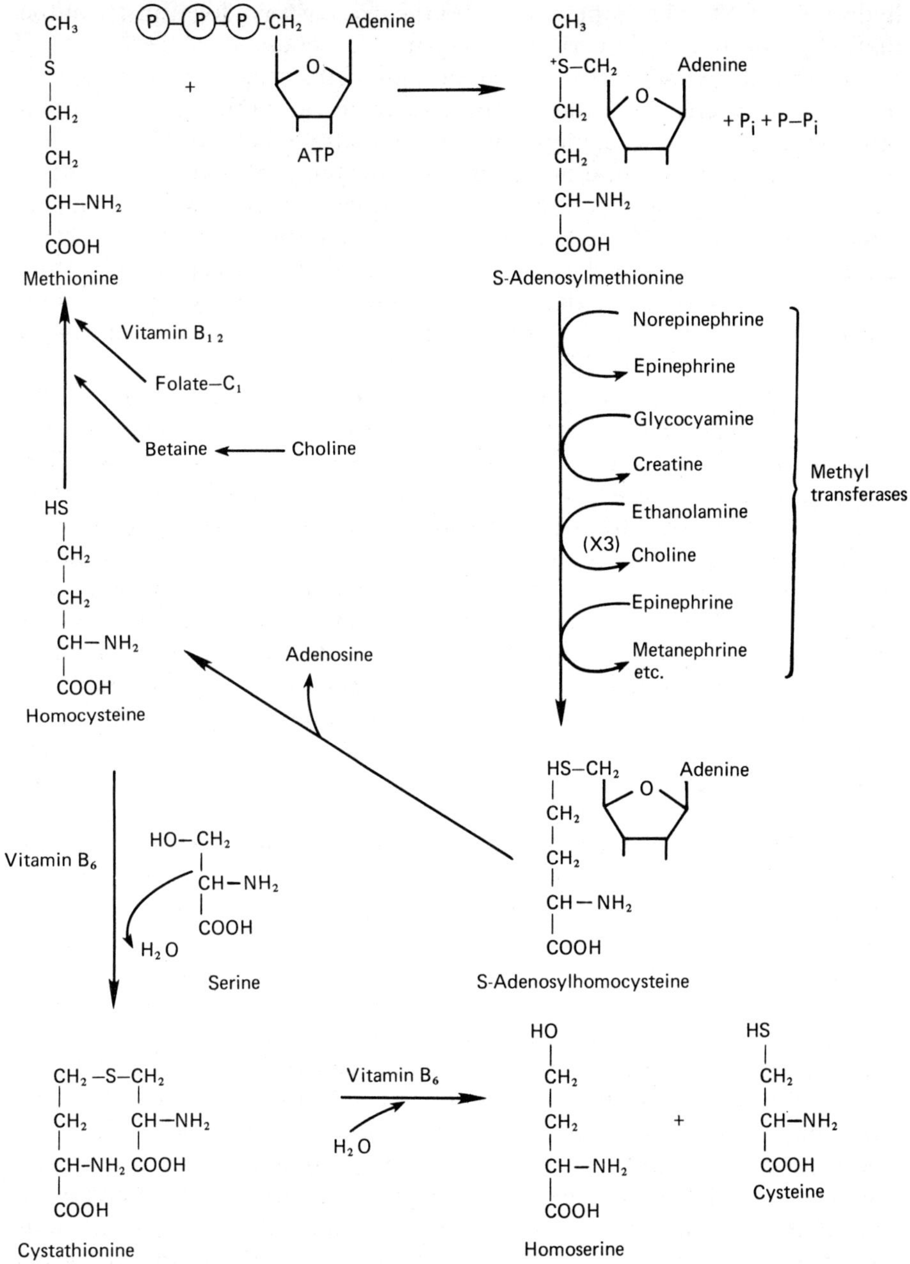

FIG. 6-16. Metabolism of the sulfur-containing amino acids, methionine and cysteine, showing their role in methyl-transfer reactions.

hydrolysis of ATP. This, then, provides the driving force for transfer of the methyl group to a wide variety of acceptor compounds.

In the case of several products of methyl transfer (*e.g.*, choline, epinephrine, and creatine), the reaction serves a biosynthetic function; in other instances (*e.g.*, metanephrine and *N*-methylnicotinamide), the end products are destined for elimination from the body. Although the transfer can occur with various atoms acting as acceptors, the methyl group is most commonly transferred to a nitrogen or an oxygen atom. In each case, a different enzyme, which is specific for the methyl-acceptor substrate, catalyzes the transfer reaction. Such enzymes are generally termed ***methyltransferases.*** They usually operate in an irreversible fashion, so that a continuous supply of methionine is required in the diet.

The other product of methyltransferase action is *S*-adenosylhomocysteine, the demethylated form of the active methyl donor. The adenosine group is released by a hydrolytic enzyme to yield free homocysteine. The latter may be reconverted to methionine at the expense of a methyl group from *betaine,* an oxidation product of the methyl-containing substance, choline. Consequently, dietary choline may spare some of the need for methionine in relation to the methyl-group requirements of the body. The conversion of homocysteine to methionine may also be carried out by employing other single-carbon donors attached to folic-acid coenzymes, as described in the next section; an obligatory agent in this pathway, which probably accounts for one tenth to one quarter of the total methyl-group utilization of the body, is vitamin B_{12}. Accordingly, the malabsorption or impaired cellular metabolism of vitamin B_{12} leads to defects in the utilization of homocysteine by remethylation.

The bulk of the homocysteine produced upon demethylation of methionine is converted in the liver to cysteine. It should be noted that the prefix *homo* refers to the homologous relation to cysteine, that is, there is one extra CH_2 in the chain of homocysteine. Thus, the conversion from the homologous amino acid should formally require only the removal of one carbon. In fact, none of the carbons in homocysteine is converted to cysteine; only the sulfur atom is transferred, which occurs in a two-reaction process known as *transsulfuration.* The sulfur is joined first by a synthetase enzyme to the amino acid, serine, to form the thio ether, *cystathionine.* In this compound, the sulfur atom acts as a bridge between the homocysteine and serine moieties. A hydrolytic enzyme then cleaves the bridge on the other side of the sulfur atom, effectively transferring the sulfur to the serine part to produce cysteine and leaving behind the homolog of serine, homoserine. Because both the synthetic and the hydrolytic enzymes function essentially in an irreversible manner, homocysteine does not accumulate to any significant degree in normal body fluids or cells. Both reactions require the coenzyme form of vitamin B_6, pyridoxal phosphate, and hence the transsulfuration pathway will depend strongly upon the presence of adequate amounts of this vitamin in the diet.

Genetic Diseases

Inherited deficiencies of either the synthetic enzyme or the hydrolytic cleavage enzyme of the transsulfuration path may be transmitted as recessive traits: in the first case, homocysteine will accumulate and be oxidized to the disulfide form, homocystine; in the second, cystathionine will build up. In both situations, these metabolities, which are not normally present in the urine to any extent, will be rapidly excreted by the kidney, hence the disorders are designated *homocystinuria* and *cystathioninuria.*

Homocystinuria is probably the second most common inborn error of amino-acid metabolism, following phenylketonuria in incidence. It is invariably associated with dislocation of the lens of the eye (ectopia lentis), skeletal changes (osteoporosis), and cardiovascular disease (thromboembolism). Frequently the patients are mentally retarded. The characteristic eye abnormality together with a positive test for disulfide (–S–S–) compounds in the urine (cyanide-nitroprusside reaction) provides a means for diagnosis. Death may occur in young adulthood due to occlusion of the large arteries by thrombi. In some cases, very large doses of vitamin B_6 will correct the biochemical disturbances, indicating that these patients have the ability to produce the synthetase apoenzyme, which, however, reacts abnormally, with a high K_m value for its coenzyme.

A similar observation has been made in several cases of cystathioninuria, in which the hydrolytic enzyme probably also has an increased dependence upon pyridoxal phosphate. The therapeutic advantages of long-term vitamin-B_6 administration, of methionine restriction, or of supplementary cysteine intake do not appear to have been completely evaluated as yet in terms of the pathology of these conditions. This is doubly difficult with cystathioninuria, which, though detected first among inmates of mental retardation institutions, shows no clear, consistent, clinical pattern.

PYRIMIDINE AND PURINE METABOLISM

These nitrogen-containing ring structures are essential components in the nucleotide coenzymes of many metabolic reactions as well as in the polynucleotides of cellular RNA and DNA. The diet appears to play a minor part in fulfilling the requirements for pyrimidines or purines in humans; amino-acid and carbohydrate precursors generally furnish the starting materials for the complete biosynthesis of the heterocyclic rings. Although virtually all tissues, especially rapidly growing ones such as the hematopoietic tissues, have the capacity for renewing the nucleic acids and nucleotides, the liver is the principal site in the body for the formation of the ring compounds when starting from scratch. Catabolic enzymes, which are utilized in the elimination of excess pyrimidines, are capable of breaking these ring structures down completely. On the other hand, the human lacks the enzymes necessary to break open the purine rings and hence must excrete any excess intact in the form of uric acid.

A complex system of controls is imposed upon pyrimidine and purine synthesis, reutilization, and degradation in order to achieve the proper balance. Defects in these balancing mechanisms are particularly crucial in relation to purine metabolism: interference with purine biosynthesis will cause a severe restriction of cell growth and multiplication, whereas its excessive catabolism to uric acid gives rise to the arthritic conditions known collectively as *gout*.

Origin of the Rings

The pyrimidine ring is derived from the condensation of two readily available precursors, aspartic acid and carbamyl phosphate. The latter is the same compound that is utilized for urea formation, but it seems to arise differently when used to produce pyrimidines (Fig. 6-17). The carbamyl-phosphate synthetase of the urea cycle uses NH_3 and is contained inside the mitochondrion where the next reaction in the pathway occurs. The synthetase involved in pyrimidine biosynthesis, however, obtains its NH_2 from the amide nitrogen of glutamine and is found outside the mitochondrion in the cytosol. The pyrimidine-nucleotide end products, UTP and CTP, act to suppress these initial reactions of the synthetic pathway in a classic example of negative-feedback control. The aspartate-carbamyl transferase enzyme, for example, is subject to allosteric inhibition by CTP, such that any excess of the latter in the cells will automatically turn off its own biosynthesis at an early stage in the pathway. Similarly, the cytosol form of carbamyl-phosphate synthetase, but not the mitochondrial enzyme, is inhibited by UTP feedback. Although competition will occur in the liver cell for the common precursors—aspartate and carbamyl phosphate—of urea and pyrimidine biosynthesis, the compartmentalization of the enzymes and the specific regulatory mechanisms allow the two pathways to operate independently and in parallel, according to the overall needs of the body.

In a series of reactions, the carbamyl-aspartate molecule is cyclized to form the ring structure and then oxidized to *orotic acid*. The latter is converted to a nucleotide with the help of the activated sugar, *phosphoribosylpyro-phosphate* (PRPP) (Fig. 6-18). This compound, which is produced from ribose-5-phosphate at the expense of two terminal phosphates from ATP, is highly energized and thus contains within itself the driving force to link the ribose carbon in position 1 to the pyrimidine position-1 nitrogen. Finally, a decarboxylation leads to formation of the product, *uridylic acid* or UMP.

Further tailoring of the pyrimidine ring by the addition of NH_2 from glutamine leads to the cytidine nucleotides, and the addition of CH_3 leads to the thymine nucleotides. The CH_3 of the thymine compounds does not

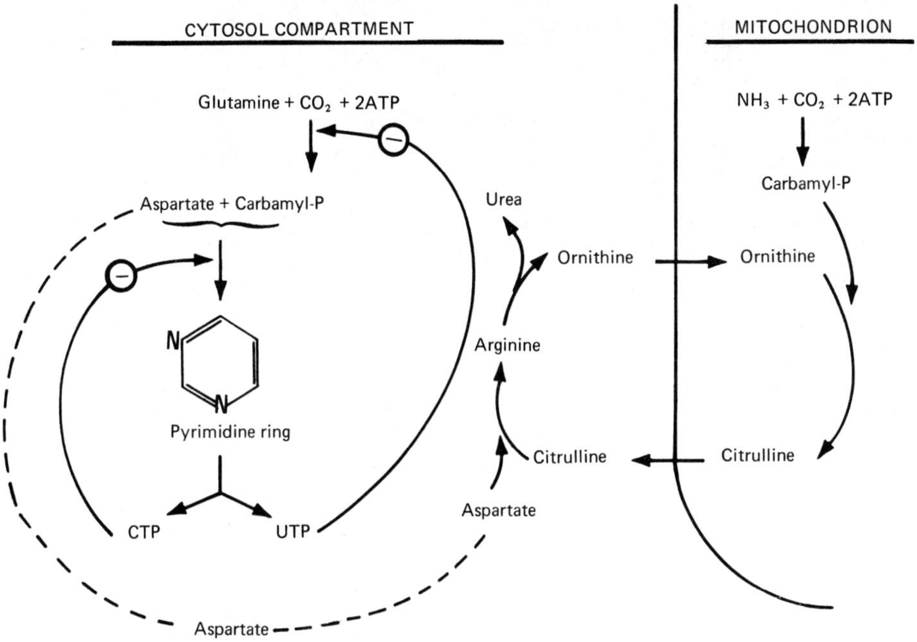

FIG. 6-17. Outline of pyrimidine ring formation, demonstrating the compartmentalization and control of the carbamyl–phosphate synthetase as distinct from that which is involved in urea synthesis.

come from the usual methyl source, *S*-adenosylmethionine, but rather from the reduction and transfer of a single-carbon fragment (formaldehyde) that is bound to the coenzyme, folic acid. This is the same type of reaction for the generation of methyl groups that was discussed earlier for the re-formation of methionine from homocysteine, and it, too, requires the participation of vitamin B_{12}. The syndrome related to the deficiency of the latter, *pernicious anemia,* is also associated with an impaired supply of thymine nucleotides for the blood-forming tissues, and the consequent inability to keep up with DNA synthesis in the differentiating blood cells leads to inadequate erythrocyte replacement in the circulation.

The synthesis of the purine ring structure is a considerably more complicated production. Its complexity and the number of intermediates involved rival those of the major metabolic pathways, such as glycolysis. A few key features of this process, however, are of interest and importance: relative to the pyrimidines, the purines are synthesized backward, with the ribose portion being added first instead of at the end of the process. There are several contributors to the purine ring atoms instead of the two simple precursors of pyrimidines: glycine is the largest single contributor, providing one nitrogen and two carbons; two of the nitrogen atoms come from glutamine and one from aspartate; one carbon arises from CO_2; and

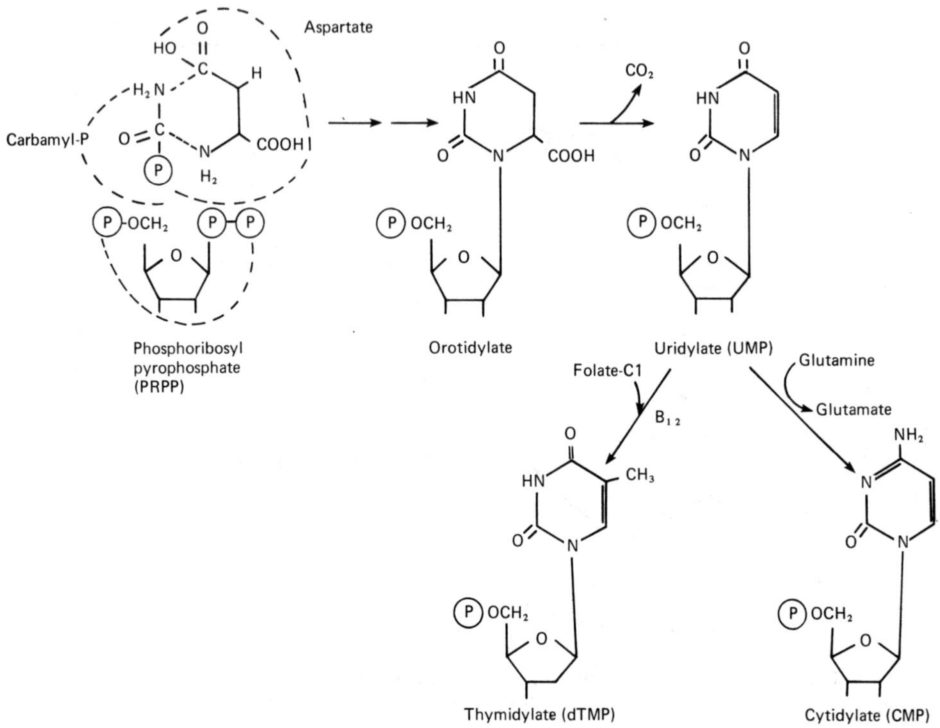

FIG. 6-18. Details of pyrimidine–nucleotide biosynthesis.

the other two carbons are added separately as one-carbon fragments joined to folic-acid coenzymes (see Figure 6-19 for the order of these additions).

The activated ribose, PRPP, acts as the starting point for purine synthesis, with an NH_2 group donated by glutamine displacing the energized pyrophosphate. Next, glycine joins on at the expense of ATP, with its carboxyl group forming an amide link to the NH_2. It is at this point, prior to the closure of the small imidazole ring, that one carbon from the folic-acid coenzyme is added to the NH_2 of the substituted glycine. In the formation of the larger, fused ring, one carbon and two nitrogen atoms are sequentially added, and again the folic-acid-dependent transfer of the last carbon atom occurs just before the ring closes. Ring closure hence produces the completed purine nucleotide, *inosinic acid* (IMP). The latter acts as the parent compound for other purine nucleotides: the addition of a NH_2 group from aspartic acid at position number 6 forms *adenylic acid* (AMP); oxidation followed by the addition of NH_2 from glutamine at position number 2 yields *guanylic acid* (GMP).

The dual requirement for folic-acid-mediated transfers at two stages leading to the completion of the purine ring renders purine biosynthesis,

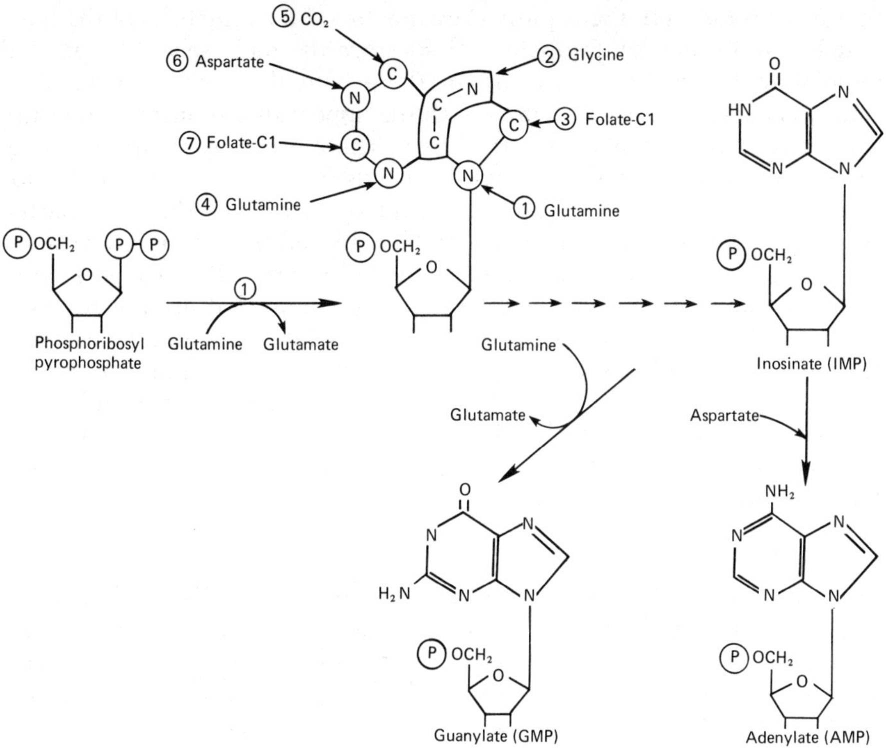

FIG. 6-19. Biosynthesis of the purine nucleotides. Additions of substituent atoms to the ring structure are indicated in order (1 to 7).

RNA and DNA biosynthesis, and hence cell proliferation acutely dependent upon the maintenance of normal folic-acid concentrations and metabolism. The coenzyme form of folic acid is the reduced derivative, *tetrahydrofolate,* and it may transfer one-carbon fragments at various oxidation levels. A major source of these one-carbon fragments is the reaction catalysed by the mitochondrial enzyme, *serine hydroxylmethyl transferase*:

$$\text{Serine} + \text{Tetrahydrofolate} \rightleftarrows \text{Glycine} + CH_2\text{–Tetrahydrofolate}$$

The bound CH_2–group, at the oxidation level of formaldehyde, may be transferred as such, oxidized to the level of formic acid or CO_2 for catabolism, or reduced to the level of CH_3 for methylations. The serine hydroxylmethyl transferase is widely distributed in cells involved in nucleic acid synthesis and in the brain, where serine serves as an important source of glycine. Folic acid is a vitamin that is required for human nutrition, and its lack in the diet constitutes one of the most prevalent deficiency syndromes in the world today. As noted previously concerning shortages in thymine production, the rapidly growing, blood-forming cells of the bone marrow and other hematopoietic tissues are the most seriously affected by folic-acid deficiency. Accordingly, the principal symptom of this deficiency is a characteristic anemia with the appearance of large, immature erythrocytes (megaloblasts) in the blood. There is a close relationship between the effects and the nutritional needs for folic acid and those for vitamin B_{12} (cyanocobalamin). A deceptive "cure" of the hemopathic symptoms in pernicious anemia may be achieved by the ingestion of excessive folic acid; however, the deficiency of vitamin B_{12} produces additional devastating effects upon the nervous system that are not reversed by folic acid. The amelioration of the megaloblastic anemia by administration of folic acid is, therefore, a dangerously misleading indicator. This is particularly critical in intestinal dysfunctions such as ileitis and sprue, where the absorption of both folic acid and vitamin B_{12} may be inadequate to serve the body's needs. Folic-acid deficiency is also commonly encountered in infancy and pregnancy.

Uric-Acid Formation

As noted previously, the pyrimidine rings may be totally catabolized in the body to CO_2 and NH_3, and the nitrogen arising from pyrimidine breakdown will thus be excreted as urea. No parallel mechanism exists, however, for purine ring degradation. Some free purines, which arise either from the digestion of dietary nucleic acids or from the catabolism of tissue nucleic acids and nucleotides, may be reutilized in the body by the so-called salvage mechanisms discussed in the next section. These reactions, however, are of main significance for the thrifty reclamation of the preformed purines in extrahepatic tissues. In the liver, free purine bases from exogenous or endogenous sources are, for the most part, deaminated to form xanthine or hypoxanthine. The oxidation of both these metabolites is catalyzed by the flavin-containing enzyme of the liver, *xanthine oxidase,* in the presence of O_2 to produce the end product, uric acid (Fig. 6-20). Since the latter cannot be further metabolized nor reutilized, like urea it finds its way into the urine as a final nitrogenous waste substance. Its presence thus provides an index of nucleic-acid metabolism in the body. Urinary uric-acid excretion reflects in part the amount of exogenous purine ingested; when a person is equilibrated on a standardized diet that is virtually purine-free, the daily uric-acid output in the urine will indicate the extent of endogenous purine turnover.

Gout and Related Diseases

Gout is a syndrome characterized by arthritis and an excess of uric acid in the blood *(hyperuricemia).* The causes, the frequency of acute inflammatory attacks, and the associated effects upon other organs are variable. Generally, the condition is more prevalent in males, and the severity of the symptoms and the rate of recurrence increase with age. The inflam-

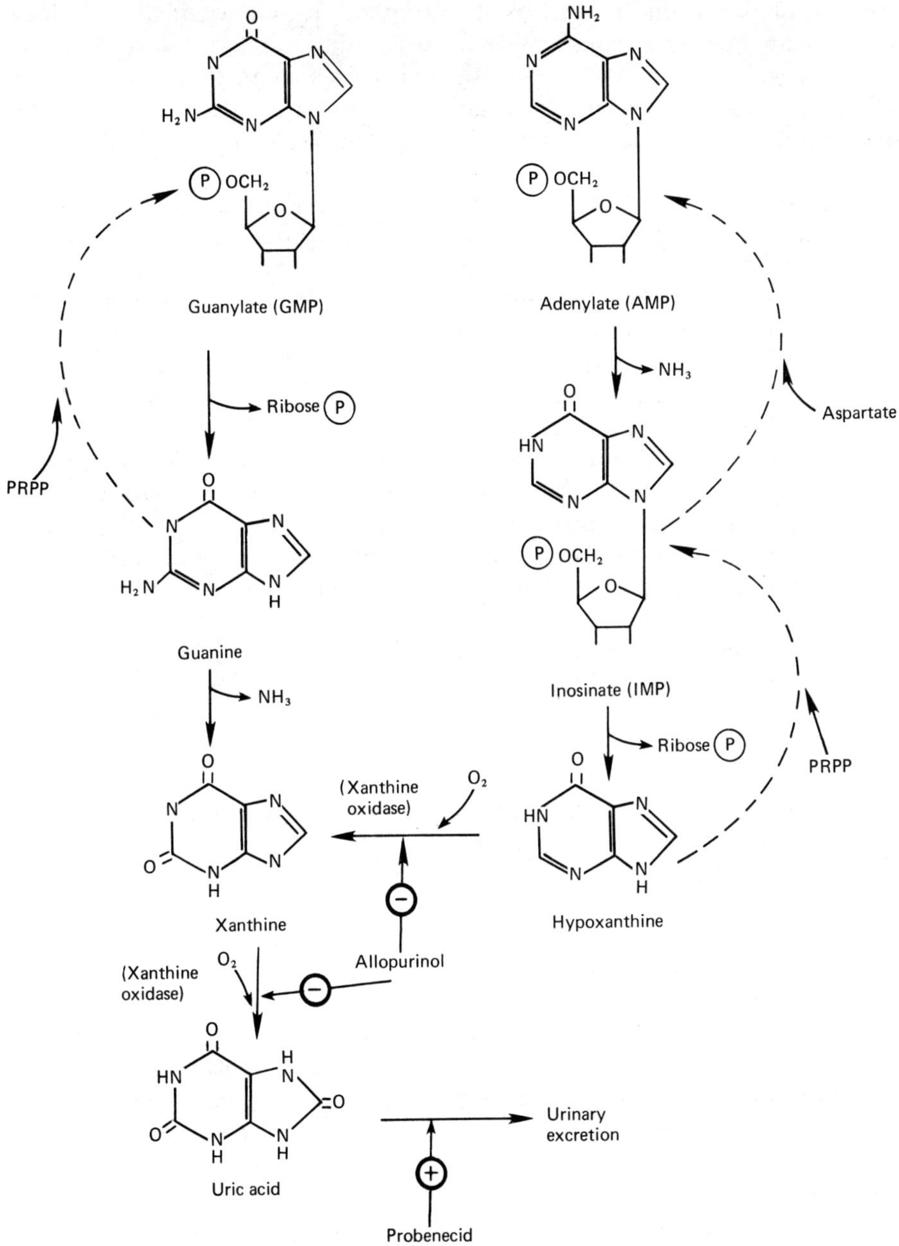

FIG. 6-20. Catabolism of the purines to form uric acid. The purine salvage reactions are shown by broken lines. *(PRPP,* phosphoribosylpyrophosphate.)

mation of the joints as well as the common accompaniment of kidney damage are clearly attributable to the precipitation of crystals of the monosodium salt of uric acid in the affected tissues. With time, the deposits of sodium urate may grow into large, visible, bulbous lesions termed *tophi*, particularly around the joints of the extremities and the calculi of the kidneys.

Even in normal humans, the concentration of urate salts in the body fluids approaches the limit of solubility. In gouty patients, the blood may contain urates in excess of the theoretical solubility, that is, in supersaturated solution. Although serum proteins will stabilize the supersaturated solutions for extensive periods, localized decreases in pH (*e.g.*, along the urinary tract) or possibly the breakdown of solubilizing factors (*e.g.*, in joint connective tissues) may lead to a localized focus of crystallization for uric acid or its salts.

The primary cause of the hyperuricemia is most usually a metabolic overproduction of uric acid; underexcretion may be a contributing factor, particularly as a secondary consequence of renal disease, alcohol ingestion, or heavy-metal poisoning of the kidneys. The metabolic disorders in gout are, like the pattern of inheritance, multifactorial. In some cases, the overproduction of uric acid is associated with a defect in the purine salvage mechanism. An enzyme, *hypoxanthine-guanine phosphoribosyl transferase,* normally can reutilize hypoxanthine or guanine by converting them to their respective nucleotides; this reaction requires the driving force provided by the other substrate, PRPP, and is strictly analogous to the process whereby free orotic acid is joined to ribose in the main pathway of pyrimidine-nucleotide biosynthesis (Fig. 6-20). In certain gouty adults, an inherited partial deficiency of this transferase enzyme results, for unexplained reasons, in more active, total purine production by the biosynthetic pathway. When the enzyme is completely absent (the so-called Lesch-Nyhan syndrome), the consequences are extremely dire, and the hyperuricemia is accompanied by severe neurologic disturbances. These are typified by retarded mental development in early childhood and compulsive self-biting. Drugs such as *allopurinol,* which reduces uric-acid formation by inhibiting xanthine-oxidase activity, and *probenecid,* which enhances the renal clearance of uric acid, have been used effectively to reverse the hyperuricemia and to relieve many of the symptoms of these disorders.

HEME METABOLISM

Heme is the bright-red complex formed between iron and the nitrogen-containing porphyrin ring. It is the functional group of the oxygen-carrying proteins, such as the hemoglobin of erythrocytes and the myoglobin of muscles, and of the oxidative enzymes, such as the oxygenases and hydroxylases of the liver and the ubiquitous cytochromes of tissue mitochondria. The maintenance and regulation of normal rates of heme for-

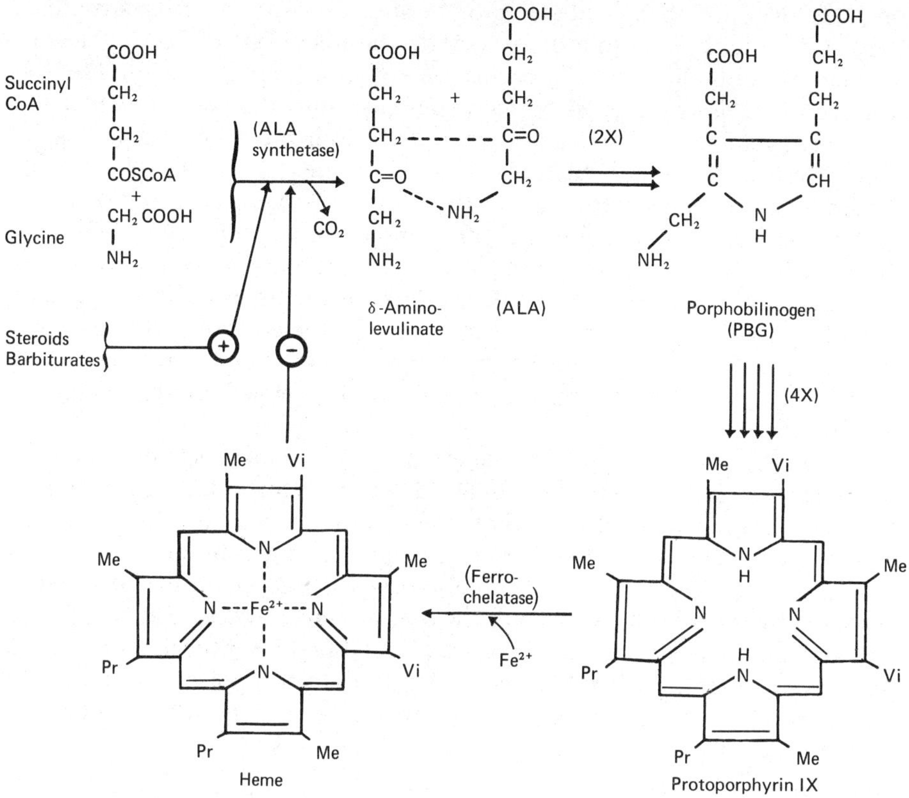

FIG. 6-21. Biosynthesis of porphyrin and heme. (*Me,* methyl; *Vi,* vinyl; *Pr,* propionic acid.)

mation and breakdown are of crucial importance, not only in relation to the essential role of hemoproteins in the body, but also because of the severe toxic effects manifested by the metabolites and by-products that may arise during disturbances of heme production or utilization. Although the hematopoietic tissues such as bone marrow, the spleen, and other components of the reticuloendothelial system play a cardinal role in heme metabolism, the liver, because of its regulatory role in porphyrin biosynthesis and its function in the elimination of the bile-pigment waste products, is the central organ of heme metabolism.

Porphyrin and Heme Synthesis

The essential features of the ***porphyrin*** ring structure and its formation may be summarized simply as follows (Fig. 6-21): first, succinyl CoA (which provides most of the carbon atoms of the porphyrin) and glycine (which provides all of the nitrogen) combine together to produce δ-*aminolevulinic acid* (ALA). Next, two molecules of ALA condense and cyclize into the substituted, five-membered ring of *porphobilinogen* (PBG). Four of the PBGs are then bridged into the larger porphyrin ring structure, which, following a couple of enzymatic modifications of the outer substituents, yields the naturally occurring isomer *protoporphyrin IX*. Finally, a ferrous ion is complexed by ferrochelatase into the very center of the flat porphyrin ring by linkages to each of the four nitrogens to yield *heme*. Both the first reaction, which is catalyzed by the enzyme *ALA synthetase*, and the last reaction, which is catalyzed by *heme synthetase*, are localized inside the mitochondria, but the other intervening steps occur in the cytosol. It is the first enzyme of the sequence that controls the operation of the entire pathway for heme synthesis. ALA synthetase is normally present at only 1/100th of the activity of the next step (the condensation reaction) and consequently is rate-limiting. Moreover, both the catalytic activity and the synthesis of this enzyme are subject to feedback inhibition by the end product, heme.

Steroid hormones, such as the estrogens and glucocorticoids, as well as certain drugs, such as the sulfonamides and barbiturates, can stimulate the induction of new ALA synthetase in the liver, whereas high concentrations of glucose will, like heme, prevent the induction of enzyme synthesis. The concentrations of available iron will modulate the conversion to heme by ferrochelatase, as will the oxygen tension (PO_2). The last factor is particularly important in the control of heme biosynthesis in the bone marrow, in keeping with the observation that erythropoiesis is adaptively accelerated by the low oxygen concentrations experienced by dwellers at high altitudes.

Porphyrias

Some of these conditions arise from genetic disorders of metabolism in the bone marrow and hence are termed *erythropoietic porphyrias;* others, which originate either from a primary, congenital liver defect or from a secondary, latent sensitivity of heme synthesis in the liver to certain toxic agents, are referred to as *hepatic porphyrias*. The characteristic feature of all these syndromes is the accumulation of various porphyrin derivatives in the tissues, although the distribution of the porphyrins will vary with their site of origin. In the erythropoietic porphyrias, for example, the accumulated porphyrins may be readily detected in the erythrocytes and the blood-forming cells by their intense fluorescence under ultraviolet irradiation. In cases of the hepatic porphyrias, the cells of the liver, but not of the bone marrow, will exhibit fluorescence under similar conditions. In many cases, the porphyrin compounds will spill out in the urine (uroporphyrias), where their deep, port-wine coloration and ultraviolet fluorescence form the basis of diagnostic tests.

In all types of porphyria but one, the afflicted individuals exhibit skin lesions of varying severity as well as photosensitivity. Another common but variable accompaniment is neurologic involvement. This is seen in the hepatic porphyrias only, and the effects range from intense pains in the abdomen or extremities to paralysis, seizures, hallucinations, psychosis, and coma. Like the hemophilias, hepatic porphyria has been referred to as the "royal malady" because of its hereditary transmission through the Houses of Hanover, Prussia, and Stuart; the madness of England's King George III has been attributed to this disorder.

The precipitating factors of acute attacks of porphyria may be either exposure to sunlight or the ingestion of drugs or hormones that act to induce the synthesis of ALA synthetase. In the hepatic porphyrias, the activity of this enzyme in the liver is elevated many times above the normal. It would seem that these syndromes differ fundamentally from other inborn errors of metabolism, in which the lesion leads to lowered enzyme activity. The increase in ALA-synthetase activity probably reflects a genetic defect in the normal control system such that the synthesis of enzyme is "hyperinducible" and not subject to the normal restraints of repression.

Bile Pigments

Like their precursors, the hemoproteins, bile pigments are brilliantly colored substances, ranging from green (biliverdin) to reddish yellow (bilirubin). As their name implies, they impart the coloration to the bile. The biliary fluid is the normal route for the elimination of the bile pigments, which are the end products of the catabolism of the porphyrin moieties of hemoproteins. Unlike the apoprotein and the iron, which are conserved in the body, the porphyrins are not reutilized at all but are quantitatively excreted as their open-chain tetrapyrrole derivatives.

When bile pigments accumulate in the blood and other body fluids either due to excessive formation or due to inadequate removal in the bile, they may produce an intense yellow coloration of the skin. This condition, termed *icterus* or *jaundice,* may result from the massive breakdown of erythrocytes (hemolysis) leading to overproduction of bilirubin from hemoglobin catabolism, but it is more commonly associated with a defect in the mechanisms whereby the liver disposes of the normal flux of bilirubin. In adults, bilirubin has few serious toxic effects; in the newborn, however, severe and permanent damage, particularly to the brain, may result from jaundice.

Heme Degradation

Several tissues are capable of hemoprotein catabolism. This is apparent from the rainbow of bile pigments produced from the extravasation and local breakdown of hemoglobin in the skin and underlying tissues, as exemplified by a "black eye" or other superficial bruises. Quantitatively, the bulk of heme degradation products arises from the hemoglobin released by the hemolysis of the erythrocyte, when it reaches the end of its normal life span of 4 months. Continuous removal of the aging cells and the subsequent conversion of hemoglobin to bile pigments occur predominantly in the spleen, with reticuloendothelial elements of the liver and bone marrow playing a secondary role. All these tissues contain a highly active enzyme system, *heme oxygenase,* which is associated with the endoplasmic reticulum (Fig. 6-22).

Before it is susceptible to attack, the heme moiety must be dissociated from the hemoglobin or other hemoproteins. In the presence of O_2 and NADPH, a membrane-bound cytochrome, designated *P450,* oxidizes one of the carbons that bridges the pyrrole rings and releases it as carbon monoxide. This effectively opens the porphyrin ring to yield the linear, tetrapyrrole chain of *biliverdin.* The globin (or other apoprotein portion) is hydrolyzed to free amino acids, which will be reconverted in part to newly synthesized protein. The iron that is released when the heme is cleaved will be almost quantitatively retained for recycling into new hemoproteins.

The biliverdin rarely accumulates or escapes from the cells of its origin into the blood, because it is acted upon as soon as it is formed by a very active, soluble, reductase enzyme that also requires NADPH. The product of this reaction is ***bilirubin,*** the principal bile pigment found in the body. Bilirubin itself is sparingly water-soluble and probably exists in cellular or extracellular fluids only as a protein complex. Because of its lipid solubility, however, free bilirubin passes readily through the hydrophobic cell membranes. Consequently, the bilirubin formed in the reticuloendothelial tissues diffuses into the bloodstream, where it forms a soluble complex with albumin in the plasma. This is the major, normal transport form of the bile pigments.

Conjugation of Bilirubin

Upon reaching the hepatic cells, the albumin–bilirubin complex dissociates, and free bilirubin is taken up across the cell membrane. It then becomes bound to a specific carrier protein in the cytosol. A major function of the liver is to convert the bilirubin, which is brought from various sites of heme breakdown in the body, into a water-soluble form that may be excreted in the bile (Fig. 6-22). This is accomplished by attaching sugar derivatives on two of the pyrrole rings, a process known as *conjugation.* The sugar derivative, *gluronic acid,* is added on in its activated form, UDP-glucuronic acid; the latter is derived in the liver by the action of a soluble dehydrogenase upon UDP-glucose, utilizing 2 mols of NAD. A specific enzyme in the endoplasmic reticulum, *glucuronyl transferase,* catalyzes the transfer of a molecule of activated glucuronic acid to each of two propionic-acid carboxyl groups in ester linkages. The resulting conjugated form of the bile pigment, *bilirubin diglucuronide,* has the hydrophilic sugar residues to offset the hydrophobic property of the rest of the molecule. It is not only highly water-soluble itself, but it facilitates the solubilization of free bilirubin by complexing with it. Together, the conjugated and unconjugated bilirubin molecules may then be passed from the hepatic cells

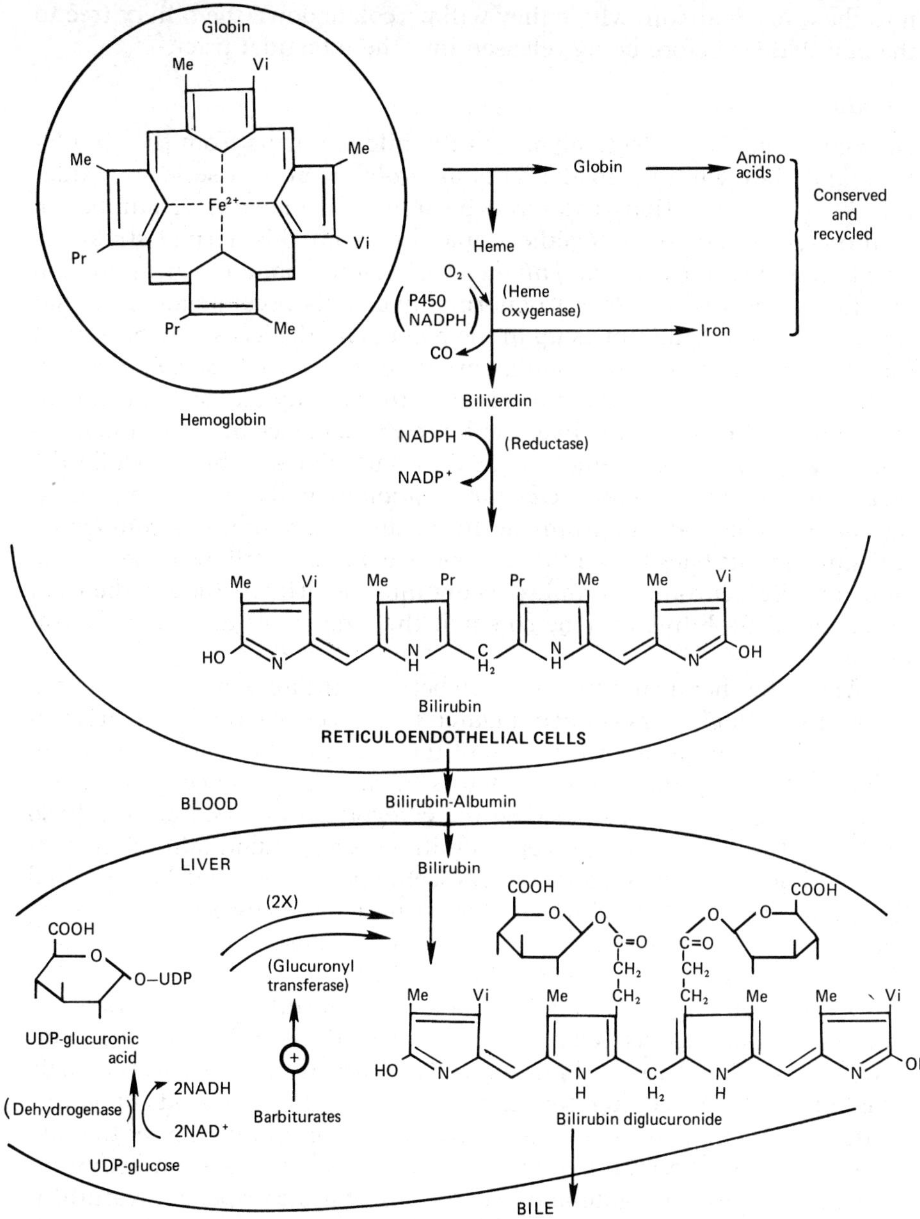

FIG. 6-22. Catabolism of heme and conjugation of the bile pigments.

into the bile canaliculi, where they will percolate down the biliary tree to the gallbladder before being released into the intestinal tract.

Jaundice

Normally, the liver effectively clears the bile pigments from the circulation, and they wind up in the feces as *urobilins* after a series of oxidation–reduction reactions catalyzed by intestinal bacteria. A number of factors—either singly or together—may interrupt this normal stream of events (Fig. 6-23). *Hemolytic jaundice* will result from an overproduction of bilirubin beyond the ability of normal liver cells to conjugate it, so that unconjugated bilirubin piles up in the blood, and the feces may be almost black from the increased urobilins excreted. *Hepatocellular jaundice* will be the outcome of diseases that impair either the uptake or the conjugation of free bilirubin by the liver, with the consequence that unconjugated bilirubin from normal hemolysis again accumulates in blood, while the fecal urobilins are decreased. *Obstructive jaundice* will occur when a blockage of the bile drainage from the liver causes a buildup of conjugated bilirubin and its back flow into the bloodstream. An overflow of the soluble bilirubin-diglucuronide complex occurs into the urine, although the fecal excretion of urobilins may be so small that the stools are virtually colorless.

Thus, the chemical differentiation between the amounts of conjugated ("direct reacting") versus unconjugated ("indirect reacting") bilirubin in the blood, as well as an estimation of the excretory bile pigments, are of value in deducing the causative factors in cases of jaundice. In practice, however, the distinctions between the categories described are rarely so clear-cut, that is, it is rare, except in initial stages, to find biliary obstruction without there being some hepatocellular damage that has resulted from regurgitation of the bile. In other situations, erythrocyte instability, intrahepatic defects (either of metabolism or transport), and cholestasis may be associated in various combinations.

A common type of jaundice due to hepatic insufficiency occurs in the newborn infant and is referred to as *neonatal hyperbilirubinemia.* This problem is related to the biochemical immaturity of the liver at birth. The fetus lacks the conjugating enzyme system, and it depends upon placental exchange with the mother's circulation for elimination of the bilirubin. At birth, the infant is cut off from this support system before his own capacity to conjugate and excrete bilirubin has ripened. Naturally, any superimposed acceleration of hemolysis—as seen in blood-group incompatibility for Rh factor *(erythroblastosis fetalis),* ABO, or other antigen—will severely aggravate the situation and intensify the hyperbilirubinemia. Moreover, in contrast with the transient "physiologic" jaundice of the full-term newborn (which normally corrects itself within 2 weeks as the hepatic conjugating enzyme develops), the retarded differentiation

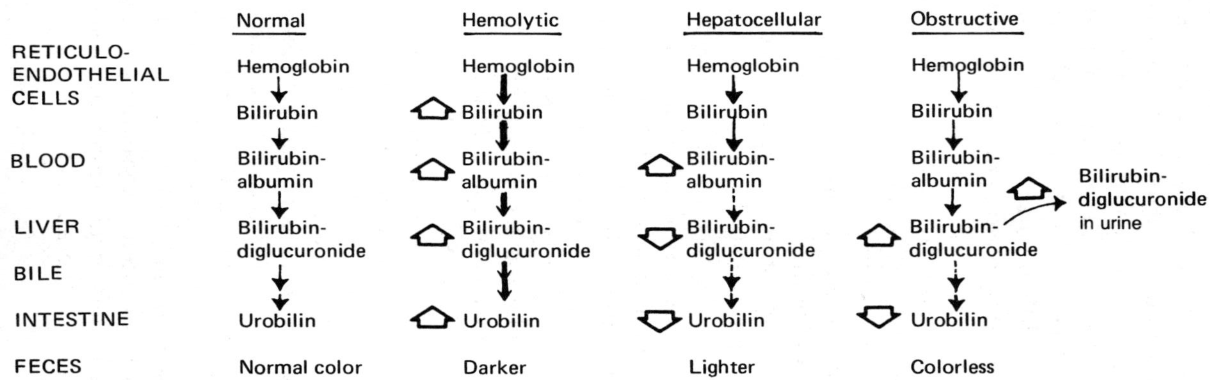

FIG. 6-23. Three categories of jaundice and the changes of bile-pigment metabolism in different tissues. The solid arrows indicate the sequence of metabolic reactions, the thickness of the arrow showing the extent of each reaction relative to normal. The open arrows indicate the changes (increase or decrease) in content of bile pigments in various tissues, body fluids and excreta.

of liver enzymes in premature infants may cause the plasma levels of unconjugated bilirubin to increase to dangerously high values.

The danger from bilirubin is related to its toxic effects upon the developing brain, which in turn is related to the ready passage across cell membranes of its unconjugated form. Neurotoxic effects from jaundice alone are rarely seen in adults, and conjugated bilirubin cannot enter the brain cells from the bloodstream. The deposition of bilirubin in the brain occurs in cases of *kernicterus.* In acute cases, the infants develop convulsions and die in the early neonatal period. Those who are less severely affected and survive through childhood may nonetheless suffer permanent neural damage that is manifested by ataxia and other motor disturbances as well as by mental retardation.

The treatment of severe hyperbilirubinemia in infancy has frequently required heroic measures: in some cases, it has proved necessary to carry out a complete exchange transfusion of the child's blood; in other instances, children were administered barbiturates, since these drugs are known to promote the induction of the hepatic glucuronyl-transferase enzyme that is required for bilirubin conjugation. More recently, the known instability of the bile pigments when subjected to ultraviolet irradiation has been exploited in what may be termed "phototherapy." In this approach, the exposure of the patient to sunlight or other ultraviolet sources breaks down the excess bilirubin to water-soluble products that can be excreted in the bile. The last is the treatment of choice for children with the inborn error of metabolism that leads to congenital absence of the bilirubin-conjugating system *(Crigler–Najjar syndrome),* since such patients often show a tendency to develop kernicterus and their livers cannot be induced to synthesize the enzyme for conjugation.

Hereditary chronic jaundice cases have been documented in which the symptoms are generally milder, the increases in circulating bile pigments intermittent in frequency and less severe in intensity, and the course of the disorder more protracted. These syndromes may be related to only a partial deficiency of the glucuronyl-transferase activity in the liver *(Gilbert's disease),* in which there is a moderate and fluctuating increase of unconjugated bilirubin in the serum. Alternatively, a defect in the ability to transport bilirubin diglucuronide from the hepatic cells into the bile *(Dubin–Johnson syndrome)* leads to variable accumulations of conjugated bilirubin in the blood. Phenobarbital has been used in patients with either disorder in efforts to induce a greater conjugation capacity and thus possibly reduce the hyperbilirubinemia. In most cases, however, the symptoms are so mild that no therapy is necessary.

THE KIDNEY

The role of the kidney is vital to the metabolism of nitrogenous substances in the body, not only from the obvious standpoint of elimination of ni-

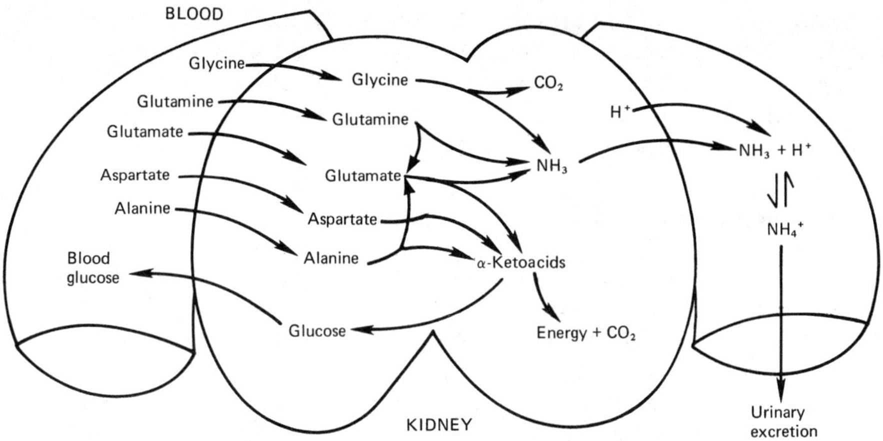

FIG. 6-24. Role of the kidney in amino-acid and ammonia metabolism.

trogenous wastes, but also in providing an alternative site for glucose synthesis from amino acids and in regulating ammonia production to compensate for excessive hydrogen ions in the bloodstream. The excretory mechanisms of the kidney cells are remarkable in terms of both their selectivity and their efficiency. To cope with the hundreds of different metabolites reaching them through the blood, the epithelial cells of the kidney possess in their membranes a host of specific transport systems that must recognize and separate each friend from foe. The friendly, essential metabolites of the circulation (*e.g.*, amino acids and glucose) must be retained in the body, and the sometimes toxic wastes (*e.g.*, ammonia and urea) must be expelled in the urine. Much of the metabolic effort and energy requirement of the kidney is directed to the transport processes involved in excretion. Gluconeogenesis and the acid–base functions of renal cells are coordinated, since both the carbon for the formation of glucose and the nitrogen for ammonia excretion are derived from the deamination of amino acids.

AMINO-ACID OXIDATION

Kidney cells exhibit a considerable ability to deaminate amino acids. Aminotransferases for many common amino acids and the glutamate-dehydrogenase enzyme are highly active in renal mitochondria. These activities allow for the oxidative deamination of the amino acids that are carried to the organ in the bloodstream (Fig. 6-24). The latter is the major source of free amino acids for renal metabolism, since the kidney cells normally carry out little proteolysis of either their own tissue proteins or those of the blood plasma.

Among the blood amino acids, those that are most readily acted upon by the kidney are glutamate, aspartate, alanine, and glycine. The first three are subject to the combined actions of aminotransferases and glutamate dehydrogenase, but the last requires a separate enzyme system, *glycine oxidase,* which acts by direct oxidative deamination. All these mechanisms function essentially irreversibly in the kidney, which has little capacity for amino-acid biosynthesis.

AMMONIUM-ION METABOLISM

One of the major jobs of the kidney is to adjust the *p*H of the bloodstream within the narrow limits required for the optimal operation of the enzyme systems throughout the body. The mechanisms of *p*H control are complex, including such factors as the production of organic acids or carbonic acid (CO_2) as tissue metabolites and the elimination of CO_2 in the lungs. The kidney provides an ideal system for disposing of hydrogen ions in the urine. This involves the production of ammonia by accelerated amino-acid breakdown. The ammonia, as uncharged NH_3, diffuses freely from the tissue, through cell membranes, and into the urine, where it may associate with a hydrogen ion to produce the ammonium cation:

$$NH_3 + H^+ \rightleftarrows NH_4^+$$

The ammonium ion does not pass freely through the cell membrane and consequently can act as a hydrogen ion sink in the urine.

The advantages of such a system to the body are several: first, ammonia is a waste product that is freely and cheaply provided to the kidney from amino-acid catabolism. Second, large amounts of hydrogen ion can be passed in the urine without decreasing the *p*H below tolerable limits for the urogenital tissues. Third, the elimination of the ammonium cation as a companion of the anions in the urine (*e.g.*, phosphate or organic-acid anions such as acetoacetate) spares the body's reserves of sodium and other needed cations. Finally, by adaptive changes in ammonia generation, the kidney can respond to fluctuations in acid production elsewhere in the body.

Glutaminases

The kidney possesses a number of enzyme systems that are capable of breaking down glutamine, but it lacks the capacity for the synthesis of glutamine, as it does for most amino-acid anabolic reactions. The source of glutamine for hydrolysis in kidney is provided by the bloodstream. Renal cells have a great avidity for the circulating glutamine, which originates in the first instance largely from the ammonia produced by liver cells. The glutamine in turn provides most of the ammonia produced within the kidney, so that glutamine may be considered as a nontoxic form of ammonia to shuttle between the liver and the kidney. The renal glutaminase functions in a manner similar to the hepatic glutaminase, releasing glutamate and NH_3 through a simple hydrolytic reaction. The kidney enzyme differs, however, in being stimulated greatly by inorganic phosphate. The glutamate produced may, of course, be further deaminated by glutamate dehydrogenase; thus glutamine may ultimately yield two molecules of NH_3.

Role in Acidosis

Under conditions of normal health, ammonium-ion excretion in the urine is a small fraction of the total output of urinary nitrogen. Adaptive increases in renal NH_3 excretion may occur in ***metabolic acidosis***—that is, when other tissues of the body increase their output of acidic metabolites to the degree that the blood *p*H drops—or, to a lesser extent, in ***respiratory acidosis***—that is, when the lungs fail to remove CO_2 rapidly enough. In some acidotic patients, for example, in untreated diabetics who produce and excrete large quantities of acetoacetic and β-hydroxybutyric acids, the urinary ammonia may increase tenfold or more, so that it becomes the major nitrogenous waste substance.

Several factors contribute to this adaptive increase of ammonia formation in response to acidosis. First, the liver is stimulated to produce more glutamine and secrete it into the bloodstream. Second, the rate of uptake of glutamine by the kidney is enhanced so that the circulating levels of glutamine are maintained at about the same steady-state concentration as in the normal condition of acid–base balance. Third, the amounts of renal glutaminase, aminotransferases, and glutamate dehydrogenase increase gradually with time. Finally, the accelerated utilization of α-ketoglutarate and other keto-acid products of amino-acid degradation is promoted by the stimulation of gluconeogenesis and citric-acid-cycle oxidation in renal mitochondria. Thus, the kidney is able to adjust its NH_3 production through increased substrate availability, the removal of other products, and the development of greater catalytic potential when the ambient *p*H falls.

AMINO-ACID TRANSPORT

As mentioned previously, the kidney has the capacity to clear the bloodstream of foreign substances quite efficiently. The foreign substance may in fact be a normal intracellular intermediate (*e.g.*, homocysteine or homogentisic acid), which appears to the kidney as an interloper since the compound does not normally pass into the bloodstream. Such substances, therefore, are removed from blood as rapidly as they enter it, and they may be actively secreted by the kidney into the urine. Such a situation is clearly pathologic, and it would present serious difficulties for the nitrogen economy of the body if the kidney were to deal indiscriminately with all the circulating amino acids in this wasteful manner. In fact, it is an important function of certain renal cells to retrieve the important substances, such as glucose, amino acids, and inorganic ions, and prevent their loss from the body in company with the waste products. This selective retention process is termed ***renal reabsorption.***

During the production of urine, the fluid and the low-molecular-weight components of blood plasma are filtered into the collecting tubules of the kidney. Throughout their length, the tubules are lined with epithelial cells of varying absorptive capacities. At the near end of the tubules are situated those cells whose function is to reabsorb the amino acids from the filtrate before it drains into the urinary tract. Under normal circumstances, close to 100% of the amino-acid content is returned to the blood in this way; this explains why the amino acids are usually only found as minor constituents of the urine.

The uptake of amino acids by renal tubular cells requires energy, since the molecules must be moved against a concentration gradient. The specificity of the process as well as other features are reminiscent of the action of an enzyme on its substrate; for example, the rate of reabsorption reaches a saturation limit analogous to the V_m of enzymatic reactions at high amino-acid levels, and amino acids of similar structure are capable of competitively interfering with the uptake of one another. All these properties indicate that the amino acid that is reabsorbed forms a carrier complex—analogous to the enzyme–substrate complex—with a protein molecule that contains a specific binding site on the outside of the cell membrane. An input of energy, either as ATP or possibly as an ionic gradient, then drives the amino acid across the membrane, where it is released inside the cell. The free carrier then switches its binding site back to the external surface to await another amino acid and thus may be said to act catalytically since the carrier itself remains unchanged. A similar carrier system acts in the gastrointestinal tract for the uptake of dietary amino acids. The carriers are specified by the genes in the same manner as the enzyme proteins are. In some instances, an inherited defect of the carrier system will affect both the gastrointestinal uptake and the renal reabsorption of a particular amino acid or a group of related amino acids (*e.g.*, see the earlier discussion of Hartnup disease). The related amino acids involved in a group-specific transport system may compete for the same carrier. Five such groups have been identified: neutral amino acids (including aromatics such as tryptophan), dibasic amino acids (including cystine), dicarboxylic amino acids, iminoacids (plus glycine), and β-amino acids.

Renal Aminoacidurias

The appearance of elevated quantities of amino acids in the urine indicates either of two general causes: a *metabolic aminoaciduria*, typified by phenylketonuria, in which an enzymatic lesion, either inherited or acquired, prevents the utilization of a particular amino acid and results in its overflow in urine; or a *renal aminoaciduria*, typified by Hartnup disease, in which the primary defect lies in the inability of the renal tubules to reabsorb certain amino acids. As in the case of the true inborn errors of metabolism, the renal-carrier deficiencies may be inherited, but they may also be acquired due to the action of toxic substances that poison the renal epithelial cells. Since a single carrier system functions for the transport of several amino acids, renal tubular defects are expressed as multiple aminoacidurias encompassing the constituents of a group-specific system, and they may be even more generalized in cases of nonspecific damage to the kidney.

Cystinuria is the best characterized and most common of the inherited renal aminoacidurias. Apart from the tendency to form "stones" composed of highly insoluble cystine in the urinary tract, the disorder is apparently without serious symptoms. To prevent this obstruction by cystine calculi, both the volume of urine and its *p*H must be kept high. Under these conditions, which may be produced by the administration of copious fluids and alkaline agents, the excreted cystine will be kept below the saturation concentration, and small stones will be dissolved. In addition to cystine, the urine contains variable quantities of the dibasic amino acids—arginine, lysine, and ornithine—which bear a superficial structural resemblance to cystine and presumably are reabsorbed normally by the same group-specific carrier system.

Other genetically determined, specific aminoacidurias of the same type as cystinuria are known, for example, Hartnup disease and iminoglycinuria. In the latter disorder, only glycine and iminoacids, proline and hydroxyproline, are excreted in excess, which again accents the dependence of tubular reabsorption upon the group-specific systems.

Generalized aminoaciduria may also be observed in certain inborn errors of metabolism. This abnormality may develop with age as, for example, in *galactosemia*, where the accumulated galactose-1-phosphate gradually poisons the renal tubules, or in *Wilson's disease* (hepatolenticular degeneration), where the retention and deposition of copper produces a similar effect. The aminoaciduria is obviously a secondary manifestation of the disease condition in such cases, and the loss of amino acids from the body nitrogen pool is rarely a significant factor in their pathology. In the generalized inhibition of tubular reabsorption, a number of other substances—for example, glucose or phosphate—will also be excreted in large amounts. Thus, one of the major pathologic sequelae of the inherited general aminoaciduria called *Fanconi's syndrome* is deficient bone mineralization, leading to consequent fractures, which is caused by the excessive loss of phosphate in the urine. In children with this disorder, massive doses of vitamin D are required to offset this phosphate excretion and allow normal bone growth. The ingestion of certain drugs (*e.g.*, nitrobenzenes) or heavy metals (*e.g.*, mercury, lead, or cadmium) that interfere with the kidney's energy production will prevent the uptake of a number of substances in the tubules and thus result in an acquired general aminoaciduria, with glycosuria and phosphaturia also being present.

NITROGENOUS WASTES

The quantitive aspects of nitrogen excretion were stressed earlier in relation to dietary intake and the discussion of nitrogen balance. Of equal importance in the assessment of health and disease is the qualitative nature of the nitrogenous waste substances. It is as if we were trying to deduce the status of household occupants from an examination of their garbage; the weight of the refuse may give us a rough idea of the level of activity of the residents, but we would have to undertake a more rigorous inspection of the containers and labels before we could arrive at any conclusions about the state of affairs within the house. The identification and analysis of the individual waste components of the urine affords us an analogous opportunity to diagnose the condition of nitrogen metabolism in the body (Fig. 6-25).

Urea is the largest single component of urinary nitrogen excretion, constituting more than three quarters of the total in normal adults. The amount excreted bears a direct relationship to dietary protein intake. Fluctuations from normal may also reflect the functional states of the liver or kidneys. In the extreme case of total renal shutdown, for example, the absence of urine production will result in elevations of blood urea *(uremia)*. At the other extreme, in complete hepatic failure no urea will be produced for the kidney to excrete. Between these extremes, the relative concentrations of urea in the blood and urine represent the balance between the synthetic capacity of the liver and the ability of the kidneys to clear this waste substance efficiently from the blood.

Creatine and *creatinine* are metabolites that arise chiefly within the muscle cells, and their excretion provides an indication of the status of the musculature of the body. Creatinine, for example, is produced in a continuous manner from muscle creatine phosphate by a nonenzymatic process. Since it cannot be converted back into creatine and since the creatinine released into the blood is actively secreted into the urine, it is rapidly and irreversibly eliminated as it is formed. The amount of creatinine excreted is independent of the dietary nitrogen intake and is quite constant from day to day in the same person, regardless of the urine volume. For this reason, it is often used as a baseline of comparison for other urinary excreta. The amount excreted by a given person is directly related to the person's size and, in particular, to the muscle mass of that person. In contrast to creatinine, creatine may be reutilized for phosphocreatine synthesis, and, in order to conserve it, it is reabsorbed by the renal tubules in the same manner as the amino acids. In young children and in pregnant women, small amounts of creatine appear in the urine, but in normal adults, little is excreted. Elevated creatine excretion is often indicative of muscle-wasting disorders, such as those that occur in starvation and various types of muscular dystrophy.

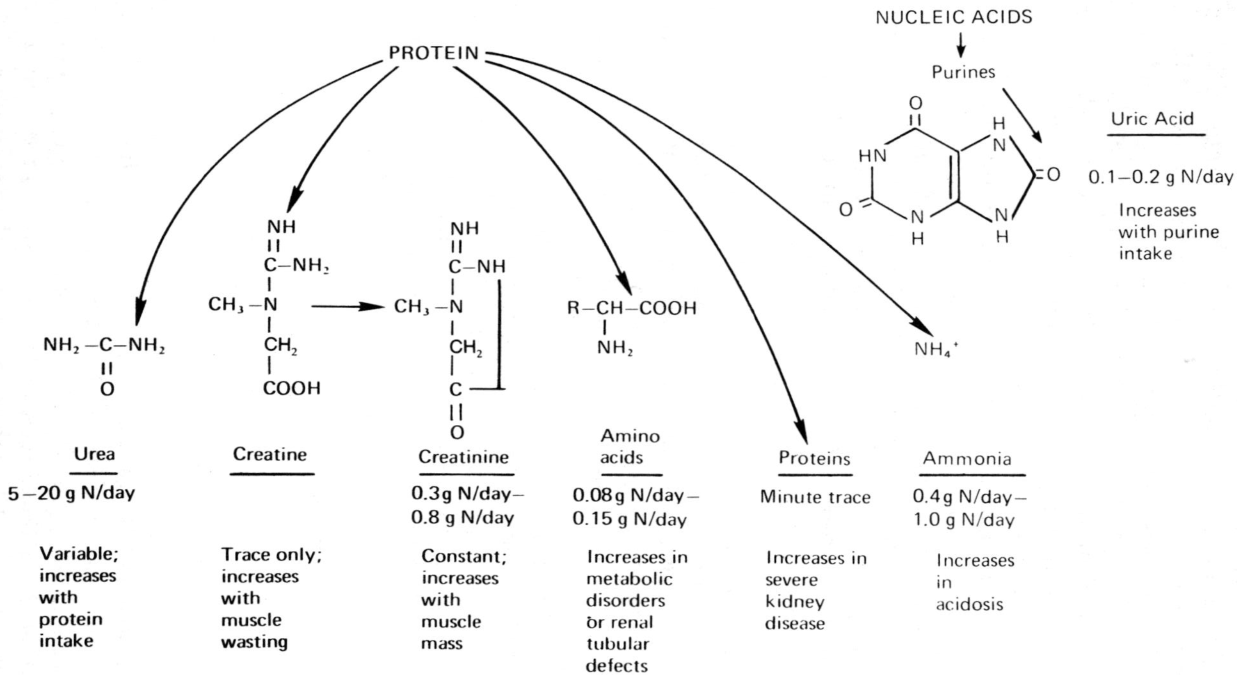

FIG. 6-25. Nitrogenous wastes in normal urine and the circumstances leading to increased excretion of each.

Uric acid excretion has been discussed in connection with purine metabolism and gout. Obviously, the quantity of uric acid in the urine of normal individuals is related to the nucleic-acid content of the diet. Since there is a high content of the latter in most high-protein foods such as meat, the excretion of uric acid follows a pattern parallel to that of urea but on a diminished scale. High-protein foods such as eggs and milk products have a very low purine content, and a diet of these may be used to lower the uric-acid production and hence its excretion. The kidneys are very effective at removing uric acid from blood plasma, employing an active secretion process in the renal tubules.

Amino acids form a small part of the normal nitrogen output in urine. Among those that are less efficiently reabsorbed are glycine, histidine, glutamine, serine, and alanine. Consequently, these amino acids appear most conspicuously in normal urine, and in conditions of generalized aminoaciduria, these components are also most notably increased. The appearance of other groups or of specific amino acids in elevated amounts in the urine is pathognomonic of metabolic disorders or renal tubular defects.

Ammonia excretion has also been discussed in some detail. It must be stressed that the ammoniacal smell of most urine samples intensifies on standing due to the decomposition of other nitrogenous wastes; consequently only determinations on fresh samples are of value. In addition to the increase in urinary ammonia produced by acidosis, increases in the levels of ammonia or amino acids in the blood plasma will increase the urinary excretion of ammonium ion.

Proteins do not normally find their way into the urine. Only smaller proteins can pass from the blood into the glomerular filtrate, and most of these are reabsorbed efficiently. Any increase above trace levels of protein in the urine is pathologic and suggestive of severe kidney damage. In such cases, for example, in *nephrotic syndrome*, the presence of high albumin levels in the urine with reciprocally lowered levels in the blood serum may be particularly notable.

THE BRAIN

The metabolism of the amino acids and their derivatives in neural tissue is important in early life for the growth and differentiation of myelin, the nerve cells, and their processes and, in the mature nervous system, for its maintenance and function. The effects upon the brain of dietary protein deprivation, such as those seen in kwashiorkor, are the most dramatic during early infancy. Although many of the other symptoms may be reversed by increased protein intake, there is good evidence now that unless such dietary supplements are instituted in the crucial first months of life when the brain is developing rapidly, permanent mental defects will ensue. Recalling the similar irreversible effects of amino-acid excesses (*e.g.*, as in phenylketonuria) during the early neonatal period will make it apparent that a correct balance of all the amino acids, in roughly equivalent proportions, is essential for normal mental development.

At all stages of life, the brain has the capacity for active transport of the essential amino acids, thus maintaining higher concentrations in the nerve cells than those which are supplied in the blood plasma. These concentration mechanisms are of particular importance for the uptake of the aromatic amino acids, which serve as the precursors of certain amines. Such amines may act as neurotransmitters or have modulating effects on nervous function, and they are produced within the nerve cells. Several nonessential amino acids—such as glutamate and aspartate (or their derivatives) as well as glycine (which can arise in nerve cells from serine by the action of serine hydroxymethyl transferase)—may also produce alterations in neural excitability and are found in high concentrations within the brain. These nonessential amino acids are synthesized inside the nerve cells, and variations in their plasma concentrations will have little effect upon the cerebral levels.

GLUTAMATE METABOLISM

Of the nonessential amino acids found in neural tissue, *glutamic acid* is the most significant from the standpoint of concentration and the interconversions to other important compounds. Among the latter are glutamine and α-ketoglutarate—which provide linkages with ammonia and carbohydrate metabolism, respectively—and the inhibitory neurotransmitter substance, γ-aminobutyrate (GABA). Together these members of the glutamate family constitute over half of the total brain amino acids. They are generated in the nerve cells using the carbon atoms provided from blood glucose, and this accounts for about one tenth of the total carbohydrate utilization in the brain (Fig. 6-26).

The glutamate of neural tissue may arise either by amino-group transfer from any of a number of amino acids to α-ketoglutarate or by the reductive amination of this latter citric-acid-cycle intermediate utilizing ammonia in the presence of glutamate dehydrogenase. This enzyme functions predominantly in the direction of glutamate synthesis in the brain, and it will be quite sensitive to ammonia levels. When excessive ammonia is generated, however, as occurs in convulsive disorders, most of it will be captured by further conversion of the glutamate to glutamine by the highly active glutamine synthetase found in neural tissue. Unlike glutamate which penetrates the cell membranes poorly, glutamine is freely diffusible and thus may escape into the bloodstream or cerebrospinal fluid, thereby removing two molecules of the highly toxic ammonia from the brain. The measurement of the glutamine concentration in cerebrospinal fluid consequently provides a good indicator of the levels of

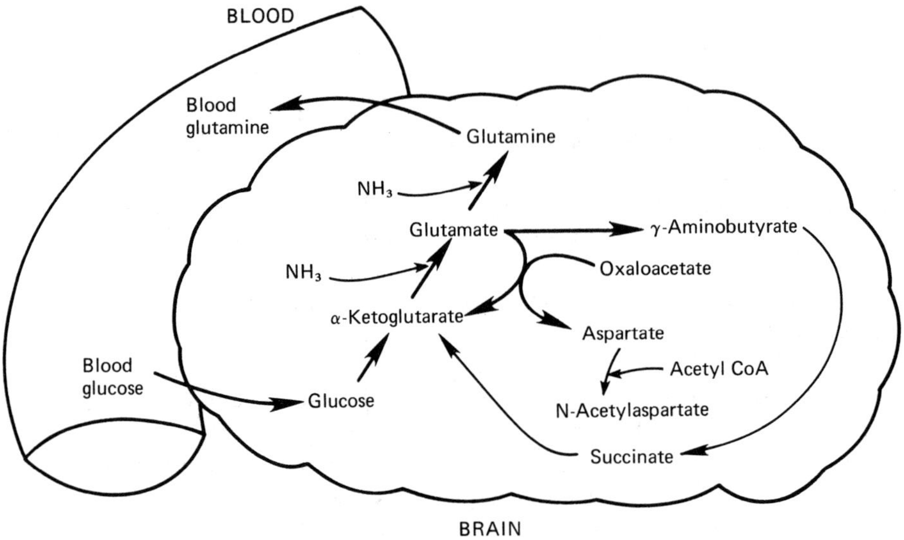

FIG. 6-26. Metabolism of amino acids in the brain.

ammonia reaching the brain in situations such as hepatic coma. The determination of the degree of interconversion of glutamine, glutamate, α-ketoglutarate, and NH_3, is complicated by the differences among different regions and cell types in brain. It seems, however, that neural tissue and compounds of the glutamate family are geared largely toward ammonia uptake and removal.

Although glutamate is not oxidized effectively by the brain and hence does not contribute significantly to its energy production, part of it may be diverted, by amino transfer, to the formation of other nonessential amino acids. Chief among these is aspartate, which is formed by the highly active glutamate-aspartate aminotransferase enzyme of neural tissue. Through reaction with acetyl CoA, aspartate can give rise to *N*-acetylaspartate, a compound whose concentrations rival those of the glutamate family and which is found in no other tissue of the body in such great amounts. The function of this unique component of the nervous system is not well understood. It may serve as a stable reservoir of acetyl groups in the brain that may be tapped for acetylcholine formation or fatty-acid synthesis. Alternatively, it may simply contribute to the fixed negative charge within the nerve cells themselves. More complex peptides containing *N*-acetylaspartate or *N*-acetylglutamate are also known to exist in the brain, but their role is obscure.

Gamma-Aminobutyric Acid

Gamma-aminobutyric acid, or *GABA* as it is frequently abbreviated in the medical literature, is another characteristic component of the brain. It is

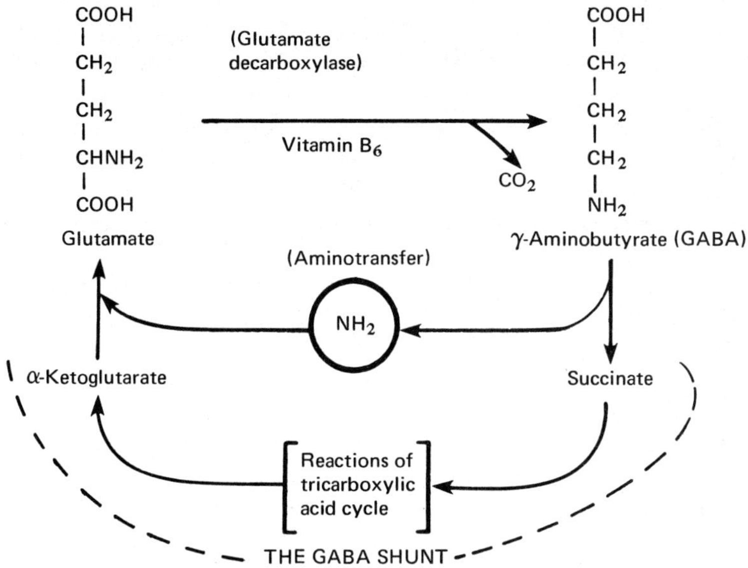

FIG. 6-27. Metabolism of γ-aminobutyrate (GABA) in the brain.

produced by the removal of the carboxyl group adjacent to the amino group of glutamate through the action of the specific enzyme *glutamate decarboxylase,* an enzyme that is mainly confined to neural tissues (Fig. 6-27). The concentrations of GABA in the brain parallel the distribution of those cells that exert suppressive or inhibitory effects upon other cells (*e.g.,* the Purkinje cells of the cerebellum), and it is considered likely that GABA acts as the neurotransmitter at the synapses of such inhibitory cells in the central nervous system (glycine appears to have a similar inhibitory neurotransmitter function in the spinal cord). These specific effects are exerted upon cells by the release of GABA on their external surface. In addition, GABA regulation inside the neurons plays a part in controlling the general excitability of the nerves; low levels of GABA are associated with convulsive states, and stable, higher levels with a reduced susceptibility to seizures. In general, the dicarboxylic amino acids, glutamate and aspartate, enhance the excitability of nerve cells, whereas the monocarboxylates, GABA and glycine, have a counteracting effect. The balance between these excitatory and inhibitory agents may therefore be significant for the modulation of the physiologic activity of the nervous system.

The metabolism of glutamate to GABA can also provide a bypass of the citric-acid cycle known as the *GABA shunt.* Only about one tenth of the total glucose utilized passes through this shunt. The GABA shunt involves the conversion of GABA to succinate and hence circumvents the α-ketoglutarate-dehydrogenase reaction.

Pyridoxine and Convulsions

As it is for many enzymes involved in amino-acid metabolism, notably the amino transferases and amino-acid decarboxylases, vitamin B_6 in the form of the coenzyme, pyridoxal phosphate, is an essential cofactor for GABA synthesis in the brain. The apoenzyme of glutamate decarboxylase binds pyridoxal phosphate poorly, so that the activity of the enzyme, and hence the concentrations of GABA in the brain, are drastically depressed in states of vitamin-B_6 deficiency.

In some infants, an inherited disorder characterized by convulsions and a high rate of mortality in the neonatal period has been shown to respond to the administration of large doses of vitamin B_6. This defect, termed *pyridoxine dependency* in untreated children, is related to an impaired conversion of glutamate to GABA, which indicates a lower than normal affinity of the glutamate decarboxylase for its coenzyme. The injection of GABA itself is ineffective as treatment, because the brain takes it up poorly from the bloodstream. The pyridoxine-dependent child, however, seems to survive and develop normally when diagnosed early and treated with massive supplements of the vitamin.

Deficiency of GABA production in the extrapyramidal centers of the brain that superimpose control upon motor functions has been demonstrated in patients with the inherited disease termed *Huntington's chorea.* This disorder is characterized by severe, progressive loss of muscle function and early death. Research into drugs that may rectify the chemical defect, for example by selectively inhibiting the GABA aminotransferase, is being actively pursued in hopes of alleviating this devastating neurologic condition.

BIOGENIC AMINES

These important metabolites of the amino acids are involved with the process of transmitting the electrical impulses from one nerve cell to another (Fig. 6-28). Upon stimulation of the first nerve cell, it releases a chemical agent into the extracellular-fluid gap between itself and the adjacent nerve cell. An increase of the ionic permeability of the second cell's membrane, which is caused by the agent in the gap or *synapse,* leads to the rapid diffusion of sodium ions into the second cell. This results in depolarization of the cell's membrane and thereby starts a wave of electrical discharges along the second nerve. The chemical agents that are produced by such stimuli and provide for transmission across the synapses of nerve cells are generally termed *neurotransmitters.* Such transmitter substances include the inhibitory agents, GABA and glycine, as noted previously. Four major types of excitatory neurotransmitters have been identified, with different distributions occurring throughout the nervous system. These include *acetylcholine* (found in cholinergic tracts), *nor-*

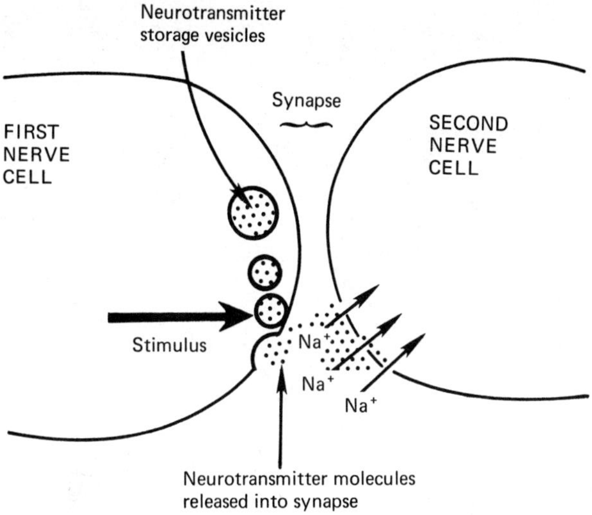

FIG. 6-28. Release and action of neurotransmitter agents at a nerve synapse.

epinephrine (found in adrenergic nerves), *serotonin* (found in serotonergic nerves), and *histamine* (which is implicated in some sensory pathways).

Acetylcholine

The action of this important substance in neural excitation may be conveniently divided according to the four phases in its life history: biosynthesis, storage, release, and degradation. Chemically, acetylcholine is a simple ester formed between acetate and the aminoalcohol derivative, choline. The latter is produced from the amino acid, serine, by decarboxylation to ethanolamine, followed by the addition of three methyl groups from *S*-adenosylmethionine (Fig. 6-29). As noted earlier, these transformations occur while the nitrogenous substances are complexed to phospholipids. Choline also plays an important additional role as a structural component of neural membranes.

The biosynthesis of acetylcholine, which is catalyzed by the specific enzyme *choline acetyltransferase,* takes place only in nerves and particularly in those regions where the nerve cells terminate at the surfaces of other nerves or of muscle cells. In the nerve endings, choline that has been either carried from the bloodstream or synthesized in the nerves reacts with the acetyl CoA that is derived principally from carbohydrate oxidation within the nervous system. The acetylcholine thus formed is rapidly taken up into a stable storage form and encapsulated within tiny, membrane-bounded vesicles (Fig. 6-28). When the nerve is stimulated, the vesicles join the surface membrane of the cell and release their contents, which consist of a concentrated packet of acetylcholine molecules, into the gap of the synapse. Here, the neurotransmitter migrates across to the

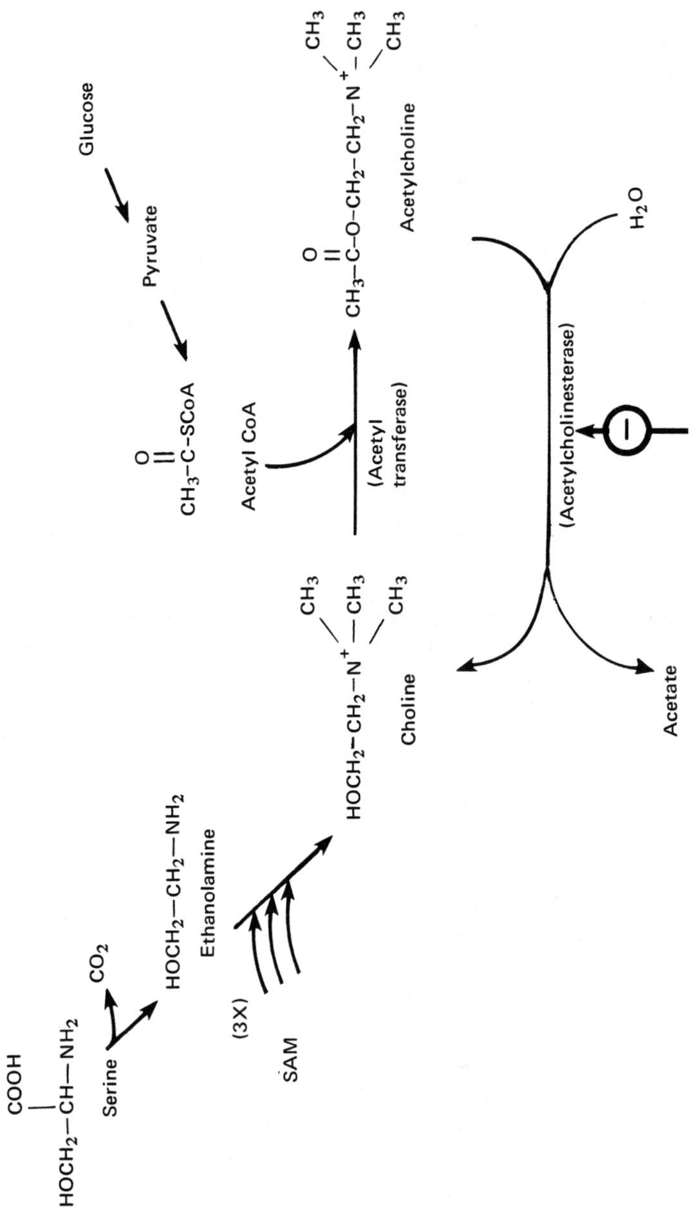

FIG. 6-29. Metabolism of acetylcholine in nerves. (*SAM, S*-adenosylmethionine.)

adjacent nerve-cell membrane to promote the sodium permeability increase that was described in Chapter 3. When its job has been done—that is, when the electrical change is produced in the second nerve cell—the remaining acetylcholine must be mopped up and removed so that its effect may be only transitory rather than sustained. This is accomplished at the membrane surface by a specific hydrolytic enzyme, *acetylcholinesterase,* which breaks the compound down to acetate and choline again (Fig. 6-29). The distribution of acetylcholine in the nervous system parallels those of the acetyltransferase and esterase enzymes, with the highest concentrations being found in the regions where the motor nerves terminate at the skeletal muscles and in the parasympathetic nerves such as the vagus nerve.

Specific inhibitors of the esterase enzyme permit the further accumulation of acetylcholine, since there is no other way to eliminate the neurotransmitter substance except through slow diffusion. Agents that produce such effects are termed *anticholinesterases* and include compounds that are useful as drugs to potentiate the effects of acetylcholine in controlled, localized situations, or as poisonous compounds. The latter include certain insecticides and nerve gases, which cause massive accumulations of acetylcholine in the nervous system and produce death in the affected organism by eliciting convulsions and respiratory failure.

The Catecholamines

The catecholamines are an important class of biogenic amines that include ***epinephrine*** (or adrenaline), the substance produced by the adrenal medulla and associated with the general systemic effects of stimulation of the sympathetic nervous system; ***norepinephrine*** (or noradrenaline), the precursor of epinephrine and the mediator of the synaptic transmission of impulses in the sympathetic nerves as well as in other classes of adrenergic fibers; and *dihydroxyphenylethylamine* or ***dopamine,*** which gives rise to the other two catecholamines and plays a role in special areas within the brain that modify motor activity. All these aromatic amines contain the dihydroxybenzene ring (catechol), and they arise in metabolism from tyrosine or phenylalanine (Fig. 6-30).

The biosynthesis of catecholamines in the adrenal medulla or in the nervous system commences with the action of *tyrosine hydroxylase,* an enzyme that is analogous to and uses the same cofactors and mechanism as the phenylalanine hydroxylase in the liver. This reaction involves the addition of a second hydroxyl group in the ring of tyrosine to form *dihydroxyphenylalanine,* which is referred to in the medical literature as *dopa.* A pyridoxine-containing decarboxylase then converts the dopa to dopamine by the removal of CO_2.

The importance of dopamine is noteworthy in the condition known as *Parkinson's disease,* which is characterized by a lack of control of motor activity and a drastic reduction in dopamine concentrations in extrapy-

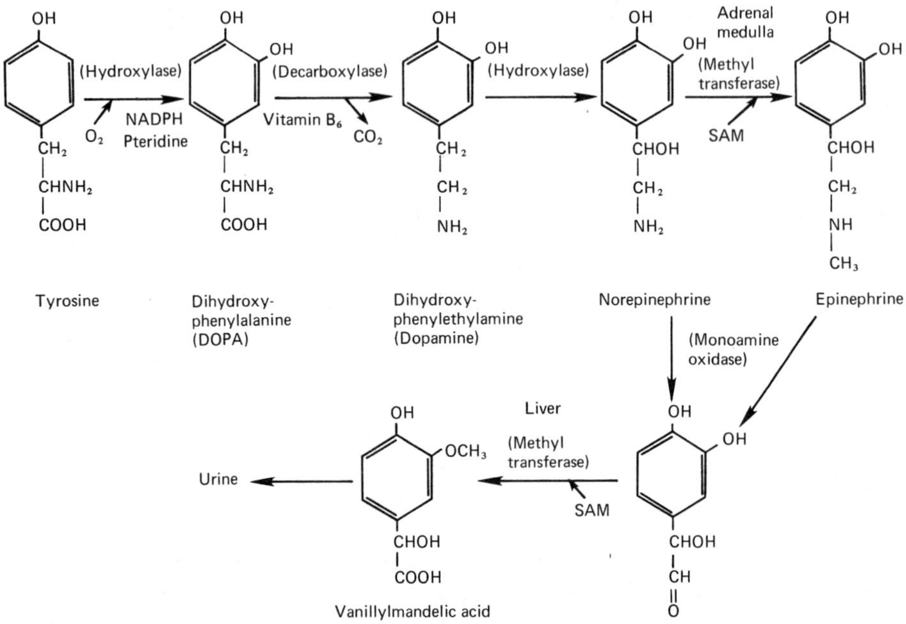

FIG. 6-30. Biosynthesis and degradation of the catecholamines.

ramidal centers of the brain that superimpose control upon the motor functions. The administration of dopamine itself has little effect since it does not seem to be able to enter the brain from the bloodstream, that is, it cannot cross the blood–brain barrier. Its precursor, dopa, does penetrate to the basal ganglia where the defect lies, however, and, as a result of the action of the intracellular decarboxylase, its administration can restore dopamine concentrations and provide short-term but dramatic relief of the symptoms of the disease in many patients.

In other regions of the nervous system, dopamine is an intermediate in the production of norepinephrine. The enzyme that catalyzes this conversion is a hydroxylase, but in this case the hydroxyl group is added in the side-chain rather than to the ring. The same reaction occurs in the adrenal medulla, where a final methylation, utilizing *S*-adenosylmethionine as the donor, generates epinephrine. Regardless of the locus of these synthetic reactions, the catecholamines are stored in the tissues in the form of granules or vesicles until such time as a stimulus arrives to release them.

After producing their effects upon cell membranes, several fates may befall the catecholamines to lead to their inactivation. First, and probably the most important from the standpoint of the rapidity of elimination, is the specific uptake of the substances by adjacent cells, effectively removing them from the membranes. Once inside the cells, a mitochondrial enzyme, *monoamine oxidase,* can rapidly strip off the amine group to leave behind the corresponding aldehyde compound. Alternatively, the catecholamines may enter the circulation and, upon reaching the liver, become inactivated biologically by the addition of a methyl group from S-adenosylmethionine to one of the ring hydroxyl groups. Eventually, a combination of both activities leads to the excretion of the catabolites resulting from catecholamine degradation as deaminated, methylated, aromatic acids such as vanillylmandelic acid (Fig. 6-30).

A number of pharmacologic agents that have inhibitory effects upon certain of the above reactions in the central nervous system may exert potent changes in behavior and mood. In general, agents such as *reserpine*, the *Rauwolfia* alkaloid, deplete the brain of norepinephrine by causing its discharge from storage vesicles and produce sedation and a tranquilizing effect. On the other hand, the monoamine-oxidase inhibitors, which prevent the elimination of norepinephrine from the brain and thereby lead to its accumulation in even higher amounts, promote excitement, wakefulness, and an antidepressant effect. The administration of the latter inhibitors may, however, provoke dangerous elevations of catecholamines in the brain if foods containing these substances are ingested simultaneously. Certain cheeses and fermented beverages, for example, contain large amounts of tyramine, the decarboxylation product of tyrosine, so that a wine-and-cheese party could very well prove fatal for an individual being treated with monoamine-oxidase-inhibiting drugs.

Serotonin

The compound ***serotonin,*** which is derived from tryptophan, functions as a neurotransmitter in certain central pathways of the brain. The serotonergic response is described as the classic protective reaction to fear, that is, withdrawal or flight, as opposed to the adrenergic response, which is associated with the aggressive tendency to fight. In addition to its localization in the brain, serotonin exists in high concentrations in the gastrointestinal tract and lung tissue.

Although the initial step in serotonin formation, the hydroxylation of the indole ring of tryptophan, may be catalyzed slowly by the nonspecific action of phenylalanine hydroxylase, a different specific hydroxylating enzyme exists in neural tissues, which produces the related compound, 5-hydroxytryptophan (Fig. 6-31). The tryptophan hydroxylase enzyme utilizes O_2 and a pteridine coenzyme, and its distribution in the brain regions coincides with the distribution of high concentrations of serotonin. The serotonin, or 5-hydroxytryptamine, is produced from 5-hydroxytryptophan by a decarboxylase enzyme.

The elimination of serotonin is chiefly as the result of the action of monoamine oxidase, so that inhibitors of this enzyme may also produce elevated concentrations of serotonin in the nervous system. The end product of serotonin catabolism, 5-hydroxyindoleacetic acid, is removed by the kidneys; hence, the urinary excretion of this metabolite provides an indicator of serotonin production and release in the body.

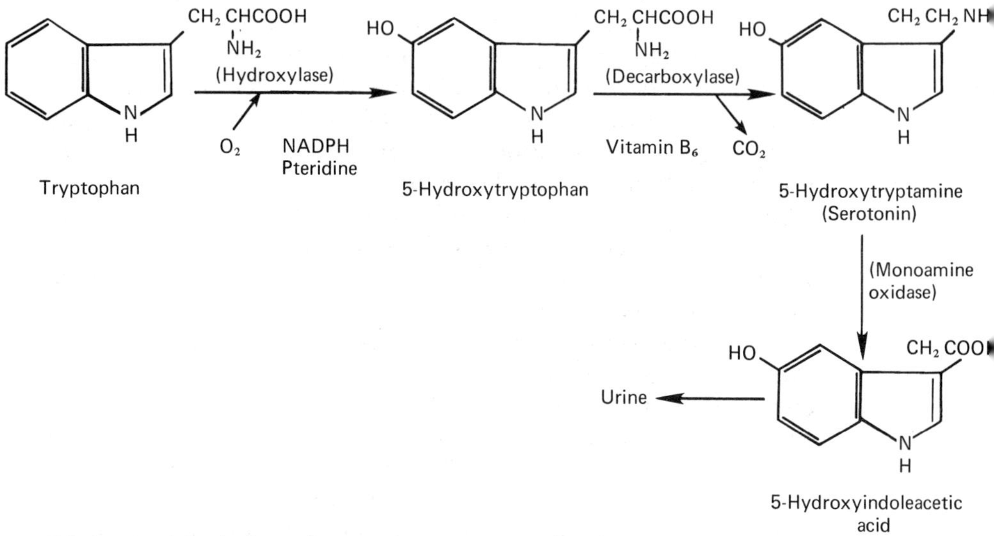

FIG. 6-31. Metabolism of serotonin.

Neuropeptides

The nervous system abounds in amino-acid polymers of varying lengths. Some have clear roles in promoting the secretion of endocrine agents (hormone-releasing factors), some have endocrine effects of their own, whereas others have less well-defined and possibly modulatory actions upon transmitter functions, mood, or behavior. Neurosecretory cells of the hypothalamus are foremost in the generation of peptides with endocrine functions, many of which act by producing effects upon the adjoining pituitary gland. Since the pituitary gland plays a major part in coordination of the various endocrine glands scattered throughout the body, the linkage of the neuropeptides between the nervous system and the pituitary is of great significance for the overall control of metabolism.

Substance P was the earliest of the neuropeptides to be suspected as a possible neurotransmitter agent. This peptide containing 11 amino acid residues is localized in sensory nerve tracts, and is released from the nerve-endings following excitation of the neurons. It has been proposed that substance P acts as a primary sensory neurotransmitter, at least in the slow conducting fibers of pain pathways.

Among the more recently detected neuropeptides that have attracted much interest are the endorphins and enkephalins. The term *endorphin* was coined to describe substances of endogenous origin, which mimicked the actions of the opium drugs such as morphine. The related *enkephalins* are smaller derivatives with similar analgesic actions, which were first detected in the brain, hence their name. Modern terminology refers to endorphins and enkephalins collectively as opioid peptides, and the membrane-binding sites that they share with morphine as opiate receptors.

The enkephalins are pentapeptides with the sequences:

Tyr–Gly–Gly–Phe–Met

or

Tyr–Gly–Gly–Phe–Leu

and are referred to as Met-enkephalin or Leu-enkephalin, respectively. Synthetic analogues of the enkephalins, with varying amino-acid substituents, have equal or greater potency and duration of action than the natural enkephalins. The goal is to develop more effective morphine-like analgesics without the addictive and other harmful side-effects of the natural opiates. The other endogenous opioid peptide, which occurs mainly in the pituitary gland and the hypothalamic regions nearby, is termed β-endorphin and contains 31 amino-acid residues with the Met-enkephalin sequence as the N terminus. The β-endorphin in turn has its origins in a larger precursor glycoprotein that also contains the peptide sequence of the adrenocorticotrophic hormone (ACTH) and is termed pro-opiocortin. Stressful stimuli to the pituitary gland lead to proportional, simultaneous release of both β-endorphin and ACTH into the bloodstream, underlining their common pro-opiocortin mode of origin. The enkephalins also seem to originate by proteolytic processing of larger polypeptide precursors, but they must be made in regions of the nervous system other than the pituitary since the levels of enkephalins in the brain are not altered by hypophysectomy.

The opioid peptides have many of the features expected of neurotransmitters. Although widely distributed in the central nervous system, the enkephalins have their highest concentration in those sensory regions of the brain and spinal cord that are involved in pain perception, and in the emotion-regulating regions of the limbic system. The opiate receptors are also highly concentrated in these enkephalin-rich areas. Excitation of these regions results in localized release of enkephalin in the vicinity of the neurons that elaborate substance P. The analgesic actions of enkephalins are postulated to result from their ability to inhibit release of substance P, the putative transmitter of the small, unmyelinated fibers in the pain pathway. Other general modulatory effects of enkephalins and β-endorphin upon memory, mood, behavior, and their possible involvement in affective disorders (*e.g.*, schizophrenia) are ongoing areas of active research.

SUGGESTED READINGS

Abood LG, Hoss W: Excitation and conduction in the neuron. In Siegel GJ, Albers RW, Katzman R et al (eds): Basic Neurochemistry, 2nd ed, pp 103–124. Boston, Little, Brown & Co, 1976

Anderson GH: Diet, neurotransmitters and brain function. Br Med Bull 37:95–100, 1981

Aoki TT, Finley RJ, Cahill GF Jr: The redox state and regulation of amino acid metabolism in man. Biochem Soc Symp 43:17–29, 1978

Axelrod J: Neurotransmitters. Sci Am 230, No. 6:59–71, 1974

Berk PD, Jones EA, Howe RB, Berlin NI: Disorders of bilirubin metabolism. In Bondy PK, Rosenberg LE (eds): Metabolic Control and Disease, 8th ed, pp 1009–1088. Philadelphia, WB Saunders, 1980

Bird ED: Chemical pathology of Huntington's disease. Annu Rev Pharmacol Toxicol 20:533–551, 1980

Boss GR, Seegmiller JE: Hyperuricemia and gout: classification, complications and management. N Engl J Med 300:1459–1468, 1979

Brown AK, McDonagh AF: Phototherapy for neonatal hyperbilirubinemia: Efficacy, mechanism and toxicity. Adv Pediatr 27:341–389, 1980

Cello JP: Diagnostic approaches to jaundice. Hosp Pract 17, No. 1:49–60, 1982

Cohen PP: The ornithine-urea cycle: Biosynthesis and regulation of carbamyl phosphate synthetase I and ornithine transcarbamylase. Curr Top Cell Regul 18:1–18, 1981

Conn HO, Lieberthal MM: The Hepatic Coma Syndromes and Lactulose. Baltimore, Williams & Wilkins, 1979

Felig P: Amino acid metabolism in man. Annu Rev Biochem 44:933–955, 1975

Freeman HJ, Kim YS: Digestion and absorption of protein. Annu Rev Med 29:99–116, 1978

Hulse JH, Spurgeon D: Triticale. Sci Am 231, No. 2:72–80, 1974

Jeejeebhoy KN: Protein nutrition in clinical practice. Br Med Bull 37:11–17, 1981

Jones ME: Pyrimidine nucleotide biosynthesis in animals: Genes, enzymes, and regulation of UMP biosynthesis. Annu Rev Biochem 49:253–279, 1980

Kelley WN, Wyngaarden JB: The Lesch–Nyhan syndrome. In Stanbury JB, Wyngaarden JB, Fredrickson DS (eds): The Metabolic Basis of Inherited Disease, 4th ed, pp 1011–1036. New York, McGraw-Hill, 1978

MacIntosh FC: Acetylcholine. In Siegel GJ, Albers RW, Katzmann R, Agranoff BW (eds): Basic Neurochemistry, 3rd ed, pp 183–204. Boston, Little, Brown & Co, 1981

McCarty DJ: The management of gout. Hosp Pract 14, No. 9:75–85, 1979

Meyer UA, Schmid R: The porphyrias. In Stanbury JB, Wyngaarden JB, Fredrickson DS (eds): The Metabolic Basis of Inherited Disease, 4th ed, pp 1166–1220. New York, McGraw-Hill, 1978

Mudd SH, Levy HL: Disorders of transsulfuration. In Stanbury JB, Wyngaarden JB, Fredrickson DS (eds): The Metabolic Basis of Inherited Disease, 4th ed, pp 458–503. New York, McGraw-Hill, 1978

Munro HN: Protein synthesis and breakdown in the liver. In Brown H, Hardwick DF (eds): Intermediary Metabolism of the Liver, pp 3–22. Springfield, Illinois, Charles C Thomas, 1973

Odell GB: Neonatal Hyperbilirubinemia. New York, Grune & Stratton, 1980

Peck EJ Jr: Receptors for amino acids. Annu Rev Physiol 42:615–627, 1980

Prusiner SB: Disorders of glutamate metabolism and neurological dysfunction. Annu Rev Med 32:521–542, 1981

Ratner S: Enzymes of arginine and urea synthesis. Adv Enzymol 39:1–90, 1973

Saheki T, Tsuda M, Takada S, Kusumi K et al: Role of arginino-succinate synthetase in the regulation of urea synthesis in the rat and argininosuccinate synthetase-associated metabolic disorder in man. Adv Enz Reg 18:221–238, 1980

Sassa S, Kappas A: Genetic, metabolic, and biochemical aspects of the porphyrias. Adv Hum Genet 11:121–231, 1981

Schildkraut JJ: Biogenic amines and affective disorders. Annu Rev Med 25: 333–348, 1974

Schmid R, McDonagh AF: Hyperbilirubinemia. In Stanbury JB, Wyngaarden JB, Fredrickson DS (eds): The Metabolic Basis of Inherited Disease, 4th ed, pp 1221–1257. New York, McGraw-Hill, 1978

Scriver CR, Clow CL: Phenylketonuria: epitome of human biochemical genetics. N Engl J Med 303:1336–1342, 1394–1400, 1980

Scriver CR, Clow CL: Phenylketonuria and other phenylalanine hydroxylation mutants in man. Annu Rev Genet 14:179–202, 1980

Seegmiller JE: Diseases of purine and pyrimidine metabolism. In Bondy PK, Rosenberg LE (eds): Metabolic Control and Disease, 8th ed, pp 777–937. Philadelphia, WB Saunders, 1980

Segal S: Disorders of renal amino acid transport. N Engl J Med 294:1044–1051, 1976

Shih VE: Urea cycle disorders and other congenital hyperammonemic syndromes. In Stanbury JB, Wyngaarden JB, Fredrickson DS (eds): The Metabolic Basis of Inherited Disease, 8th ed, pp 362–386. New York, McGraw-Hill, 1978

Snyder SH: Catecholamines, serotonin, and histamine. In Siegel GJ, Albers RW, Katzmann R et al (eds): Basic Neurochemistry, 2nd ed, pp 203–217. Boston, Little, Brown & Co, 1976

Snyder SH, Innis RB: Peptide neurotransmitters. Annu Rev Biochem 48:755–782, 1979

Stoff JS, Epstein FH, Narins R, Relman AS: Recent advances in renal tubular biochemistry. Annu Rev Med 38:46–68, 1976

Tschudy DP, Lamon JM: Porphyrin metabolism and the porphyrins. In Bondy PK, Rosenberg LE (eds): Metabolic Control and Disease, 8th ed, pp 939–1007. Philadelphia, WB Saunders, 1980

Veale AMO: Screening for phenylketonuria. In Bickel H, Guthrie R, Hammersen G (eds): Neonatal Screening for Inborn Errors of Metabolism, pp 7–18. New York, Springer–Verlag, 1980

Visek WJ: Ammonia metabolism, urea cycle capacity and their biochemical assessment. Nutr Rev 37:273–282, 1979

Waterlow JC, Jackson AA: Nutrition and protein turnover in man. Br Med Bull 37:5–10, 1981

Watkins JC, Evans RH: Excitatory amino acid transmitters. Annu Rev Pharmacol Toxicol 21:165–204, 1981

Wellner D, Meister A: A survey of inborn errors of amino acid metabolism and transport in man. Annu Rev Biochem 50:911–968, 1981

Wurtman RJ, Growden JH: Dietary enhancement of CNS neurotransmitters. Hosp Pract 13, No. 3:71–77, 1978

Wyngaarden JB, Kelley WN: Gout. In Stanbury JB, Wyngaarden JB, Fredrickson DS (eds): The Metabolic Basis of Inherited Disease, 4th ed, pp 916–1010. New York, McGraw-Hill, 1978

Zieve L, Nicoloff DM: Pathogenesis of hepatic coma. Annu Rev Med 26:143–157, 1975

Chapter Seven Integration of Metabolism

The average student of biochemistry may be excused if he emerges from his first exposure to the subject with a series of disconnected impressions of metabolism occurring in isolated compartments. As with the chapters and subsections of this book, a preliminary look at the subject requires some breakdown of the tangled complexity of the body's chemistry into tighter groupings of closely related compounds and events. At some point, however, it is necessary to try to interweave these individual reactions and to try to understand how the whole fabric comes to constitute a purposive living system.

INTEGRATIVE SYSTEMS

It is apparent to anyone who has reflected upon the matter at all that the human body is infinitely more than a conglomerate or colony of various cell types. The cells, differentiated for specific metabolic and physiologic purposes, are associated together as tissues, and the latter are assembled structurally as organs. The arrangement provides for a neat division of labor, but it requires supervisory agencies to oversee the activities of the various parts and to ensure that they are in harmony with the goals of the whole organism (Fig. 7-1).

There are three major systems that perform this integrative role. Foremost is the *nervous system,* the communications center and primary computer, with its receiving mechanisms to assess the body's status relative to the environment (*e.g.,* lack of oxygen, hunger, thirst, pain, and other danger warnings) and its transmitting processes to relay appropriate commands so that other body organs may adjust to these external changes. Second is the *endocrine system,* the factory and depot for chemical messengers, with its wide-ranging influence upon growth, reproduction, and

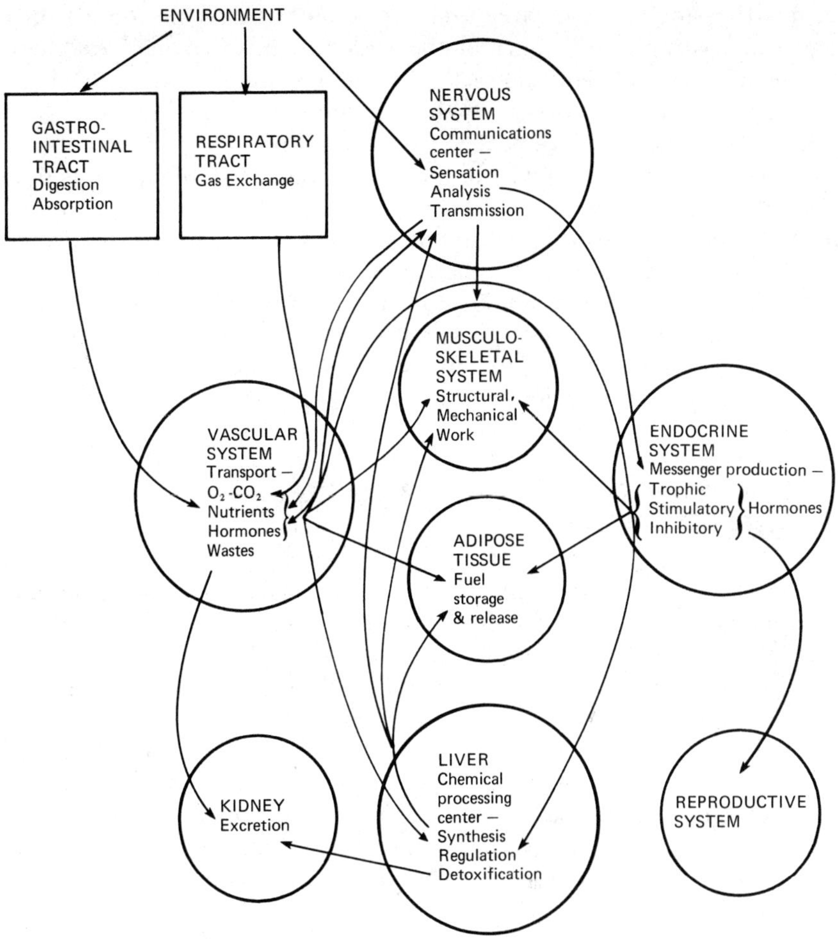

FIG. 7-1. Integrative functions of the nervous system, endocrine system, and vascular system upon body metabolism.

development, as well as other long-range goals of the organism. Its more rapid adaptations enable this system to maintain the homeostasis of blood components, such as glucose, lipids, and calcium, and to ensure the proper balance between the synthesis and the breakdown of tissue components in relation to energy needs. Third is the *vascular system,* the vehicle for the transport of all chemical substances in the body, with its pipes and mobile carriers for the delivery of oxygen, organic nutrients, and minerals from the external environment to cells deep within the body, for the relay of building materials, fuels, and chemical messengers between the tissues, and for the elimination of invading organisms, toxins, and the end products of metabolism.

In healthy states, these three systems smoothly intermesh and complement one another. Hormones of the endocrine system mediate their effects through the bloodstream of the vascular system, and the levels of circulating hormones returning to their glands of origin regulate their own release by negative feedback. Superimposed upon this is the control of hormone-releasing factors within the hypothalamus of the nervous system, which in turn exerts its effects upon the master endocrine gland, the pituitary. Epinephrine from the adrenal medulla of the endocrine system, as well as norepinephrine from the nervous system, will intensify the blood flow and pressure throughout the vascular system. These and many more examples of physiologic controls illustrate the tight interdependence of this troika of integrative systems in the coordination of the multifarious migrations and transformations of molecules throughout the body.

INTERRELATIONS OF CARBOHYDRATES, FATS, AND PROTEINS

Just as the three main integrative agencies must harmonize, so also must there be consonance among the three major regions of metabolism. There are several ways in which these different classes of metabolites interact, overlap, and converge. These ways include, among others, their common energetics, their shared precursors and intermediates, and their common final pathways for carbon and hydrogen oxidation (Fig. 7-2). The levels of dietary intake of each substance in relation to activity will play the major role in determining how a balance may be struck, as noted in the following sections. Superimposed upon these gross considerations of supply and demand are also the fine controls of the integrative systems in relation to physiologic or pathologic states.

COMMON ENERGY SUPPLIES

The common energy supplies, in the form of either phosphorylative energy as ATP or reductive energy as NADH or NADPH, are equally available for the biosyntheses of a wide range of simple or complex organic compounds and macromolecules. The direction toward the anabolism of any of these substances has as its first requirement the provision of chemical energy from one of these mobile sources. It follows that when the cell places a priority upon the production of a particular class of substances, it must be at the expense of the catabolism of another substance. The common energy pool within a cell may therefore be viewed as a finite amount of driving power for which all the anabolic and other endergonic processes must compete.

When the liver, for example, is called upon to manufacture more glu-

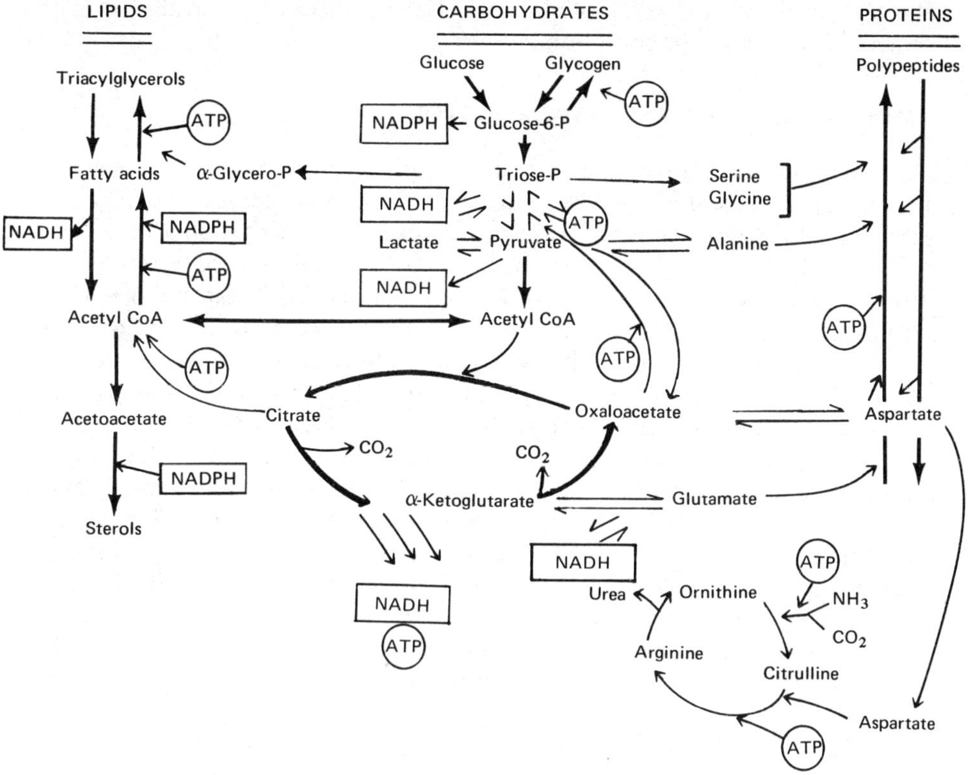

FIG. 7-2. Interrelations of lipids, carbohydrates, and proteins in metabolism.

cose by reversing glycolysis, it cannot simultaneously synthesize proteins and fats; on the contrary, it will be necessary to hydrolyze some existing protein or fat and to oxidize some of the resulting amino acids or fatty acids in order to generate the NADH and ATP needed for gluconeogenesis. It may be—as in the case of glucocorticoid stimulation of the induction of certain enzymes required for maximal rates of gluconeogenesis—that while a specific protein is being synthesized in greater amounts, a number of others are simultaneously being catabolized to fulfill the resultant energy requirements. Similarly, in any cell that is producing more membranes, an accelerated breakdown of carbohydrate will be required to drive the biosynthesis of the necessary proteins and lipids. In the hepatic and adipose-tissue cells, the process of lipogenesis is particularly dependent upon glucose utilization to generate the NADPH and ATP necessary for the conversion of acetyl CoA to fatty acids. The polymerization reactions of glycogenesis, DNA and RNA duplication, and polypeptide chain elongation all require a coupled input of energy from the parallel breakdown of other organic compounds.

SHARED PRECURSORS AND INTERMEDIATES

Shared precursors and intermediates provide another level of overlap among the metabolic pathways where competitive interactions may occur. The total stock of carbon, especially in the form of carbohydrate, will influence the degree of synthesis of lipids and proteins. Precursors of the three-carbon skeletons of glycerol for lipogenesis or of alanine or serine for protein synthesis, for example, will derive directly from the trioses in the glycolytic pathway of carbohydrate metabolism.

Possibly the most important single intermediate—which is shared in all categories of metabolism and which hence provides a pivotal linkage—is acetyl CoA. The two-carbon fragment may be derived from the catabolism of alkyl chains and sterols among the lipids, of certain aromatic and aliphatic amino acids, or of glucose and other carbohydrates. It may be used for the synthesis of important esters, such as acetylcholine, or for the biogenesis of fatty-acyl chains or steroid ring compounds. Recall, too, the essential regulatory role of acetyl CoA as the positive effector of pyruvate carboxylase, providing oxaloacetate either for gluconeogenesis, for aspartate production, or for the maintenance of flux through the citric-acid cycle.

It is important to note that there is a one-way flux toward lipogenesis from carbohydrate or protein sources through acetyl CoA. Since no mechanism exists in the human for converting the two-carbon fragment to the three-carbon compounds required for gluconeogenesis or for the formation of nonessential amino acids such as alanine, there is no way that the fatty-acid carbon can be reconverted to glucose or protein. In the catabolism of lipids, small amounts of three-carbon intermediates are regenerated from the glycerol moiety, but these are not significant in relation to the large amounts of carbon in the side-chains of the triacylglycerols, for example.

THE FINAL COMMON PATHWAY

The final common pathway for all the metabolic systems is, of course, the citric-acid cycle and the reactions of the respiratory chain. These mitochondrial processes serve to coordinate a number of metabolic sequences at several levels. The citric-acid cycle is the principal source of cellular carbon dioxide for the carboxylation reactions that prime fatty-acid synthesis and gluconeogenesis; this carbon dioxide also supplies the carbon of urea as well as of certain portions of the purine and pyrimidine rings. Intermediates of the cycle also lead to cross-linkages between carbohydrate and nitrogen metabolism (α-ketoglutarate and glutamate, oxaloacetate and aspartate, or succinate and heme). The role of aspartate in the metabolic sequence from citrulline to arginine with the formation of fumarate provides a direct linkage between the fumarate-to-oxaloacetate segment of the citric-acid cycle and the urea cycle.

There are also several ways in which the citric-acid-cycle intermediates intermesh with lipogenesis. Citrate cleavage gives rise to the acetyl-CoA precursor for fatty-acid biosynthesis, and it acts as a positive modifier for the initial acetyl-CoA carboxylase reaction as well. In the presence of NADP-linked decarboxylating enzymes, both isocitrate and malate may provide the NADPH required for subsequent reductive stages of fat synthesis.

Breakdown products from the catabolism of carbohydrates and fats funnel into the cycle as acetyl CoA; most of thc amino-acid catabolites enter it as the α-keto acids. In either case, the total oxidative degradation of the compounds in the mitochondria transfers the hydrogen from these substances to the flavoproteins that are fixed in the membranes or to the mobile molecules of NAD. The former, $FADH_2$, must be reoxidized *in situ* by the cytochrome system; the latter, NADH, may also transfer its reducing equivalents to the respiratory chain or it may be used to drive reductive syntheses elsewhere in the cell following its transport out of the mitochondria by one of the substrate shuttle carriers. A key factor in determining the direction to be taken by NADH will be the adenine–nucleotide state: high ADP and low ATP levels, which signify low energy reserves, will draw the NADH into the respiratory chain by stimulating the energy coupling of oxidative phosphorylation; low ADP and high ATP levels will exert the opposite effect and, in restraining the electron-transport system, they will divert NADH for alternative reductive reactions, such as those of glutamate synthesis or gluconeogenesis.

INTERRELATIONS OF TISSUES AND ORGANS

In the first parts of this book, I have attempted to show how the metabolic responsibilities of various tissues have been assigned. In some cases, the biochemical capacities of the cells have become extremely specialized and limited, as, for example, are those of the erythrocytes with their solely anaerobic glucose catabolism. In others, the cells are virtually plenipotent from the standpoint of their capacity to manifest a broad range of enzymatic activities, for example, the hepatocytes, with their involvement in the anabolism, catabolism, and interconversions of carbohydrates, fats, and proteins, as well as in other metabolic duties.

All the cells in a given individual possess the same DNA molecules, and at some point in its life, each cell has the capacity for expressing that DNA by replication when it divides or by transcription when it synthesizes proteins during growth. To variable degrees, as the cells differentiate toward their final specialized forms in the body's organs, they lose these features of gene expression, and the DNA becomes progressively silent. In the extreme case, the DNA may vanish entirely, as in the erythrocyte; alternatively, it may lose the capacity for replication, as it does in the mature nerve cells, which thus are incapable of dividing in the adult brain. In most cell types, only a fraction of the total DNA may be transcribed to give rise to the messenger RNA that corresponds to the enzymes and other proteins characteristic of that particular cell. In others, the DNA cannot be read for a particular messenger sequence unless a repressor substance is removed. The extent and the timing of these intricate controlling factors determine the course of biochemical specification in tissue and organ development. In giving up their full metabolic capacities in the interests of specialization, the differentiated tissues have become reliant upon other cell types to supplement them with needed metabolites and to eliminate their waste substances.

In the previous chapters, I have presented the metabolic fate of molecules starting from their point of entry: their digestion and absorption in the gastrointestinal (GI) tract. What naturally follows then is the passage of these nutrients through the main gate, aptly named the portal vein, into the liver. Thus, it seemed logical to describe the major metabolic events at the point where they first occur in the body: in the hepatic cells. This emphasis upon the role of the liver in metabolism is equally appropriate in another context, that is, in relation to the great metabolic activity of this primary chemical factory (Fig. 7-1). In terms of adaptability and versatility, no other organ possesses such an immense array of enzymatic transformations affecting the whole body (Fig. 7-3). Let us try to summarize these activities as they pertain to other organ systems in the human.

THE LIVER

The liver acts as a gland with both exocrine and endocrine functions. The *exocrine secretion* into the GI tract is the bile, a solution that contains the waste products, such as the bile pigments, as well as the important promoters of fat digestion, the bile salts. The latter provide the main route for sterol oxidation, and the reabsorption of bile salts from the GI tract constitutes an important self-regulating feedback loop, since the oxidation of cholesterol in hepatic cells is inhibited by the bile salts returned from the gut. The amount of cholesterol biosynthesis in the GI tract is substantial, and it is also subject to control by this feedback loop, since it is significantly inhibited by the bile salts taken up by the mucosal cells.

The *endocrine secretion* from liver into the systemic circulation does not consist of hormones but rather of metabolites that are dispatched by the bloodstream for the use of other cells, and which thus alter the activity of these cells. These include (1) *glucose,* which is secreted mainly during

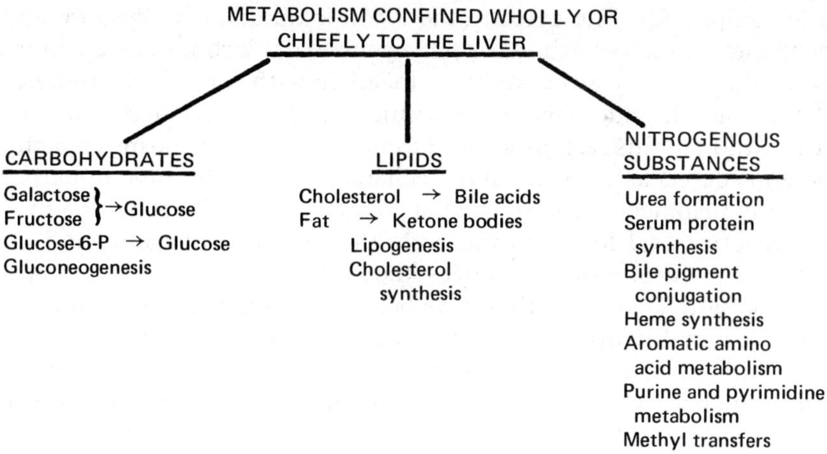

FIG. 7-3. Metabolic processes primarily carried out by the liver. (Functions also performed in other organs or tissues are glucose dephosphorylation and gluconeogenesis, which occur in the kidney to a minor extent, lipogenesis in adipose tissue, cholesterol synthesis in the gastrointestinal mucosa, heme synthesis in the reticuloendothelial system, aromatic amino-acid metabolism in neural tissue, as well as purine and pyrimidine metabolism and methyl transfers, which occur in several other tissues.)

fasting, glucocorticoid or glucagon stimulation, and vigorous exercise, and which provides requirements for brain and muscle glycolysis; (2) *triacylglycerols,* which are released after carbohydrate ingestion or insulin stimulation, and which contribute mainly to adipose-tissue lipogenesis; and (3) *ketone bodies,* which are produced in excess following starvation or in high-fat and low-carbohydrate diets, and which generate alternative energy sources for muscle and neural tissue. These, of course, are but a few of the general ways in which liver metabolites affect the reactions in other tissues. In addition, the liver is responsible for the synthesis and secretion of albumin, serum lipoproteins, and the clotting factors of the blood, as well as other important nitrogenous metabolites to be used by other tissues.

The liver bears a heavy responsibility for the metabolism of a broad spectrum of foreign substances in the body, including drugs, environmental pollutants, and ethyl alcohol. Reactions catalyzed by liver enzymes of detoxification include conjugation with amino acids, sulfate, or glucuronic acid to render lipophilic agents more water-soluble for excretion. Oxidative conversion of exogenous compounds into more polar derivatives is another common detoxifying process, catalyzed in the liver endoplasmic reticulum by cytochrome P-450 in the presence of NADPH and oxygen. This versatile hydroxylating system is capable of catalyzing demethylations, deaminations, and hydroxylations of a wide range of aliphatic and aromatic compounds. The cytochrome P-450 is an adaptive system, and administration of phenobarbital or polycyclic hydrocarbons causes an induction of the hydroxylating activity, together with a marked proliferation of the smooth endoplasmic reticulum membranes that contain the enzymes. Although this response and the enhanced hydroxylating capacity may be considered as a protective mechanism, it has become apparent that in many instances the end-products of cytochrome-P-450 action may be more reactive and lethal to the body than the parent compound. In particular, many aromatic substances known to cause cancer in humans will not react with DNA until they have been converted to active mutagens by the microsomal hydroxylation system, by the addition or unmasking of electrophilic groups. Thus they may be ingested as inactive precarcinogens, and the body carries out a type of suicidal metabolism to generate the carcinogens *in situ.*

A foreign substance whose metabolism also brings about pathologic effects is ethyl alcohol. The liver is the major instrument for conversion of ethanol to its non-toxic metabolite, acetate, which may be used as an energy source by other tissues such as heart muscle, which actively oxidizes acetate by the citric acid cycle. Although microsomal membranes can oxidize ethanol to a small extent, two soluble NAD-linked enzymes, *alcohol dehydrogenase* and *acetaldehyde dehydrogenase,* carry out the bulk of ethanol oxidation in liver. In the process acetaldehyde is generated as an intermediate compound and produces toxic effects if it is allowed to accumulate. This is the basis for the use of disulfiram (Antabuse), an irreversible inhibitor of acetaldehyde dehydrogenase, in aversion therapy of alcoholism. The ingestion of ethanol following disulfiram causes accumulation of acetaldehyde, with most unpleasant consequences for the alcoholic. Chronic ingestion of excessive amounts of alcohol has three notable effects upon liver metabolism: (1) an increase in the NADH:NAD ratio; (2) an accumulation of triacylglycerol (fatty liver); (3) a proliferation of smooth endoplasmic reticulum. Effects upon the microsomal electron transport system can lead to potentially catastrophic sensitivity or tolerance to other drugs that interact with the liver detoxifying membranes. Eventually, the inevitable result from the cumulative effects of ethanol toxicity is cirrhosis of the liver, derangement of nervous system functions, and death. Abilities of different individuals to handle varying levels of consumption of ethanol are apparently related to genetic variations of the liver dehydrogenases. Current research is actively pursuing the linkages between the inherited susceptibility for development of neurologic and hepatic pathology from ethanol or metabolites such as acetaldehyde, and the kinetics and other biochemical parameters of the dehydrogenase variants.

The extrahepatic tissues, however, have some metabolic autonomy. The main reactions of energy production (glycolysis, the citric-acid-cycle reactions, oxidative phosphorylation, and fatty-acid oxidation) as well as the major biosynthetic processes (protein synthesis, nucleic-acid synthesis, lipogenesis, and glucogenesis) are present in other cell types, although at various levels of activity. Nonetheless, the dependence of these tissues upon metabolism in the liver is frequently of the greatest significance for normal function and homeostasis. Let us therefore summarize and review some of the prinicpal examples of these relations among the liver and other tissues (Fig. 7-4).

ADIPOSE TISSUE AND THE LIVER

Adipose-tissue and hepatic metabolism of fat is closely integrated. Triacylglycerol produced in the hepatic cells is an important source of the fat laid down in the adipose depots. Another is the newly formed triacylglycerol produced in the adipocytes themselves. Since glucose will provide the carbon for both the fatty acids and the glycerol moiety that are formed in the adipose tissue, and since the liver plays a major role in controlling blood glucose levels, therefore lipogenesis in the depots will be doubly dependent on hepatic metabolism.

This flow of carbon from hepatic to adipose cells in times of plenty will be reversed when the energy intake of the body drops; then the activation of hormone-sensitive lipase in the adipocytes leads to the release of both glycerol and fatty acids into the bloodstream. These breakdown products of triacylglycerols can provide for the energy requirements of the liver during fasting, with fatty-acid oxidation generating the NADH and ATP needed for gluconeogenesis. Alternatively, the fatty acids may be utilized directly by extrahepatic tissues, such as the muscles, in lieu of glucose.

MUSCLE AND THE LIVER

A direct relationship between muscle and hepatic metabolism is noted at several levels. In the well-nourished state, both tissues will remove glucose from the circulation and deposit excess amounts as glycogen. Increased physical activity, which provides the major stimulus for glycogenolysis in muscle, promotes the breakdown of glycogen as far as lactate, and the latter diffuses into the blood. Decreased food intake, which is the chief activator of hepatic glycogen breakdown, leads to its conversion only as far as glucose, which also passes into the blood. A similar contrast pertains to the balance between glycolysis and gluconeogenesis in the two tissues. The muscle—with its high phosphofructokinase and pyruvate-kinase activity but low fructosediphosphatase, pyruvate-carboxylase, and PEP-carboxykinase levels—is poised in the direction of glycolysis of the incoming glucose to generate ATP and lactate. The liver—with its opposite distribution of enzymes—is poised toward gluconeogenesis, and, at the expense of other intracellular sources of ATP (*e.g.*, fatty-acid oxidation), it can

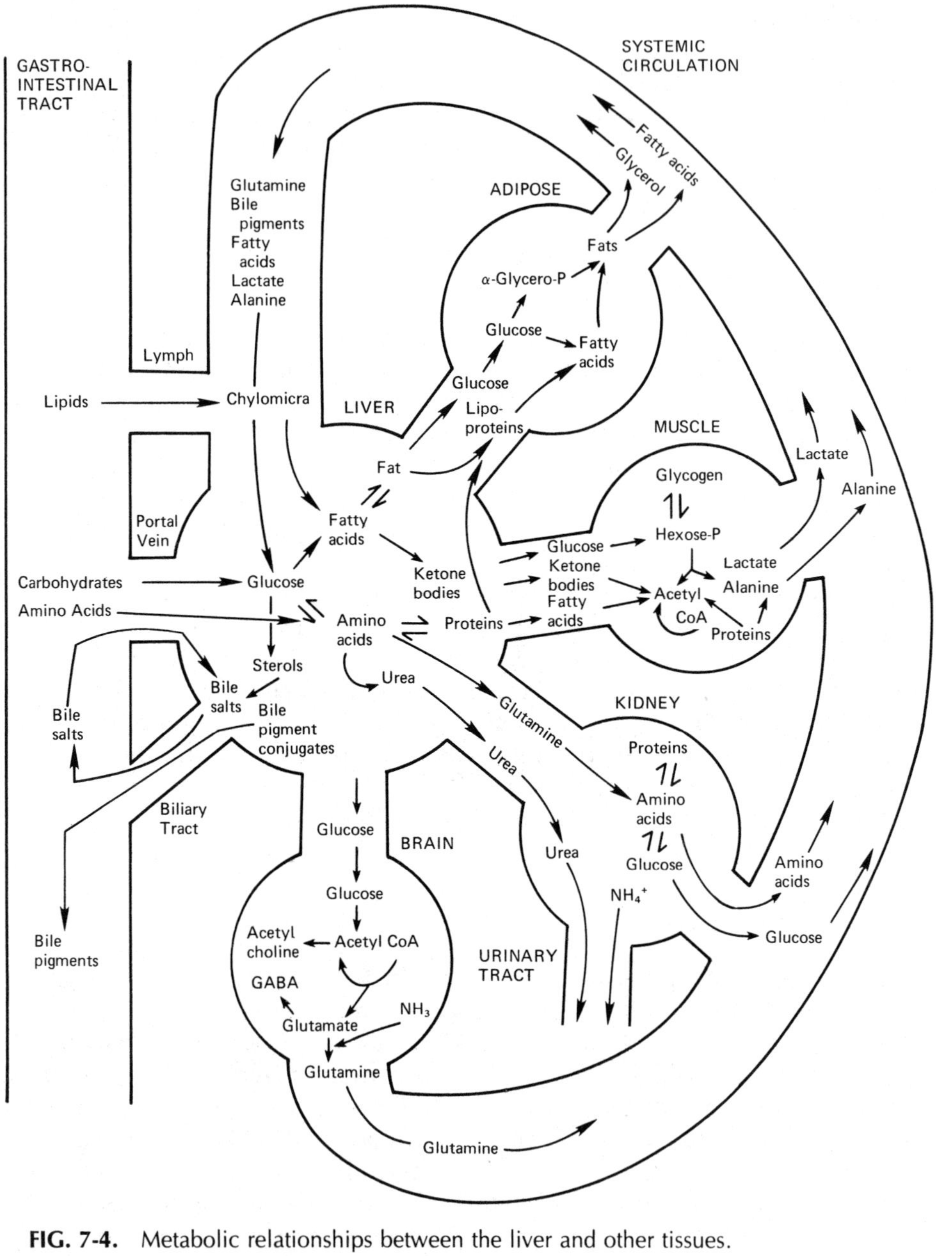

FIG. 7-4. Metabolic relationships between the liver and other tissues.

regenerate glucose for the muscle's requirements from the incoming lactate.

In conditions of prolonged food deprivation when muscle protein is catabolized, the skeletal muscles will convert significant quantities of pyruvate by amino transfer to alanine, and the latter can then be used as the carbon source for hepatic gluconeogenesis instead of lactate. The lactate–glucose cycle involves the liver and the skeletal muscles; heart muscle, it should be remembered, is an aerobic tissue, well-endowed with mitochondria and a lactate-dehydrogenase enzyme system that tends to consume lactate rather than to produce it.

One way in which both skeletal and cardiac muscles function similarly at the expense of the liver is their high utilization of the ketone bodies, β-hydroxybutyrate and acetoacetate. Since the liver lacks the enzyme for activating acetoacetate to its CoA derivative, large amounts of these ketone bodies pass into the bloodstream under conditions of low-calorie, low-carbohydrate, or high-fat intake, when hepatic fatty-acid oxidation outstrips its capacity for acetyl-CoA removal. Circulating ketone bodies serve as signals that glucose supplies are limited in these physiologic states, and convert the muscles from carbohydrate utilization to the use of acetoacetate as an alternative fuel. The muscles, as well as the kidneys and some other extrahepatic tissues, can activate acetoacetate at the expense of succinyl CoA and derive energy from the oxidation of the resultant acetyl CoA. Even under basal conditions, there is a constant flow of acetoacetate and other ketone bodies from the liver to the peripheral tissues, where they generate a significant proportion of metabolic energy, particularly in the heart. Both cardiac and skeletal muscle, however, despite their dependence upon hepatic metabolites, derive a large measure of their normal ATP requirements from the oxidation of fatty acids derived by mobilization from fat depots.

THE KIDNEY AND THE LIVER

The kidney and the liver share the capacity for gluconeogenesis, but the kidney will usually generate only a small fraction (about one tenth) of the total glucose output. Under circumstances in which liver function is impaired or when acidotic conditions stimulate the breakdown of amino acids to the α-keto-acid precursors of glucose in the kidney, this organ will become a more significant contributor to gluconeogenesis.

The kidney depends upon the output of glutamine from the liver to provide ammonia in order to balance hydrogen-ion excretion. The liver, on the other hand, requires the excretory functions of the kidney to eliminate urea and other waste products from the circulation, while retaining the glucose, amino acids, and proteins that the liver produces.

THE BRAIN AND THE LIVER

Finally, the metabolic functions of the brain and the liver are interconnected, principally by the absolute dependence of neural tissue upon the steady supply of incoming glucose that is maintained by hepatic regulation. The catabolism of glucose provides (1) the energy requirements of the brain cells for maintaining the active transport of the ions involved in nerve excitation, (2) the acetyl CoA for lipid synthesis and maintenance of the myelin and for acetylcholine synthesis, and (3) the material for the formation of the carbon chains of glutamate, γ-aminobutyrate (GABA), and other cerebral amino acids.

Since brain tissue actively metabolizes amino acids but is very susceptible to ammonia intoxication, it has developed protective measures for NH_3 fixation, mainly by producing glutamate and glutamine. Ultimately, the liver must eliminate the NH_3 arising from the brain and other peripheral tissues by forming urea. Liver failure or other metabolic disorders that result in defective urea-cycle activity have profound ramifications for mental function and development as a consequence of this dependence of the brain upon the liver for removal of ammonia and other neurotoxic end products of metabolism.

MAINTENANCE AND THE STEADY-STATE

The mature, healthy individual tends to remain in a state of balance with the environment. This involves a suppression of the growth processes that is no less remarkable than the ability of cells to grow and divide and is just as vital to survival. Failure to apply the normal braking mechanisms may lead to gigantism, obesity, or the runaway anarchy of malignancy. Undoubtedly, the future understanding of cancer may come not so much from a consideration of why the tumor cells are malignant in their growth and invasiveness as from an understanding of why normal cells are not.

In the adult, the main factor determining the normal metabolic balance is the food intake in relation to energy expenditure. Deficient intake leads to reversible mobilization of the energy stores of the body in the short run, but over long periods, deprivation or starvation causes irreversible tissue wasting and degeneration. Chronic, excessive intake of food can also lead to the pathologic consequences of obesity, which are due to overstocking of the storage depots. Before discussing these two extremes of imbalance, let us consider the means of body maintenance and review the regulators that affect the enzymes of anabolism and catabolism.

ANABOLIC HORMONES

These agents reinforce the growth-promoting effects of increased food intake. *Growth hormone,* a polypeptide secreted by the anterior pituitary gland, stimulates the biosynthesis of RNA and proteins in virtually all types of cells. This general stimulation of nitrogen retention is accompanied by an increased amino-acid uptake from the circulation. One factor that enhances the release of growth hormone is an elevated level of amino acids in the bloodstream, so the maximal stimulation of tissue protein anabolism is coordinated with a plentiful supply of precursors.

Insulin is secreted by the pancreas in response to high blood levels of either glucose or amino acids, and it enhances the uptake of both classes of metabolites into the tissues. Insulin also promotes an increased utilization of glucose for glycogen synthesis, lipogenesis, and glycolysis, while restraining gluconeogenesis in the liver and inhibiting lipolysis in the fat depots. Anabolism of tissue proteins is stimulated by this hormone, and the catabolism of amino acids is simultaneously depressed.

Thyroxine, which is secreted by the thyroid gland in response to thyroid-stimulating hormone (TSH) from the anterior pituitary, produces a general stimulation of growth and tissue differentiation. This effect is expressed mainly by an enhancement of protein synthesis and, in particular, by an increased development of mitochondrial oxidative enzymes. Thus, the thyroid hormone exerts a controlling influence on the oxygen consumption and general metabolic rate of the tissues.

Other hormones with more specific anabolic functions are the steroid *estrogens,* which are produced by the ovaries and which promote RNA and protein syntheses in selected target cells in the uterus of females; and the steroid hormone counterparts in males, the *androgens,* which are produced by the testes cause similar accelerations of RNA and protein anabolism in many tissues, including those of the musculoskeletal system.

CATABOLIC HORMONES

A number of endocrine factors act to compensate for the increased energy demands in states of stress or food deficiency. *Glucagon,* a polypeptide hormone, is secreted by the pancreas in response to depleted blood glucose and thus counteracts the hypoglycemic effect of insulin. The primary action of glucagon is to cause a specific increase in hepatic glycogenolysis with resultant release of glucose into the circulation. Other changes in the liver that may be produced by glucagon include the activation of gluconeogenesis and lipolysis. The free fatty acids released by lipase action serve as an alternative energy source and favor the process of gluconeogenesis.

The catecholamine hormones, *norepinephrine,* which is released by the stimulation of sympathetic nerves, and *epinephrine,* which is secreted by the adrenal medulla, promote similar effects to those of glucagon but their action extends to other tissues besides the liver. Thus, glycogenolysis is activated in the muscles as well as in the liver and lipolysis is stimulated in the adipose tissue, thereby raising both plasma glucose and fatty-acid levels.

The glucocorticoid steroid hormones such as *cortisol,* are secreted by the adrenal cortex in response to adrenocorticotropic hormone (ACTH) from the anterior pituitary. Cortisol opposes many of the actions of insulin, inhibiting the uptake of glucose into cells and its conversion to fats, and blocking protein synthesis in peripheral tissues. Simultaneously, the formation of the enzyme proteins involved in amino-acid catabolism and gluconeogenesis in the liver is stimulated. Consequently, the proteins in the extrahepatic tissues will be depleted, since protein hydrolysis will continue while their synthesis is depressed. The result is a flow of amino acids from the periphery to the liver, with increased amounts of carbon being channeled into glucose production and increased amount of nitrogen into urea production.

STARVATION

Undernutrition and hunger capture our attention periodically when wars, crop failures, or climatic disasters precipitate famine in various regions of the world. Biafra, Bangladesh, or Ethiopa momentarily hit the headlines but soon recede from international attention. The view that starvation is a regional and periodic occurrence, however, is far from the truth. In reality, this is a chronic, global problem, associated with poverty and other types of social deprivation in all countries. The Food and Agricultural Organization of the United Nations estimates that one fifth of the population in the developing nations is undernourished and that fully one half of all those in the most sensitive category, children under six, suffer from protein and calorie malnutrition in the Third World. Recent surveys in the United States and Canada have revealed significant nutritional inadequacies, particularly among lower income groups. The situation is insidious because of its self-perpetuating nature: inadequate means for food production or procurement lead to defective physical and mental development of the child, which lead in turn to the inability to raise the standards in the next generation.

Undernutrition is frequently subdivided into two classifications: *kwashiorkor,* which was discussed previously in relation to nitrogen metabolism is largely due to protein deficiency, and *marasmus,* a condition that has its roots in a general shortage of energy and protein in the diet. In classic cases, kwashiorkor appears in later infancy at 1 to 3 years of age and is characterized by edema and the retention of fat (particularly in the liver, which may be grossly enlarged), flaky skin, poor hair texture and pigmentation, and an apathetic, irritable disposition. In marasmus, the growth failure is much more evident in early infancy, being seen at 6 months to 1 year of age, and, in contrast with the edema of infants with kwashiorkor, who often appear puffy and moon-faced, the children with marasmus appear shrunken due to extreme wasting of the muscles and the absence of body fat. The hepatic, skin, and hair changes are not seen in typical cases of marasmus, and the infants are alert and responsive, not irritable. These two extremes represent the opposite poles of a continuum. In practice there will be overlaps, but it is obviously important from the standpoint of treatment to determine to what extent the previous diet has lacked protein, calories, or both.

The chemical changes and the metabolic adaptations of the body in states of starvation have been outlined in several previous sections. The basic principle that unifies all these alterations is based on the energy equivalence of the different classes of foodstuffs and the body tissue components. When the total caloric intake from the major fuels—carbohydrates and fats—is less than the energy expenditure by the body, it will catabolize both incoming protein and its own cellular protein, which leads to the general wasting of muscles and other organs noted in marasmus. Starvation may therefore be considered as a condition in which the individual must cannibalize his own tissues to serve his energy requirements. Before this drastic, often fatal eventuality is reached, a series of shifts in metabolism occur, depending of course upon the nutritional state at the onset of the condition.

Let us assume that we are considering a normal, well-nourished adult in good health, who is suddenly deprived of adequate caloric intake. In the short term, food deprivation provokes a lowering of the blood glucose levels, which induces the pancreas to release more glucagon and less insulin. An acute activation of glycogenolysis occurs in the liver to restore blood glucose levels toward normal. If food is not eaten within 24 hr, most of the hepatic glycogen will be depleted, and, after the first day or so, other mechanisms must be called into play. Continued glucagon release, for example, will activate hormone-sensitive lipase to produce more fatty acids for oxidation within the liver. Similarly, ACTH, catecholamines, and other hormones, which are released in response to the stressful stimuli to regulatory centers, will promote the activation of adipose-tissue lipase, and mobilization of the resulting fatty acids from fat depots will provide an energy source for peripheral tissue, the liver, and other visceral organs. In contrast with the short-term energy supply provided by hepatic glycogen storage, the triacylglycerols of adipose tissue can provide sufficient ATP for body functions for several weeks.

As the body shifts toward its long-term adaptation to starvation, there will occur, in the first weeks, a significant increase in the amount of ketone bodies produced by the liver, because of the predominance of fatty-acid oxidation and the decreased provision of pyruvate or oxaloacetate, which fails to maintain sufficient flux through the citric-acid cycle. Peripheral tissues, such as heart and eventually the brain, adapt to derive a good deal of their energy from these ketone bodies after several weeks. If this production of ketone bodies exceeds their utilization, however, metabolic acidosis will ensue, resulting in both respiratory compensation (loss of CO_2) and renal compensation (excretion of NH_4^+). Simultaneous with the shift to fatty-acid mobilization and oxidation, the enzymes involved in fatty-acid biosynthesis—such as acetyl-CoA carboxylase, fatty-acid synthetase, and the citrate-cleavage enzyme—are progressively lost from both adipose tissue and the liver.

The proteins in tissues are mobilized to varying degrees as energy sources as the state of starvation continues. The labile proteins of the GI tract and the digestive enzymes of the pancreas are the first to go, paradoxically, to impair the utilization of the small amounts of food that might be ingested. Subsequently, proteins of the visceral organs, such as those of the liver and spleen, are catabolized, followed by the functional proteins of muscle and finally those of the nervous system. The order of utilization parallels the rates of turnover of the proteins in the dynamic steady-state described previously. Amino-acid catabolism, accentuated by the secretion of glucocorticoids in greater amounts and of insulin in reduced amounts, gives rise to direct sources of ATP for the peripheral tissues and to a carbon supply for glucose formation by the liver and kidney. The marked increase of the cortisol: insulin ratio in blood serves as a useful diagnostic marker of marasmus, and acts as the primary regulator for release of endogenous energy metabolites from peripheral tissues. Hepatic enzymes of gluconeogenesis—such as the aminotransferases, phosphatases, and those involved in the conversion of pyruvate to PEP—are considerably increased in amount in the first weeks of starvation, despite the overall stimulus for protein catabolism. As starvation progresses and the brain derives proportionately more energy from β-hydroxybutyrate than from glucose, the rate of gluconeogenesis will gradually decline. Coincidentally, the muscle mass, the whole body activity, and hence the energy expenditure will also decline. Finally, when all the lipid stores are depleted, the body will begin to break down even the essential proteins of the heart, lungs, blood cells, and so forth, and death then ensues from cardiorespiratory collapse.

OBESITY

It is one of the tragic ironies of modern civilization that while roughly one third of the people in underdeveloped countries are subject to various forms of food shortage, the most serious nutritional problem in the more affluent countries of the western world is overnutrition. The prevalence of obesity is rather difficult to assess, since there is no objective criterion to define the limits of body fat content. Recent surveys in the United States and Canada suggest, however, that more than one third of the adult population may exceed the ideal weight figures by more than 20%. This grossly overweight group suffers a markedly greater mortality from medical causes in the sudden-death category. At the other extreme, an obsession with reducing diets among non-obese individuals and the fashion of the "lean and hungry look" may lead to undue emaciation or anorexia. Recent mortality figures indicate that it is probably safer to be somewhat (5%–10%) overweight rather than underweight, especially with increased age.

There is currently much debate concerning the causes of adult obesity, its relation to the differentiation of adipose tissue in childhood, and the role of genetic factors. Studies with families and with identical twins have shown concordance of weight gain and heritability during various stages of development, although environmental factors also play a major part in the growth of adipose tissue cells. The adipocyte cell population can grow in two ways; (1) by increased numbers of cells *(hyperplasia)*, or (2) by increased size of existing cells *(hypertrophy)*. The conventional wisdom based upon studies in rats was that the numbers of fat cells were determined during a "critical period" of infancy, when overfeeding could induce the differentiation of precursor cells *(preadipocytes)*. In later life eating in excess of energy requirements would result in enlargement of cells within a fixed population of adipocytes. Thus, *childhood obesity* was viewed chiefly as hyperplastic, while *adult obesity* was hypertrophic. A corollary assumption was that obesity in early life predisposes one to obesity in adulthood.

Recently improved methods of counting and measuring adipose cells in the human have revealed that although adipocyte proliferation may be most rapid between years 1 and 20, there is significant differentiation of new fat cells in the adult. Moreover, obesity at all ages will be accompanied by cellular hypertrophy; hyperplastic obesity may occur in adults as well as in children. Thus it would seem equally important to prevent the development of obesity, with its attendant proliferation of fat cells, throughout life.

The cruel paradox of severe hunger in the midst of plenty is well known to fat persons, and this suggests that the appestat—the mechanism in the hypothalamus that controls food intake—is set too high in these individuals. In other instances, obese persons may eat no more than lean individuals, but their activity levels are also markedly less. Despite the increased work load involved in moving their extra weight, such sufferers from obesity usually seem to get much less exercise and may also consume less energy in maintaining normal physiologic functions. As a machine, the obese body is unfortunately much too efficient in its energy coupling.

A net gain of fat in the storage depots occurs when the lipogenesis due to excess dietary carbohydrates or fats outstrips the mobilization of the fatty acids and their oxidation to provide the energy requirements of the peripheral tissues. Increased food intake and increased insulin secretion favor lipogenesis; increased exercise and increased secretion of the lipolytic hormones (the catecholamines and glucagon) promote fat utilization. Because of the interrelations among many factors and organ functions, it is erroneous to focus unduly upon the adipose tissue as the site of the lesion in obesity. The fatty deposits are simply the last link in a chain of metabolic events.

Ingested fats contribute directly to the triacylglycerols of adipose tissue, as witnessed by the fact that the unsaturated fatty-acid composition of the depots tends with time to reflect that of the exogenous fat in the diet. Dietary carbohydrate, however, is a key contributor to lipogenesis in obesity. In the first place, absorbed glucose in excess of energy needs or glycogen-storage capacity is channeled into fatty-acid synthesis. The hepatic cells form the most important locus for the endogenous formation of fatty acids, which will be predominantly of the saturated variety. The extra glucose and triacylglycerols exiting from the liver are taken up by the adipose cells, where some additional fatty-acid synthesis occurs. The esterification of the fatty acids entering the cells, as well as of those generated in the adipose tissue itself, will depend upon conversion of the incoming glucose to the α-glycerophosphate acceptor. In the adipose tissue of obese individuals, an enhanced conversion of glucose to α-glycerophosphate is coupled with decreased oxidation of α-glycerophosphate by the mitochondrial dehydrogenase. The consequence is that glucose is diverted to the glycerol of glycerides and a predominance of esterification over free fatty-acid release from the cells will result.

Another way in which the obese individual may be super-efficient is in the metabolism of the brown fat. As noted in Chapter 5 this type of adipose tissue is of greatest consequence in the newborn in whom it functions to maintain body temperature through the oxidation of fatty acids uncoupled from ATP formation. Brown adipose tissue is present in adults in a number of locations, particularly in the interscapular and perirenal areas. The tissue is thermogenic as indicated by increased oxygen consumption following cold exposure or infusion with catecholamines. There are also indications that enhanced metabolism of brown adipose tissue occurs in normal individuals on hypercaloric diets with high fat content. By such a device, lean individuals may resist fat deposition by burning the excess in an energy-inefficient mode. Obese individuals, or children of obese parents, who seem to have a greater risk of becoming obese, show a much lower metabolic response and hence retain the excess energy intake as enhanced deposits in the regular white adipose tissue.

Superimposed upon the imbalances created by dietary excess are the effects produced by insulin. A considerable increase in the circulating level of insulin is observed in the obese individual, as is an exaggerated response to the ingestion of glucose. The combination of greater caloric intake plus hyperinsulinism not only promotes glucose uptake into the adipose tissue and activates pyruvate dehydrogenase to produce more acetyl CoA, but also induces the syntheses of a number of enzymes—citrate-cleavage enzyme, acetyl-CoA carboxylase, and fatty-acid synthetase—in both the liver and adipose cells that lead to accelerated lipogenesis. The increase in insulin concentration will also depress the activity of hormone-sensitive lipase, further tilting the imbalance toward fat deposition.

The treatment of obese persons is one of the most difficult and frustrating enterprises, particularly from the long-range standpoint. An extreme application of will power may lead to a substantial weight loss over a period of months, only to be followed by some change of habits or emotional crisis that completely reverses the picture in a short period. The biochemical basis for this "yo-yo" phenomenon is readily appreciated by consideration of the fact that the adaptive increases of enzymes of lipogenesis occur within hours of feeding, but the de-adaptation of the same enzymes requires several weeks of fasting. Frequent cruel jokes and discriminatory practices, coupled with the popular attitude that fat people are food addicts who exert no effort to control their eating habits, further add to the feeling of frustration.

Complete starvation under hospital supervision has been used for drastic weight reduction, but a review of the preceding section will underline the dangers of such draconian measures. The major problem, of course, is the fact that an even greater loss of lean body mass than of depot fat may occur in total starvation. Since the obese person generally has gained fat at the expense of muscle tissue in the first place, he is in a poor position to accept further tissue wasting. In treatment, then, it seems more reasonable to reduce the total caloric intake to a point below the basal amount needed for the maintenance of activity and to ensure that enough protein is present to provide for the synthesis of muscle and other lean tissues as well as to offset any negative nitrogen balance due to increased gluconeogenesis. A reduction of carbohydrate intake will obviously prevent the tendency to lipogenesis that is induced by high blood glucose levels and hyperinsulinism. This will be most effective if the high sucrose content of most refined, modern diets is replaced by the less rapidly absorbed natural carbohydrates, such as the starches. The great surges of blood glucose that follow the heavy, single meal of the day that is a common eating pattern in obesity are major contributors to the adaptive increases of lipogenesis. Consequently, it is more useful to apportion the dietary intake among many, small, equal meals at intervals throughout the day.

The simultaneous adoption of a graduated, well-supervised exercise program will aid in the weight-loss process. First, the increased energy expenditure contributes to catabolism rather than anabolism of fat. Second, the stimulus to muscle development ensures the loss of depot tissue while promoting a compensatory buildup of lean body mass. Third, as the exercise becomes progressively more strenuous, the accompanying release of catecholamines accentuates lipolysis and the utilization of fatty acids by the muscles. Fourth, the strengthening and increased capacity of the heart and respiratory system (training effect) will reduce the likelihood of cardiovascular illness. Finally, the increased mobility and stamina that he gains will allow the individual to participate more fully in the pleasurable activities of life and enhance that immeasurable but most important attribute, his feeling of well-being.

SUGGESTED READINGS

Coleman JE: Metabolic interrelationships between carbohydrates, lipids and proteins. In Bondy PK, Rosenberg LE (eds): Metabolic Control and Disease 8th ed, pp 161–274. Philadelphia, WB Saunders

Coward WA, Lunn PG: The biochemistry and physiology of kwashiorkor and marasmus. Br Med Bul 37:19–24, 1981

Craddock D: Obesity and its Management, 3rd ed. Edinburgh, Churchill Livingstone 1978

Cravioto J, Hambraeus L, Vahlquist B (eds): Early Malnutrition and Mental Development. Stockholm, Almqvist and Wiksell, 1974

De Pierre JW, Seidegard J, Morgenstern R, et al: Induction of drug-metabolizing enzymes: A status report. In Lee CP, Schatz G, Dallner G (eds): Mitochondria and Microsomes, pp 585–610. Reading, MA, Addison–Wesley Publishing Co, 1981

Dobbing J, Smart JL: Vulnerability of developing brain and behavior. Br Med Bull 30:164–168, 1974

Epstein FH: Metabolic requirements for renal function. Hosp Pract 14, No. 6:93–102, 1979

Felig P, Wahren J: Protein turnover and amino acid metabolism in the regulation of gluconeogenesis. Fed Proc 33:1092–1097, 1974

Guroff G: Molecular Neurobiology. New York, Marcel Dekker, 1980

Häger A: Adipose tissue cellularity in childhood in relation to the development of obesity. Br Med Bull 37:287–290, 1981

Hems DA: Carbohydrate metabolism in relation to kidney function. In Schmidt U, Dubach UC (eds): Renal Metabolism in Relation to Renal Function, pp 26–39. Bern, Hans Huber, 1976

Himms–Hagen J: Cellular thermogenesis. Annu Rev Med 38:315–351, 1976

Holloszy JO, Booth FW: Biochemical adaptations to endurance exercise in muscle. Annu Rev Med 38:273–291, 1976

Iverson LL: The chemistry of the brain. Sci Am 241, No. 3:134–149, 1979

James WPT, Trayhurn P: Thermogenesis and obesity. Br Med Bull 37:43–48, 1981

Kappas A, Alvares AP: How the liver metabolizes foreign substances. Sci Am 232, No. 6:22–31, 1975

Keys A: Overweight, obesity, coronary heart disease and mortality. Nutr Rev 38:297–307, 1980

Li T–K: Enzymology of human alcohol metabolism. Adv Enzymol 45:428–483, 1977

Neely JR, Morgan HE: Relationship between carbohydrate and lipid metabolism and the energy balance of heart muscle. Annu Rev Physiol 36:413–459, 1974

Orrenius S, Thor H, Jernström B: The influence of inducers on drug-metabolizing enzyme activity and on formation of reactive drug metabolites in the liver. In Ciba Foundation Symposium 76 (new series): Environmental Chemicals, Enzyme Function and Human Disease, pp 25–39. Amsterdam, Excerpta Medica, 1980

Paine AJ: Hepatic cytochrome P-450. Essays Biochem 17:85–126, 1981

Peters TJ: Ethanol metabolism. Br Med Bull 38:17–20, 1982

Ruderman NB: Muscle amino acid metabolism and gluconeogenesis. Annu Rev Med 26:245–258, 1975

Salans LB: Obesity and the adipose cell. In Bondy PK, Rosenberg LE (eds): Metabolic Control and Disease 8th ed, pp 495–521. Philadelphia, WB Saunders, 1980

Simpson DP: Regulation of renal metabolism by changes in acid–base balance.

In Angielski S, Dubach U (eds): Biochemical Aspects of Renal Function, pp 58–64. Bern, Hans Huber, 1975

Sjöström L: Fat cells and body weight. In Stunkard AJ (ed): Obesity, pp 72–100. Philadelphia, WB Saunders, 1980

Sokoloff L: Metabolism of ketone bodies by the brain. Annu Rev Med 24:271–280, 1973

Tizard J: Early malnutrition, growth, and mental development in man. Br Med Bull 30:169–174, 1974

Vary TC, Reibel DK, Neely JR: Control of energy metabolism of heart muscle. Annu Rev Physiol 43:419–430, 1981

Glossary

Active Center—that region of an enzyme molecule where the substrate combines and is activated to react during catalysis (syn. *catalytic site*)

Active Transport—the process whereby energy is utilized to carry a substance across a membrane, from a solution of lower to higher concentration *(i.e.,* against the concentration gradient of the substance)

Acyl Carrier Protein—a polypeptide whose prosthetic group, pantetheine, carries the intermediates of fatty acid biosynthesis in a thioester linkage during the reaction sequence catalysed by fatty-acid synthetase

Adenosine Triphosphate (ATP)—the nucleotide formed by joining the purine base, adenine, through ribose to three phosphate groups linked together. It is the principal mediator of energy-driven reactions in cells and is derived by phosphorylation of *adenosine diphosphate (ADP)*

Adenylate Cyclase—an enzyme in the plasma membrane of cells, with a regulatory subunit on the outer surface whose interaction with appropriate extra cellular hormones will activate or suppress the catalytic subunit on the inner surface which converts ATP to intracellular cyclic AMP (syn. *adenyl cyclase)* see also **Cyclic AMP)**

Allosteric Modifier—a substance which combines with a **regulatory enzyme** at one region *(allosteric site)* and modifies activity at the catalytic site; it may either enhance the activity *(positive modifier)* or inhibit it *(negative modifier)*

Amino Acid—an organic compound with a hydrocarbon residue attached to an amino (NH_2) group and a carboxyl (COOH) group, that serves as the monomeric unit of the proteins

Aminotransferase—one of a group of enzymes that catalyzes removal of the amino group from an amino acid to produce the α-keto acid and then reversibly transfers the $-NH_2$ to a second α-keto acid to yield a new amino acid (syn. *transaminase)*

Amphipathic Property—the characteristic of a molecule possessing two regions of opposing solubility such that one end of the molecule dissolves readily in water **(hydrophilic),** while the other end does not **(hydrophobic)**

Anabolic Hormone—an endocrine substance whose metabolic effects upon the target tissue are primarily to promote the biosynthesis of proteins, fats, or polysaccharides, or to inhibit their degradation (ant. *catabolic hormone)*

Anabolism—biosynthetic transformations requiring energy input *(e.g.,* production of macromolecules from small precursors) (ant. *catabolism)*

Antibody—a serum protein that is synthesized in response to the entry of a foreign substance **(antigen);** it combines selectively with the antigenic agent to neutralize its toxic or infectious effects (see also **Immunoprotein)**

Anticodon—a sequence of three bases in transfer RNA that is complementary to the **codon** triplet of messenger RNA (see also **Codon, Transfer RNA**)

Antigen—A substance, either free or combined at the surface of a bacterial or tissue cell, that elicits the formation of a specific **antibody** protein and combines avidly and selectively with the induced antibody (see also **Immunoprotein**)

Antimetabolite—a substance that bears a structural resemblance to a normal substrate or coenzyme, thus competing with it in metabolism and antagonizing its actions

Apoenzyme—the inactive protein portion remaining when an essential bound cofactor or prosthetic group is removed; it may regain catalytic activity when the prosthetic group is restored to form the holoenzyme (see also **Prosthetic group, Holoenzyme**)

Beta (β) Oxidation—the enzymatic process in mitochondria whereby fatty acids are oxidatively degraded at the second carbon atom (β-position) from the carboxyl group, such that two-carbon fragments are successively released as acetyl CoA, with successive shortening of the fatty-acid carbon chain

Bilayer—a bimolecular sheet formed by the associations of hydrophobic surfaces of two lipid monolayers to produce a sandwich two molecules thick with only the hydrophilic surfaces exposed on either outer side (see also **Monolayer)**

Bile Pigment—a breakdown product of hemoglobin (see **Bilirubin)**

Bile Salt—the salt form of a steroid acid, usually conjugated to an amino acid such as glycine; it is produced in liver by oxidation of cholesterol and secreted in the bile into the intestine to help emulsify fats and aid in their digestion and absorption

Bilirubin—a degradation product formed by opening the porphyrin ring derived from heme degradation, consisting of four pyrrole rings linked in a straight chain (syn. **bile pigment)**

Biologic Half-Life—a measure of the time required for the body either to renew one half of its normal components or to eliminate one half of a foreign substance taken into the body

Carbohydrate—a polyhydroxylic compound of carbon, hydrogen and oxygen; it may be a simple sugar **(monosaccharide)** or contain a few **(oligosaccharide)** or many **(polysaccharide)** sugar units joined together

Carnitine—a carboxylic hydroxy acid related to choline found in high concentrations in muscle; it plays a role in transporting fatty-acyl groups across the mitochondrial inner membrane

Catabolic Hormone—an endocrine substance whose metabolic effects upon the target tissue are primarily to promote the degradation of proteins, fats, or polysaccharides or to inhibit their biosynthesis (ant. **anabolic hormone)**

Catabolism—degradative reactions releasing free energy and heat *(e.g.,* oxidation of organic compounds to carbon dioxide and water) (ant. **anabolism)**

Catecholamine—a family of aromatic substances with sympathomimetic activity, which possess an *o*-dihydroxybenzene (catechol) ring attached to an aliphatic amine portion

Cholesterol—the major steroid derivative in the body; it exists as the free alcohol in cell membranes and as the fatty-acid ester in blood plasma lipoproteins and serves as precursor of other important steroid derivatives, such as the sex hormones, bile salts, and so forth

Chromaffin Granule—a secretory vesicle containing high concentrations of catecholamines, particularly located in the adrenal medulla and sympathetic neurons

Chromosome—one of the structures found in the nucleus of animal cells that contains the DNA of the genes, associated with histones and other proteins and with RNA

Chylomicron—a small lipid droplet composed mainly of fat with small amounts of cholesterol and phospholipids and a thin coating of protein; it is synthesized in the intestinal mucosa cells from products of lipid digestion and released into the lymphatic vessels

Citric Acid Cycle—the series of mitochondrial reactions whereby a two-carbon moiety (acetyl CoA) is completely oxidized by combining with oxaloacetic acid to form citric acid, the latter then being catabolized to 2 mols of CO_2, and regenerating the oxaloacetic acid (syn. *Krebs cycle; tricarboxylic acid cycle)*

Codon—a sequence of three bases (triplet) in messenger RNA which specifies the ordering of a particular amino acid through the complementary **anticodon** triplet of its attached transfer RNA during protein synthesis. *Initiator* and *terminator codons* start and stop translation of the message

Coenzyme—an organic cofactor that functions in enzyme catalysis; it generally contains a **vitamin** as the reactive group

Coenzyme A (CoA)—a nucleotide derivative containing the vitamin, pantothenic acid; its functional thiol (—SH) group forms thioesters with acyl groups of important metabolic intermediates *(e.g., acetyl-SCoA)*

Coenzyme Q (CoQ)—a fat-soluble, substituted benzoquinone that functions as an electron carrier in the respiratory chain between flavoproteins and cytochromes

Cofactor—a nonprotein substance that is required for the activity of an enzyme; it may be inorganic (*e.g.*, a metal ion) or organic *(e.g.*, a **coenzyme)**

Competitive Inhibition—the action of an inhibitor that so closely resembles the substrate that it competes with it for binding sites at the enzyme's active center. (V_{max} is unaffected but K_m is increased in the presence of inhibitor.)

Complementary Base-Pairs—the specific pairs formed by hydrogen bonds between a purine base of one nucleic acid strand with a pyrimidine of another strand; adenine (A) always pairs with thymine (T) or uridine (U), guanine (G) with cytosine (C)

Constitutive Enzyme—an enzyme whose concentration is fixed and not subject to stimulation **(induction)** or inhibition **(repression)** of its biosynthesis by regulatory agents

Cori Cycle—the process in carbohydrate metabolism whereby lactic acid that is derived from muscle glycolysis is carried through the bloodstream and is recycled by liver gluconeogenesis to glucose which is returned in the bloodstream to the muscle (syn. *glucose-lactate cycle)*

Coupling Inhibitor—an agent that prevents the interaction of ADP with the energy-coupling steps in oxidative phosphorylation and, hence, blocks both ATP synthesis and respiration

Coupling Site—one of the oxidative reactions of the respiratory chain in which there is a large release of free energy that is capable of being coupled to the formation of ATP from ADP and inorganic P (see also **Oxidative phosphorylation)**

Cyclic AMP—a nucleotide that acts as the intracellular mediator in the target cells after their interaction with nonpenetrating hormones (syn. *adenosine 3', 5'-cyclic phosphate;* **second messenger)**

Cytochrome—one of a class of heme-containing proteins, localized to the membranes of mitochondria and endoplasmic reticulum, whose iron atoms undergo cyclic reduction and oxidation during the sequential electron transfers of respiration

Cytochrome Oxidase—the terminal carrier of the respiratory chain; it is the only cytochrome component capable of transferring electrons directly to oxygen (syn. *cytochrome* a-a_3)

Dehydrogenation—an oxidation reaction in which hydrogen atoms are abstracted from the substrate

Denaturation of Proteins—unravelling of the natural folded conformation of a polypeptide; it is generally accompanied by irreversible decrease of solubility and loss of biologic activity of the protein

Deoxyribonucleic Acid (DNA)—the polymer formed of nucleotides containing deoxyribose (dAMP, dCMP, dGMP, dTMP); it is the chemical basis of the gene and is formed of two interwined, complementary strands *(double helix)*

Derepression—removal or neutralization of action of a repressor molecule that blocks enzyme synthesis (see **Induction, Repression)**

Dynamic State of Body Constituents—the condition of turnover in tissue proteins and other components, which may appear to exist in a static equilibrium in the absence of growth, but nonetheless are in a steady state of synthesis and degradation, undergoing continual exchanges with dietary substances (see also **Steady state)**

Electron Carrier—a compound which, by alternating between oxidized and reduced states, serves as an intermediary in the transfer of electrons from substrates to oxygen along the respiratory chain (see also **cytochrome)**

Emulsion—a fine, milky dispersion of one liquid in another liquid with which it is not capable of producing a solution

End-Product Inhibition—a form of metabolite control in which the final products of a metabolic sequence exert **negative feedback** on the whole pathway by inhibiting a pacemaker regulatory enzyme, usually near the beginning of the sequence

Endoplasmic Reticulum—an interconnected network of flattened tubules in the cytoplasm. Regions with ribosomes attached to the cytoplasmic surfaces of the tubules are termed *rough endoplasmic reticulum* and are involved in the synthesis of proteins, especially those for secretion; regions lacking particles are termed *smooth endoplasmic reticulum* and play roles in the processing of secretory proteins and in drug detoxification.

Energy of Activation—a measure of the energy required to raise a molecule to the excited transition state in order to permit a reaction to proceed

Enzyme—a protein molecule which combines selectively with and acts specifically upon another molecule, the **substrate,** to catalyze the conversion of substrate into particular **product** molecules

Enzyme Inhibitor—a substance that selectively interacts either to block or to modify the conformation of the active center of an enzyme, thereby slowing or preventing its catalytic action upon the substrate

Enzyme–Substrate Complex—the transient, activated intermediate formed when the substrate is bound in the active center of an enzyme during catalysis

Epinephrine—a hormone produced by the adrenal medulla, which produces effects on target tissues by raising cyclic AMP concentrations in the cells, which promotes glycogenolysis in muscle and liver, raising blood glucose levels, and activates lipolysis in adipose tissue, raising blood fatty acid levels. (see **Catecholamine**) (syn. *adrenaline)*

Essential Amino Acids—those that are not synthesized at all or rapidly enough in the body and, hence, must be provided in the diet. (For humans they are *arginine, histidine, isoleucine, leucine, lysine, methionine, phenylalanine, threonine, tryptophan, valine.)*

Essential Fatty Acids—those polyunsaturated long-chain acids that may not be synthesized in the body and, therefore, are essential in the diet. (They include *linoleic acid* (C18:2), *linolenic acid* (C18:3), *arachidonic acid* (C20:4.)

Excitation of Neurons and Muscle—the alteration of membrane permeability in response to external stimuli that produces a change of the membrane potential due to movements of cations (K^+ and Na^+) across the cell membrane (syn. *depolarization)*

Exon—expressed sequence of the gene that codes for a polypeptide sequence (see **Intron**)

Fatty Acid—an aliphatic carboxylic acid, usually with a single carboxyl group attached to a long, straight hydrocarbon chain, with an even number of carbons between 14 and 24 and a variable number of double bonds from 0 to 6 in the carbon chain

Fatty Acid Synthetase—the multienzyme system that catalyzes the stepwise condensation of one molecule of acetyl CoA and seven molecules of malonyl CoA and the reduction of intermediates with NADPH to produce palmitoyl CoA

Flavoprotein—a protein which contains a riboflavin coenzyme (FMN or FAD) as its prosthetic group

Free Energy (F)—the energy that is available to perform work as a system proceeds toward equilibrium. A process will occur spontaneously in nature only if its free energy change (ΔF) is less than zero (syn. *Gibb's free energy* [*G*]).

Gene—the unit hereditary factor; it is the segment of a DNA molecule that specifies one polypeptide chain (syn. *cistron*)

Genotype—the entire genetic constitution of an individual

Globulin—one of the group of proteins in blood plasma that is precipitated by half saturation with ammonium sulfate. The various globulin proteins differ in their lipid and carbohydrate components and may be separated by ultracentrifugation and electrophoresis

Glucagon—a polypeptide hormone secreted by the α-cells in the pancreatic islet tissue. It specifically activates the formation of cyclic AMP in liver cells, thus raising the blood glucose by stimulating liver glycogenolysis and gluconeogenesis

Glucocorticoid—any of the steroid hormones produced by the adrenal cortex which promote the synthesis of glucose by gluconeogenesis and, hence, elevate the concentration of liver glycogen and blood sugar

Gluconeogenesis—biosynthesis of glucose from *lactate, pyruvate,* or *oxaloacetate* or from noncarbohydrates which give rise to these compounds or to other intermediates of glycolysis

Glucose—an aldehyde-containing, six-carbon sugar (aldohexose), which is the most important and abundant monosaccharide in body fluids and is the monomer of the storage polysaccharide, glycogen, in the tissues

Glutamic Acid—the five-carbon dicarboxylic amino acid, α-aminoglutaric acid. It is derived in metabolism by aminotransfer or reductive amination of the citric acid cycle intermediate, α-ketoglutaric acid

Glycogen—the storage carbohydrate of liver, muscle, and other body tissues; it is a highly branched polysaccharide of glucose monomers

Glycogen Synthetase—the enzyme that catalyzes the elongation of the linear chains of glycogen by sequential addition of glucose units from UDP glucose to the terminal 4-hydroxyl groups of growing polysaccharide chains

Glycogenesis—the synthesis of glycogen by polymerization of glucose monomers

Glycogenolysis—the splitting of the glycogen polymer chain to yield free glucose or glucose phosphate esters

Glycogenosis—one of a group of inborn errors of metabolism which exhibit the common characteristic of excessive storage of glycogen, though the tissues affected and structures of the polysaccharides may vary (syn. *glycogen storage disease*)

Glycolipid—a fatty substance containing one or more sugars or carbohydrate derivatives; it will usually also contain the long-chain base, sphingosine (see also **sphingolipid**)

Glycolysis—the breakdown of hexose sugars, such as glucose, to three-carbon acids, such as *pyruvate* and *lactate,* yielding energy as ATP. In the absence of oxygen, lactate is the predominant carbon product *(anaerobic* glycolysis)

Glycolytic Pathway—the overall sequence whereby glucose is converted to pyruvate or lactate (see **Glycolysis**)

Glycoprotein—a polypeptide containing one or more sugars or carbohydrate derivatives as its prosthetic group; it frequently is a component of cell membranes, where the carbohydrate residues may act as antigenic determinants or other recognition sites on the exterior surface

Golgi Apparatus—a system of stacked, flattened tubules usually lying between the nucleus and plasma membrane whose cisternal spaces contain secretory proteins undergoing modification prior to being exported from the cell

Heme—the complex that is formed when ferrous iron is chelated in the center of the protophorphyrin ring; it is the prosthetic group of oxygen-carrying proteins (**hemoglobin,** *myoglobin*) and respiratory proteins **(cytochromes)**

Hemoglobin—the pigmented protein of red blood cells; it reversibly binds oxygen to its heme prosthetic group and carries it from the lungs to the peripheral tissues

High-Energy Compound—a substance whose hydrolysis results in the release of a large amount of energy, *i.e.,* a negative free energy change of more than 5 Cal/mol under standard conditions

Histone—a strongly basic protein, rich in arginine and lysine residues, that forms a complex with DNA in the chromosomes of eucaryotic cells

Holoenzyme—a conjugated protein that contains a prosthetic group which is necessary for the catalytic action of the enzyme (see also **Prosthetic group, Apoenzyme**)

Hormone Receptor—a protein molecule with a specific, high-affinity binding site for the hormone, located on the outer surface of a responsive target cell

Hydrophilic—a description for molecules or portions of molecules which are *polar* by nature and associate readily with water molecules (ant. **hydrophobic**)

Hydrophobic—a description for molecules or portions of molecules that are *nonpolar* by nature and prefer to associate away from water molecules (ant. **hydrophilic**)

Immunoprotein—a protein, usually found in the γ-globulin fraction of blood plasma, that confers resistance to disease or eliminates another extrinsic molecule through a specific binding in response to an antigenic challenge (syn. *antibody, immunoglobulin*)

Inborn Error of Metabolism—an inherited disease in which a defect in a gene results in the failure to produce a normal catalytically active enzyme

Inducible Enzyme—an enzyme that is normally present in low amounts and will be induced to undergo accelerated biosynthesis by the increased concentration of a metabolite or hormone inducer molecule (syn. *adaptive enzyme*)

Induction—the promotion of synthesis of an enzyme by a metabolite, the *inducer* molecule, which is often a substrate for the enzyme (syn. *derepression*)

Insulin—a polypeptide hormone secreted by the β-cells in the pancreatic islet tissue, which lowers the blood glucose by enhancing its uptake and utilization by tissues and by suppressing its synthesis in gluconeogenesis

Intermediate—substance that is formed at one of the stages during conversion of starting material to end-products by a sequence of enzyme reactions (see **Precursor**)

Intron—intervening sequence of a gene that does not code for polypeptide, but is located between the expressed sequences in the gene (see **Exon**)

Ionophore—a substance that forms a lipid-soluble complex with a specific ion and thus is capable of carrying the ion through the hydrophobic bilayer of biologic membranes

Isoelectric Point—the pH of the solution in which a protein will possess an equal number of positively and negatively charged groups, that is, no net charge

Isoenzyme—one of the variant forms in which an enzyme may be present in the same species; it will perform the same catalytic reaction as the other isoenzymes but may be distinguished by its different kinetic properties and electrophoretic migration (syn. *isozyme*)

Ketone Body—any of a group of products of the incomplete oxidation of fatty acids, namely, *acetoacetic acid, β-hydroxybutyric acid* or *acetone*

Kinase—an enzyme that catalyzes transfer of the terminal phosphate group from a nucleoside triphosphate (usually ATP) to an acceptor molecule (usually an alcohol), forming a phosphate ester (syn. *phosphotransferase*)

Krebs Cycle (see **Citric acid cycle**)

Krebs–Henseleit Cycle (see **Urea cycle**)

Lipase—any of the group of esterase enzymes found in pancreatic secretion into the intestine, inside the fat cells, or at the cell membrane of liver and other tissues, that catalyze the hydrolysis of ester linkages between glycerol and the fatty acids of fats or phospholipids

Lipid—an organic compound of natural origin, which is soluble in non-polar reagents such as chloroform, ether, and alcohol (the so-called fat solvents) but is insoluble in water; it usually contains a long hydrocarbon chain or a sterol

Lipidosis—one of a group of inborn errors of metabolism characterized by abnormal accumulations of lipids in the tissues

Lipogenesis—the synthesis of fats by condensation of acetyl-CoA molecules and reduction to fatty acids and esterification of the latter to form the triacylglycerols

Lipolysis—the hydrolytic decomposition of lipid, usually pertaining to fat hydrolysis into glycerol or partly acylated glycerol, plus free fatty acids

Lipoprotein—a water-soluble combination of a lipid or a number of different lipids with a protein. (Combinations of lipids with protein which are fat-soluble are generally termed *proteolipids*)

Lipoprotein Lipase—an esterase present in many tissues, especially muscle and adipose tissue, that attacks the chylomicrons hydrolysing the triacylglycerols and, hence, clears the milky turbidity of blood plasma after a fatty meal (syn. *clearing factor*)

Lysosome—a membrane vesicle containing hydrolytic enzymes, which carries out intracellular digestive functions

Macromolecule—a very large molecule with a polymeric structure, such as a **protein** or **nucleic acid**

Messenger RNA—the single-stranded ribonucleic acid molecule transcribed from the segment of DNA that codes for a protein molecule. It carries the genetic message from the nucleus to the ribosome and orders the amino acids into the correct sequence during protein synthesis

Metabolic Acidosis—accumulation of abnormal quantities of organic acids *(lactate, acetoacetate)* that arise from excess tissue metabolism

Metabolic Antagonism—opposition to the actions of a substance by an analogue whose structure is similar enough to permit it to compete in metabolism (see also **antimetabolite**)

Metabolic Debt (of Muscle)—the condition prevailing after prolonged activity of muscle, resulting in depletion of high-energy reserves that must be restored later when the muscle is at rest

Metabolism—the chemical transformations of molecules, brought about in living organisms by the actions of enzymes, under the control of genes, hormones, energy state, and the end-products

Methyltransferase—one of a group of enzymes that catalyzes the transfer of a methyl group from the activated methyl donor, S-adenosyl-methionine, to a specific acceptor (syn. *transmethylase*)

Micelle—a very fine, colloidal dispersion of molecules, usually as a parallel array, within a liquid phase, which fails to form a true solution of the molecules

Michaelis–Menten Constant (K_m)—the concentration of a substrate that results in one half of the maximal rate for an enzymatic reaction; it is a characteristic value for each enzyme–substrate combination and reflects the binding affinity

Michaelis–Menten Equation—the kinetic expression for the dependence of enzyme rate *(v)* upon substrate concentration [*S*]:

$$v = V_{max} [S]/[S] + K_m$$

where V_{max} is the maximal rate with saturating substrate concentration and K_m is the **Michaelis–Menten constant**

Microfilament—a fine threadlike contractile structure which is particularly abundant in the cytoplasm of motile cells and participates in cellular movements and changes of shape

Microtubule—a hollow cylindrical structure that plays a role in maintaining the shape and rigidity of cells and forms the mitotic spindle during cell division

Mitochondrion—the double-membrane enveloped organelle of eucaryotic cells that is responsible for cellular respiration, oxidation of pyruvate and fatty acids, the citric-acid cycle, and the aerobic generation of ATP by oxidative phosphorylation

Monoamine Oxidase—a flavoprotein enzyme that catalyzes the oxidation of amines into their corresponding aldehydes plus ammonia

Monolayer—a monomolecular surface film produced by packing lipid molecules side-by-side at an interface between two liquids (*e.g.*, oil–water) or liquid and gas (*e.g.*, air–water)

Monomer—the single unit of a polymeric molecule

Mutation—chemical modification of a gene such that one or more of the bases in the DNA are deleted or altered or a new base is inserted, thus changing the base sequence of the genetic code, and hence the protein coded by the gene either is produced with an altered amino acid sequence or is not produced at all

Myelin—a multilamellar system of membranes, particularly rich in sphingolipids, that forms a sheath around certain nerve fibers

Na^+, K^+-ATPase—an active transport system that uses the energy from hydrolysis of ATP to drive sodium ions outward across the plasma membrane, with an opposite migration of potassium ions into the cell

Negative Feedback—an autoregulatory system whereby an increased concentration of an end-product of metabolism inhibits one of the enzymes in the sequence leading to its formation (see **End-product inhibition**)

Neurotransmitter—a chemical mediator that carries a nerve impulse across a synapse or nerve–muscle junction. Included in this group of agents are acetylcholine and a number of aromatic amine substances (syn. *synaptic transmitter*)

Nicotinamide Adenine Dinucleotide (NAD)—a coenzyme of oxidation–reduction reactions containing the vitamin, nicotinamide, as its functional group. The oxidized form (NAD^+) is reduced to form NADH by accepting two electrons as a hydride ion (H^-)

Nicotinamide Adenine Dinucleotide Phosphate (NADP)—This coenzyme differs from NAD in possessing an additional phosphate group. It has analogous oxidized ($NADP^+$) and reduced (NADPH) forms but has a different specificity for dehydrogenases

Nitrogen Balance—the nutritional state of an individual in relation to his protein metabolism; it is a measure of the differential between ingested and excreted nitrogen

Nitrogen-Ring Compounds—a group of heterocyclic substances including **pyridine** (C_5H_5N), found in nicotinamide coenzymes; **pyrimidine** ($C_4H_4N_2$) and **purine** ($C_5H_4N_4$), found in the nucleic acids and nucleotides

Noncompetitive Inhibition—the action of an inhibitor that interferes with groups involved with catalysis at the enzyme's active center without affecting substrate binding. (V_{max} is decreased but K_m is not altered in the presence of inhibitor.)

Nonessential Amino Acids—those which may be synthesized in the body rapidly enough to meet the needs for protein synthesis. (In man they include *alanine, aspartate, asparagine, cysteine, glutamate, glutamine, glycine, proline, serine, tyrosine.*)

Norepinephrine—a neurotransmitter produced in sympathetic nerves and released in response to nerve stimulation; it produces similar metabolic effects on peripheral tissues to those evoked by epinephrine (see also **catecholamine**) (syn. *noradrenaline, 3, 4-dihydroxyphenylethanolamine*)

Nucleic Acid—a polymer of nucleotides joined together by phosphodiester linkages (syn. *polynucleotide*) (see also **Deoxyribonucleic acid, Ribonucleic acid**)

Nucleolus—the region inside the nucleus that contains much of the nuclear RNA and is involved in the synthesis of ribosomal RNA

Nucleoprotein—a complex formed between a nucleic acid and a protein, usually a basic protein such as a **histone**

Nucleoside—a compound consisting of a nitrogenous base (purine or pyrimidine) joined to a pentose sugar. (A **nucleotide** has one or more phosphate groups attached to the sugar of a nucleoside.)

Nucleotide—a compound consisting of a nitrogenous base (purine or pyrimidine) joined through a pentose sugar to one or more phosphate groups; it may function as a coenzyme or polymerize to form the **nucleic acids** (DNA and RNA)

Nucleus—the largest organelle in eucaryotic cells; it contains within its membrane the DNA of the genes bound to histones in the chromosomes and an RNA-rich structure, the nucleolus

Organic Acid—a compound with a hydrocarbon residue attached to an acidic group (usually COOH = *carboxylic acid*)

Ornithine Cycle (see **urea cycle**)

Oxidative Deamination—the removal of an amino group from a compound and its release as ammonia, accompanied by transfer of hydrogens from the compound to a coenzyme, NAD or FAD (ant. **reductive amination**)

Oxidative Phosphorylation—the mitochondrial process whereby the energy released during the passage of electrons along the respiratory chain is partially conserved by the coupled synthesis of ATP from ADP and inorganic phosphate

Oxidation–Reduction Reaction—a chemical process whereby one or more electrons are transferred from a substance being oxidized to a substance being reduced

Pentose–Phosphate Pathway—the reaction sequence that oxidizes glucose-6-phosphate to produce NADPH and either pentose sugars or carbon dioxide (syns. *hexose monophosphate shunt; phosphogluconate pathway*)

Peptidase—one of the group of hydrolytic enzymes that attacks the peptide linkages of the small peptide fragments which arise from digestion of proteins

Phenotype—the characteristics of an organism that are the product of inherited factors interacting with the environment

Phosphagen—a compound that serves as a reservoir of high-energy phosphate for the regeneration of ATP from ADP *(e.g., phosphocreatine)*

Phosphatase—an enzyme that catalyzes the hydrolytic cleavage of a phosphate ester to release inorganic phosphate (syn. *phosphohydrolase*)

Phosphatidic Acid—the phospholipid produced by acylation of α-glycerophosphate with 2 mols of fatty-acyl CoA; it is the precursor of other phosphoglycerides and triacylglycerols

Phosphoglyceride—a lipid containing glycerol esterified to a phosphate group and one or two fatty-acyl groups (syn. *glycerophosphatide*)

Phospholipid—one of the family of lipid substances containing phosphate and either glycerol or sphingosine linked to long-chain fatty acids (syn. *phosphatide*)

Phosphorylase—the enzyme that catalyzes the degradation of the linear chains of glycogen by addition of inorganic phosphate to cleave the glycosidic links sequentially from the terminal 4-hydroxyl ends yielding glucose-1-phosphate

Phosphorylation—the transfer of a phosphoryl ($-PO_3$) group from one compound to another (syn. *phosphoryl group transfer*)

Plasma Membrane—the outer, limiting envelope of animal cells; it consists of a bilayer of lipids with a similar amount of proteins either embedded in or attached to the bilayer

P:O ratio—a measure of the stoichiometry of the phosphorylations coupled to electron flow in the respiratory chain; it is the quotient of moles of phosphate esterified to form ATP divided by the atoms of oxygen consumed during respiration

Polymer—a large molecule produced by linking single molecular units **(monomers)** into an elongated, linear chain

Polymerization—the head and tail attachment of multiple single molecules **(monomers)** to form long chains **(polymers):** with sugars as monomers the polymers are termed **polysaccharides;** with amino acids, **polypeptides** or **proteins;** with nucleotides, **polynucleotides** or **nucleic acids**

Polynucleotide—a long chain or polymer of nucleotides (see also **Nucleic acid**)

Polypeptide—a long chain or polymer of amino acids (see also **Protein**)

Polyribosome—a number of ribosomes joined to a strand of messenger RNA (syn. *polysome*) (see also **Ribosome**)

Polysaccharide—a long chain or polymer of sugar molecules *(monosaccharides)*

Porphyrin—one of a class of compounds with four pyrrole rings joined in a flat cyclic structure by methyne bridges (syn. *cyclic tetrapyrrole*)

Positive Feedforward—an autoregulatory system whereby a precursor of a metabolic sequence stimulates the activity of one of the enzymes responsible for its utilization (ant. **negative feedback**)

Precursor—substance that serves to provide atoms for the synthesis of metabolic end-product

Product—the resultant substance arising from catalytic action of an enzyme upon its substrate

Prostaglandin—one of a family of 20-carbon hydroxy fatty acids with a cyclopentane ring in the center of the chain. The prostaglandins have variable effects upon metabolism in different tissues by modulation of cyclic AMP action

Prosthetic Group—a nonprotein component attached to a protein that is generally essential for its biologic activity. (Removal of the prosthetic group leaves the *apoprotein.*)

Protease—one of the family of hydrolytic enzymes that attacks certain peptide linkages of large polypeptides, that is, intact proteins

Protein—a polymer of amino acids joined together by peptide (CO–NH) linkages (syn. *polypeptide*)

Protein Kinase—an enzyme that catalyzes transfer of a phosphate group from ATP to the serine or tyrosine hydroxyl group of a protein acceptor

Proteolysis—the hydrolytic cleavage of proteins at certain peptide linkages to yield smaller polypeptide fragments and, when complete, free amino acids

Protonmotive Force—the electrochemical gradient of hydrogen ions that is generated across the mitochondrial inner membrane during respiration and is proposed by the chemiosmotic theory of energy coupling to provide the driving force for ATP formation or cation movements by oxidative phosphorylation

Purine—a nine-membered heterocyclic ring system with imidazole condensed to a pyrimidine ring

Pyrimidine—a six-membered heterocyclic ring with two nitrogen atoms in the one and three positions (syn. *1,3-diazine ring*)

Recombinant DNA—a molecule of DNA from one cell, which is attached to a carrier molecule of DNA for transfer into another cell of the same or a different species

Reduction Potential (E'_0)—a measure of the affinity of a substance for electrons; under standard conditions it is the electromotive force in volts when reductant and oxidant forms of the substance are present in 1.0 M concentrations

Reductive Amination—the addition of an ammonia molecule to an α-keto acid forming an α-amino acid at the expense of the reducing energy furnished by NADH or NADPH (ant. **oxidative deamination**)

Regulatory Enzyme—an enzyme whose activity may be controlled by its interaction with an **allosteric modifier** or by *covalent modification* or the enzyme protein. It generally catalyzes an early step of a metabolic sequence.

Renal Reabsorption—the uptake of substances that have passed into the glomerular filtrate by active transport mechanisms of the kidney tubules for retention in the bloodstream

Replication—the production of an identical copy of the double-stranded DNA molecule when the gene is duplicated during cell division

Repression—the inhibition of the synthesis of an enzyme, often resulting from accumulation of a biosynthetic end-product of the pathway in which the enzyme participates

Respiratory Acidosis—a decrease in *p*H of body fluids occasioned by excessive retention of carbon dioxide in the body (syn. *hypercapnic acidosis*)

Respiratory Chain—a series of flavoproteins, hemoproteins (cytochromes), and nonheme iron proteins in the mitochondrion, which act to transfer electrons from substrates to oxygen

Respiratory Control—the condition pertaining to oxidative phosphorylation, when energy demands upon the mitochondrion are low and the rate of electron transfer through the respiratory chain is strictly dependent upon the ADP concentration (syn. *acceptor control*)

Respiratory Inhibitor—a compound that specifically blocks one of the electron transferring reaction involved with substrate oxidation or the respiratory chain

Ribonucleic Acid (RNA)—the polymer formed of ribose-containing nucleotides (AMP, CMP, GMP, UMP); it performs key roles in the biosynthesis of proteins in the ribosomes of cells

Ribosome—an assembly of RNA and protein molecules that acts to join together the amino acids aligned by messenger RNA during protein synthesis. Several ribosomes joined to one messenger RNA strand are termed a *polyribosome* or *polysome*.

Second Messenger—a cyclic nucleotide (cyclic AMP or cyclic GMP), produced in the target cell as the result of interaction of a nonpenetrating hormone at the outer cell surface, which promotes the intracellular response of the hormone (first messenger)

Sphingolipid—any of the class of lipids containing the long-chain amino alcohol, sphingosine, including the phospholipid, *sphingomyelin*, and the glycolipids, *cerebrosides* or *gangliosides*

Steady State—the dynamic condition resulting when substances in the body are being synthesized and are simultaneously undergoing an equivalent rate of breakdown, so that no net change in content ensues (see also **Dynamic state of body constituents**)

Steroid—any of the family of substances containing the complex ring structure, cyclopentano-perhydrophenanthrene

Sterol—any of a group of alcohol derivatives of the steroid ring structure. The main sterol in body tissues and fluids is **cholesterol**

Substrate—a substance which is acted upon catalytically by an enzyme

Sugar—a polyhydroxylic conpound with an aldehyde or ketone group on one of the carbons (syn. *monosaccharide*)

Synaptic Vesicle—a membrane-bounded body which contains a high concentration of a neurotransmitter substance in the presynaptic nerve cytoplasm

Transcription—the production of a strand of RNA that is the complement of one of the strands of a DNA molecule

Transfer RNA—a small ribose nucleic acid molecule which selectively binds a particular amino acid and, by specific attachment of complementary base pairs through three of its bases **(anticodon)** to three bases on messenger RNA **(codon)**, aligns that amino acid at the appropriate position during protein synthesis

Translation—the production of a protein at the ribosome, whereby the genetic message carried in the nucleotide sequence of the messenger RNA is decoded into the amino acid sequence of the polypeptide

Tricarboxylic Acid Cycle (see **Citric acid cycle**)

Uncoupling Agent—a compound that prevents the coupling of ATP formation with respiration but allows the transfer of electrons to proceed unabated or at an accelerated rate

Urea Cycle—the series of reactions whereby carbamyl phosphate and aspartate combine sequentially with ornithine, producing first citrulline, then arginine, with the latter undergoing hydrolysis to urea and regenerating the ornithine (syn. *Krebs–Henseleit cycle; ornithine cycle*)

Vitamin—an organic substance present in trace amounts in foods, which is not synthesized rapidly enough in the body and must be provided in the diet. Vitamins often form the functional groups of **coenzymes**

Zymogen—an inactive precursor form of an enzyme, which requires covalent modification by removal of a blocking group from the active center region before acquiring catalytic activity

Zymogen Granule—a safe storage form of the inactive precursors of certain degradative enzymes, within the cells responsible for their production and secretion

Acronyms and Abbreviations

A adenine
ACP acyl carrier protein
ACP-SH acyl carrier protein, sulfhydryl form
ACTH adrenocorticotrophic hormone
ADP adenosine diphosphate
ALA δ-aminolevulinic acid
AMP adenosine monophosphate = adenylic acid
ATP adenosine triphosphate

C cytosine
CDP cytidine diphosphate
CHO carbohydrate
CMP cytidine monophosphate = cytidylic acid
CoA coenzyme A
CoASH coenzyme A, sulfhydryl form
CoQ coenzyme Q
$CoQH_2$ coenzyme Q reduced form
CTP cytidine triphosphate

dATP, dCTP, dGTP, etc. deoxy forms of nucleotides
DNA deoxyribonucleic acid
DNP 2,4-dinitrophenol
dopa dihydroxyphenylalanine
dopamine dihydroxyphenylethylamine
DPG 1,3-diphosphoglycerate

FA fatty acid
FAD flavin-adenine dinucleotide
$FADH_2$ flavin-adenine dinucleotide, reduced form
FMN flavin mononucleotide = riboflavine phosphate

G guanine
GABA γ-aminobutyric acid
GDP guanosine diphosphate
GMP guanosine monophosphate = guanylic acid
GTP guanosine triphosphate

HDL high-density lipoprotein
HMG-CoA β-hydroxy-β-methylglutaryl CoA

IMP inosine monophosphate = inosinic acid

LDL low-density lipoprotein

NAD nicotinamide-adenine dinucleotide
NADH nicotinamide-adenine dinucleotide, reduced form
NADP nicotinamide-adenine dinucleotide phosphate
NADPH nicotinamide-adenine dinucleotide phosphate, reduced form
Na^+,K^+-ATPase sodium-potassium adenosinetriphosphatase

PBG porphobilinogen
PEP phosphoenolpyruvate
PGF-2α prostaglandin, type F-2α
P_i inorganic phosphate
$P–P_i$ inorganic pyrophosphate
PRPP phosphoribosylpyrophosphate

RNA ribonucleic acid

SAM *S*-adenosylmethionine

T thymine
TPP thiamine pyrophosphate
TSH thyroid-stimulating hormone
TTP thymine triphosphate

U uracil
UDP uridine diphosphate
UMP uridine monophosphate = uridylic acid
UTP uridine triphosphate

VHDL very-high-density lipoprotein
VLDL very-low-density lipoprotein

Index